A BOOLE ANTHOLOGY

SYNTHESE LIBRARY

STUDIES IN EPISTEMOLOGY,

LOGIC, METHODOLOGY, AND PHILOSOPHY OF SCIENCE

VOLUME 291

A BOOLE ANTHOLOGY

Recent and Classical Studies in the Logic of George Boole

Edited by

JAMES GASSER

University of Lausanne, Switzerland

KLUWER ACADEMIC PUBLISHERS

DORDRECHT / BOSTON / LONDON

A C.I.P. Catalogue record for this book is available from the Library of Congress.

ISBN 0-7923-6380-9

Published by Kluwer Academic Publishers,
P.O. Box 17, 3300 AA Dordrecht, The Netherlands.

Sold and distributed in North, Central and South America
by Kluwer Academic Publishers,
101 Philip Drive, Norwell, MA 02061, U.S.A.

In all other countries, sold and distributed
by Kluwer Academic Publishers,
P.O. Box 322, 3300 AH Dordrecht, The Netherlands.

Printed on acid-free paper

Printed in the Netherlands.

TABLE OF CONTENTS

RECENT STUDIES

1. BACKGROUND

2. MATHEMATICAL ASPECTS

3. PHILOSOPHICAL ASPECTS

4. CONSEQUENCES

PREFACE

Centuries of stagnation in the study of logic were followed by an explosion of progress in the late nineteenth century. It was George Boole's first book on logic, published in 1847 as *The Mathematical Analysis of Logic*, that provided the spark. Boole showed not only that algebraic formulas can be used to express logical relations but also that an entirely algebraic logic can be devised as a pure calculus admitting various interpretations. His logic was not merely developed from mathematics but went on to become a tool that mathematicians actually make use of. It was Boole, then, who realised Leibniz's project of creating a truly mathematical logic that would involve replacing verbal reasoning with a symbolic calculus.

While few would deny Boole's place as a key figure in the history of logic, there are many who do not consider him a truly great thinker on a level with, say, Frege. As Donald Gillies has put it, Boole's research programme 'led to important, but not revolutionary advances'.[1] A similarly cautious appraisal is made by Michael Dummett in a review that first appeared in the *Journal of Symbolic Logic* and is reprinted in this volume (pp. 79–85). Other authors expressing negative opinions have showed less concern for fairness and accuracy. Until recently, the prevailing view was that since Boole had not had the advantage of reading the *Begriffsschrift* he therefore amounted to little more than a pre-modern curiosity. After a long period of finding fault with Boole wherever possible and comparing him unfavourably with other early modern logicians, a new attitude developed following studies by Hailperin, Laita, Van Evra and others. The 'reassessment' brought about by these authors led to abundant new research that has confirmed Boole's importance and perhaps even true greatness. It is no accident that the adjective 'Boolean' is now in everyday use, at least by anyone who has ever used a computer search programme.

Nor is it an accident that interest in Boole's work has probably never been greater than it is today. Two new editions of *The Mathematical Analysis of Logic* have appeared in the past five years;[2] in addition, Boole's *Selected Manuscripts on Logic and Its Philosophy,* which had never been published, appeared in 1997.[3] There has also been a resurgence of interest in the man behind the adjective, George Boole himself, especially since the publication in 1985 of Desmond MacHale's full-length biography.[4] As articles on Boole continue to proliferate, the present anthology constitutes an attempt to capture some of the 'buzz about Boole'.

The idea of the present collection arose during the planning of a conference organised by the Swiss Society for Logic and Philosophy of Science. One of the society's more active members, Gérard Bornet, sent me a note in the mid-1990s which I quote in full: '1847 + 150 = 1997. Why not a meeting?'. (I was well acquainted with Gérard's research on Boole, but it did come as a surprise that he would adopt the language of equations in his own communications!) It was not long before our society decided to prepare a major international conference to mark the 1997 sesquicentenary of the 1847 publication of Boole's *The Mathematical Analysis of Logic*. The meeting took place at Lausanne on 26–27 September 1997. Generous financial support from the Swiss National Science Foundation, the Swiss Academy of Sciences, and the University of Lausanne made it possible to invite the leading specialists in the field, almost all of whom were able to accept. As the conference programme began to take shape and the prestige of the invited speakers began to sink in, the idea of an ordinary volume of proceedings seemed inadequate. At the suggestion of my friend and teacher John Corcoran, I undertook instead to prepare an *anthology*, combining under one cover the recent work presented at the conference and classic texts frequently referred to. From Ivor Grattan-Guinness I later learned of the existence of little-known but useful studies by Samuel Neil and G.P. Young published in 1865, the year of Boole's death, which had never been reprinted. Hence this anthology, hence its two major divisions. The 'Classical Studies' have been previously published; the 'Recent Studies', based on papers presented at the conference, have not.

The 'Classical Studies' have been re-set for this volume but an attempt has been made to remain faithful to the original presentation of each, so far as was consistent with the Kluwer style adopted for the volume as a whole. Theodore Hailperin's classic 'Boole's Algebra Isn't Boolean Algebra' is reprinted here with the author's corrections. At the suggestion of an anonymous referee, the 'Recent Studies' have been grouped together under the four headings that appear in the table of contents: 'Background', 'Mathematical Aspects', 'Philosophical Aspects', and 'Consequences'. These subject categories, like those in the 'Classical Studies', give only an approximate indication of content.

Institute of Applied Mathematics, University of Lausanne J.G.

---

[1] Gillies, Donald, ed. 1992. *Revolutions in mathematics*. Oxford: Clarendon Press, p. 267.

[2] In *From Kant to Hilbert: A source book in the foundations of mathematics*, edited by W.B. Ewald. Vol. I, pp. 451–509 (Oxford: Clarendon Press, 1996), and as a monograph (Bristol: Thoemmes Press, 1998).

[3] Grattan-Guinness, Ivor, and Gérard Bornet, eds. 1997. *George Boole: Selected manuscripts on logic and its philosophy*. Basel: Birkhäuser.

[4] MacHale, Desmond. 1985. *George Boole: His life and work*. Dublin: Boole Press.

ACKNOWLEDGEMENTS

Throughout the preparation of this volume, I have enjoyed the support of my colleagues in the Institute of Applied Mathematics of the University of Lausanne, whom I thank for their interest, encouragement, and helpful suggestions. I am particularly indebted to Henri Volken and Marek Błaszczyk.

It is with sincere pleasure that I thank Nathalie Janz, Evelyne Thommen and Helmut Linneweber-Lammerskitten for their help with the Boole 1997 conference.

Gérard Bornet deserves warm thanks for suggesting the conference, as does John Corcoran for suggesting the anthology and for helping me in too many ways to be mentioned here.

Appreciation must be expressed to Jolanda Voogd and Rudolf Rijgersberg of Kluwer Academic Publishers for expert professional advice and patience with a pernickety editor.

I am deeply grateful to the contributors for their confidence and support. They met my various requests cheerfully and conscientiously, and showed considerable forbearance and understanding as the project seemed to eternalise. Special thanks are due to Ivor Grattan-Guinness, who spared no effort to obtain complete and legible copies of the Neil and Young studies, and to Maria Panteki and Shahid Rahman, who were particularly eager to help in every way that they could.

My greatest debt is to my wife Susan and son Marc, true Booleans both, who were happy to tolerate the uninterpretable in the firm belief that everything would make sense in the end.

I am grateful for permission to reproduce copyrighted materials as follows:

Luis M. Laita: 'The Influence of Boole's Search for a Universal Method in Analysis on the Creation of His Logic' (*Annals of Science*, vol. 34, 1977, pp. 163–176); copyright © 1977 Taylor & Francis Ltd; reproduced here by kind permission of Taylor & Francis Ltd and Luis M. Laita. http://www.tandf.co.uk/journals/asc.htm

James W. Van Evra: 'A Reassessment of George Boole's Theory of Logic' (*Notre Dame Journal of Formal Logic*, vol. 18, no 3, July 1977, pp. 363–377); copyright © 1977 The Notre Dame Journal of Formal Logic; reproduced here by kind permission of The Notre Dame Journal of Formal Logic and James W. Van Evra.

TRUE
FALSE
ON
OFF
GEORGE
BOOLE
IN
OUT
UP
DOWN
OPEN
CLOSE
S. Harris

SAMUEL NEIL

THE LATE GEORGE BOOLE, LL.D., D.C.L.,

Professor of Mathematics, Queen's College, Cork; Author of 'The Laws of Thought', etc.

I

'The unity of the forms of thought in all the applications of reason, however remotely separated, will one day be matter of notoriety and common wonder; and Boole's name will be remembered in connection with one of the most important steps towards the attainment of that knowledge'.—*Professor Augustus De Morgan.*

'It has often been observed that the biography of eminent literary men presents but few objects of general interest. The progress of their lives seems only measured by the order of their attainments and productions. From these we estimate the gradual rise and advance of mind and character, through every successive change from the nursery to the grave'. So, upwards of thirty years ago, George Boole expressed himself in a lecture of the genius and discoveries of Sir Isaac Newton. The remark is as correct regarding himself as of any other of those gifted spirits who have given new thoughts to their age, or flashed the fresh light of vigorous and original minds upon the mysteries of life and nature. Few lives have been passed so uneventful and incidentless, and fewer still have been so successful in shedding the subtle radiance of intelligence down into the deep places of the soul, or more prolific in casting off those seeds of thought which multiply as the years increase, and grow with the ages. Of the possibilities enclosed within his own spirit, the best uses, we may be sure, have been made. Comparatively early as was the age at which 'Death touched him, and he fell asleep', we dare not call his demise premature, however inscrutable the purpose of it, since Heaven willed it, and it came to pass.

'The man who, for his race, might supersede
The work of ages, dies worn out—not used;—
And in his track disciples onward strive,
Some hair's-breadths only from his starting-point.
Yet lives he not in vain; for if his soul
Hath entered others, though imperfectly,
The circle widens as the world spins round,—
His soul works on, though he sleeps 'neath the grass'!

Among modern lives that of George Boole's is noteworthy on many grounds. As an instance of persistent industry ultimately winning success; of a modest and unobtrusive career broadening into European fame; of dutifulness being crowned with

J. Gasser (ed.), A Boole Anthology, 1–25.

honour; of self-culture attaining the higher reaches of professional elevation; of humble worth gaining lofty recognition; of the possibility of energy, earnestness, courage, and thought leading to reputation, usefulness, and a foremost place among the men who occupy the vanguard of the world's progress, the narrative of the life and life-work of George Boole is second to none in interest and instructiveness. Of the noisy celebrity given by applauding crowds he was never covetous, for he knew it was of short continuance. He aspired to the nobler renown of acknowledged merit, borne witness to by his own works and the unsolicited testimony of those who best understood the value of his achievements; but still more he desired to attain that highest glory which consists in 'a conscience void of offence toward God and toward man'. Of a man so wise and good, so thoughtful yet so simple, who gave forth the hardly-earned learning and the elaborately-wrought products of his genius so unpretendingly, so free from self-assertion; of one who taught so lovingly, thought so intensely, laboured so assiduously, and lived so blamelessly, whose insight into truth was so clear, whose heart was so noble, true, and dutiful, one may well write; but who shall write worthily? 'As a *man*', an old friend of his writes to us, 'I have scarcely ever known George Boole's equal. Looking back through the forty years—and I am nearly threescore and ten—that I have known him, I cannot recall an act or word of his which I could wish to have been otherwise. A perfect and blameless man is, we know, not to be found on earth; and doubtless, my friend had his faults; yet, in honestly and searchingly revising the past, I see only a life of unfailing duty and self-sacrifice, incapable of a mean or low thought, with not an iota of vanity—even when he had become one of the landmarks of science, and must have known it'. Such is the voice of friendship regarding his personal character; and in the sequel it will appear that praise as high and as unreserved has been given—and given deservedly—of his intellectual endowments, and of the grandly original results in the forth-bringing of which he employed them. We prefer, however, that an 'unvarnished tale' of his life, labours, studies, and investigations should prove his worth, to the exhaustion—if that were possible—of the language of panegyric in commendation of the earnest student, the successful toiler, and the nobler thinker, who has so recently 'entered into his rest'.

George Boole was born in Lincoln, 2nd November, 1815. He was the son of respectable though not of wealthy parents, engaged in trade in that Witham-washed and cathedral-crowned city. Though not rich in this world's goods, his relatives were held in general esteem. His father took great interest in scientific pursuits, and was an eager worker for the diffusion of useful knowledge among the people. In the early days of Mechanics' Institutes he laboured earnestly with others in getting one established in Lincoln. It was a very flourishing concern, well patronized and supported; and was not a little indebted to its intelligent and scientific librarian and museum-keeper, Mr. Boole. It is hardly to be doubted that the influence of this love of knowledge, indulged in at a home fire-side, where daily bread was sweetened and sanctified by daily toil, and the honest earnings of labour were spent, not on the dainties or luxuries of sense, in the gratification of whim or folly, but in the ennobling acquisitions of mental usefulness and delight, was especially powerful in its effects on the susceptible boyhood of George Boole. Here he learned at once the dignity of self-reliance, the sacredness of human duty, however lowly, and the worth of knowledge, even in the humble levels of common life.

His early home-training combined example and precept in promoting a healthy and innocent Christian life. Strict without being stern, and sedate without being sour, the management to which Mr. Boole subjected his family was well fitted to develop alike the heart, the mind, and the activities of hte inmates of his household. George, after a childhood passed within sound of the tones of Great Tom, and during a boyhood spent in rambles by the Witham's banks, receiving as into a mirror the image of Nature, and in strolls beside august ruins, or in the shadow of Roman gateways, mediæval castles, and cathedrals reckoned among the noblest in the empire, and among many other objects of interest in some of the narrow, irregular streets which exist in the old historic *Lindum* of the Romans, received the common lessons given to boys in the day schools of such towns. Though George Boole did not disdain the sports of boyhood, he loved better to spend his holiday time in the healthy, rational, and heart-charming pursuits afforded in country excursions on the Lincoln heaths, through which the Witham flows, before and after it passes under the bridges of 'Merry Lincoln'. The limited means of his parents, eager though they were to forward the scholarship of their son, prevented them from being able to afford him more than an ordinary education. In the National School he was first a pupil and then a monitor. But the class routine of tasks neither satisfied him nor his father. The latter supplemented the school work by courses of reading, training in observation, and some scientific instruction. The boy profited from all, but his thirst for knowledge remained unslaked and unsatisfied. Oh, how he envied the grammar school boys their privilege of attaining the means of proficient advancement in the knowledge of the ancient classical languages! A kindly neighbour, Mr. William Brooke, printer and bookseller, supplied what the eager-hearted boy, when about ten years old, desired—initiation into the mysteries of Latin grammar. Thus furnished, his own unspurred diligence enabled him to acquire fluency and facility in translation, and subsequently to reach the inner spirit of the authors of the Roman classics. To this he subsequently added, by self-tuition, an acquaintance with the original literature of Greece. His soul entered into communion with the writers of the olden times, and difficulties vanished before his invincible perseverance. The self-reliant energy of genius is best shown in incident. The elder Boole was an enthusiast in seeking converse with intelligent spirits, and in asking knowledge from every available source; and he loved above all to hold intercourse with the learned—often, indeed, to the detriment of his more material interests, we believe. He was naturally proud of the progress of his son under circumstances somewhat adverse to profitable progress. On one occasion a stranger from one of the universities, probably Cambridge, on a visit to Lincoln, was induced by George's father to examine the young student. He found his grammatical and lexicographical acquaintance with the ancient languages very considerable indeed; but he revealed to him at once the existence and the importance of prosody—the science of pronunciation and versification. This circumstance, we believe, suggested to the young scholar the propriety of systematically re-studying the entire course he had gone through, in following out which resolve, George Boole expended two full years of mind-tasking, and conscientious industry. The late Rev. George Stephens Dickson, the learned and accomplished incumbent of St. Swithin's, was useful at this stage of his progress too, in providing him with the opportunity of perusing books, and of consulting him upon points of doubt or difficulty: kindnesses these, which the recipient was able to repay in after years, when, at a somewhat late period of life, that reverend gentleman betook himself to mathematical studies, and found in his youthful friend one of the ablest instructors the county could furnish. This belongs, however, to the after-time of a friendship which lasted until death, and to studies of which we have as yet made no record.

His father's fondness for mathematical and optical studies had begun to touch the inner springs of the mind of the aspiring youth. He could, he thought, as readily master the elements of such investigations as those of the languages of Greece and Rome. When about seventeen years of age, the study of mathematics, under the supervision of his father, was fervently pursued; and he found such invigoration and excitement in the new discipline, that he speedily perused, studied, and mastered all the books he could procure. Very eagerly did father and son alike pore over the problems which by their knottiness puzzled, or by their newness bewildered the earnest pair of ardent hearts that beat with the ambition of success. By dint of diligent experimenting and sedulous thought they managed to conquer or elude the difficulties their course presented, and found it possible to overmaster them. Besides this, George Boole read copiously on all possible subjects, exhausting the library of the mechanics' institute of which he was a member, and finding the weekly provision of reading contained in the publications of the Society for the Diffusion of Useful Knowledge, the *Penny Magazine*, *Chambers' Edinburgh Journal*, &c., too little for the gratification of his capacious appetite for information. He was one of the most enthusiastic members of the mechanics' institute, and devoted much time and interest to the welfare of that city foundling. He soon became known, even in Lincoln, 'his own country', as one worthy of honour for the self-containedness of his mind and the earnestness of his endeavours after self-improvement.

In the unaided toils to which he afterwards committed himself, George Boole resolutely worked his way. With a dauntless heart he re-essayed every stronghold of difficulty until he vanquished the obstacles to his progress. It is true that in this method of study he was often brought to a halt by very slight impediments, and frequently grudged the time lost and energy spent in overcoming trivial hindrances which a helping hand could speedily have removed. But then he was training his mind to self-reliant effort, his eye to look unappalled at the apparently pathless places which he came upon in his progress, his intellect to keen scrutiny and experimental dexterity, and his moral nature to steadiness and earnestness. He was educating himself to be a discoverer even while only traversing the open fields of knowledge; and he was strengthening the sinews of his soul by the struggles he was compelled to make to surmount the difficulties or vanquish the obstacles which opposed his forth-going. The self-trainer is at a disadvantage in comparison with the school-taught in speed and perhaps accuracy; but it is not all disadvantage—strength, resoluteness, and ingenuity are developed in him of which the other knows little, and for original investigations these are the true preparations.

In May, 1829, the Society for the Diffusion of Useful Knowledge issued a 'Life of Sir Isaac Newton', which was substantially a translation of the sketch of that man of rare scientific insight which J. B. Biot had supplied to the *Biographie Universelle*.

The extra importance thus given to this memoir, and the questions which it raised, led to the production, in 1831, of a 'Life of Newton' by Sir David Brewster, and a general discussion among scientific men of the merits of the expounder of celestial mechanics. Stirred by the agitation of the times, which had a peculiarly local interest to all Lincolnshire men, the Right Hon. Lord (afterwards Earl) Yarborough, Patron of the Lincoln and Lincolnshire Mechanics' Institute, commissioned a marble bust of Isaac Newton, which he presented to the members of that association. It was felt by the committee of management that advantage should be taken of this circumstance, and the interest it excited, to have a word in season spoken to those who were present at the inauguration festival, regarding the lessons to be derived from the life and labours of this illustrious Lincolnshire worthy; and so high was their appreciation of the energy, ability and competency of George Boole, that he was unanimously requested to deliver an address on the 'Genius and Discoveries of Sir Isaac Newton'. To this he gave a reluctant consent, having stated as his objections his own youth and the magnitude of the topic.

On Thursday, Feb. 5th, 1835, in the presence of Lord Yarborough, the donor, and an intelligent and eager audience of the citizens of Lincoln, George Boole rose, in the chapel of the ancient Grey Friars, granted by the corporation for use as a lecture hall, to implement his engagement to the committee. The youthful appearance of the lecturer, who was then only a shade above nineteen years of age, excited surprise, and when he gradually unfolded the mighty theme of his discourse surprise was heightened into admiration, and hearty plaudits from all parties in the building welcomed the *début* of the young and self-taught expositor of the worth and wisdom of the distinguished natural philosopher and mathematician whose memory the assembly had met to honour. The address was published at the request of the meeting, given expression to by Lord Yarborough, and through the combined kindness of one of George Boole's oldest friends, and of one of his former—I dare not in this case say old—pupils, whose reverence for her master's memory is great, the writer of this paper has been enabled to peruse that thirty years old lecture, which has now become scarce and rare. It is—even in memory of the magnificent eloge delivered on the same topic by the Nestor of the orators of our day, Lord Brougham, 21st Sept., 1858, at the inauguration of a statue to Newton on St. Peter's Hill, Grantham—an able and excellent address, a large-minded and thoughtful discourse.

He announces, in the outset, his intention to bestow his thoughts less on the events of the life than on an exposition of the labours of 'the *mind* of Newton'. Glancing hastily at the incidents of his birth and upbringing, he passes on to epitomize and explain the 'Treatise on Optics',—'one of the most elaborate and original of his works, and one which carries on every page the traces of a powerful and comprehensive mind'. Of this Boole says,—

> 'It is true that his theory has been left imperfect; admit that in some of the applications it has failed, but at the same time we must acknowledge that in what he failed he did not fail as a common mortal, and that the marshalled intellect of Europe has vainly endeavoured to fill up the chasm. There is in the very idea of light something so vague and intangible, that our imagination can with difficulty attribute to it an independent and material existence. Yet, granting this, and assuming as our *data* that, under certain known circumstances, known impressions are received which we designate colour, the analysis of its primitive elements, and of the laws and effects of their combinations,

would still remain a mighty problem. It is singular, that of all the subtle and mysterious agencies,—light, heat, and electricity, attraction, connected by one general link, and commissioned by their author to confer upon dead matter the life and beauty of the universe,—light is the only one that has yet thoroughly unfolded the harmony of its laws, and submitted itself to human scrutiny. That genius which stands foremost in the triumph was the calm, patient, all-surmounting genius of Newton'.

After another instalment of narrative the lecturer proceeds to give an account of the work on which 'Newton's claim to the discovery of universal gravitation must rest',—prior to the issue of which 'he found it an unsupported hypothesis', by the publication of which 'he left it an established truth. This was the 'Principia, or Mathematical Principles of Natural Philosophy'. After analyzing this book, recording the main incidents of his middle life, and indicating 'the religious impressions' which changed the object of the great thinker's life, and led him to employ himself thenceforth chiefly on 'theological inquiries', the lecturer proceeds to give this excellent estimate of his state of feeling:—

'We cannot but consider this latter portion of Newton's life as highly and singularly happy: all that can make old age honourable he possessed, with scarce a shadow of its dotage and infirmities. He was not a father, but the natural affections were in him expanded into the broad principle of universal philanthropy. Subdued passions, moderate wealth, and the much-loved blessing of peace, all tended to smooth and to illumine the rugged path of declining life. If desire of fame had been the meteor of his youth, it could not now disturb his repose, for he had long been at the summit of all earthly ambition. If the recollection of 'the single talent well employed' be attended with pleasure, surely that pleasure must have been felt in its keenest relish by him, who had received from his Maker ten golden talents, and well employed them all. These are the materials of happiness, and all these were possessed by Newton. But more than these, the support and solace of his faith, the prospect of future happiness which grows brighter as all other prospects decay, these were inalienably his. And though genius has often been a wandering star, the minister of licentiousness or the associate of scepticism, in his life we have ample testimony that such is not a natural or necessary alliance. Nor, perhaps, is there less to admire in the high excellence and unblemished purity of his moral character, than in that halo of philosophical glory which has gathered around his name'.

The story of Newton's death being told, the inscription on his monument in Westminster Abbey being translated, he passes on to a general estimate of his genius and influence, in which occurs the following outflash of rhetoric:—

'The very pertinacity with which error retains its hold is one of the strongest arguments for the final and eternal establishment of truth; it results from a natural fear that, in the wreck of received opinions, the very foundations of credibility should be destroyed, and mankind a second time involved in darkness and obscurity. For this reason, perhaps, there have been few ancient kingdoms which have not bequeathed to other times a faith or a philosophy more enduring than themselves. The Coliseum and the Acropolis are in ruins, but the philosophies which sprung up beneath their shadows are yet deep in the tide of human opinions, still influencing, with an unseen but mighty influence, the character of this age. The dark creed of the ancient Persian is yet descending from sire to son in the sacred annals of the Guebres; and the faith and fame of Zoroaster are yet triumphant against the desolation of his country, and the sword and the Koran of its Mahometan oppressors. But more especially is this truth to be observed in the records and remains of more ancient dynasties—in patriarchal Assyria, in sepulchral Egypt. The long succession of their kings and warriors is now doubtful or forgotten; the colossal relics of their primeval architecture are daily mouldering; but their sombre religion, their wild astrology, originating in the days of their greatness, are perpetuated when the very shadow of that greatness is no more. If, from the history of false and discarded systems, we pass by analogy to the more enduring influence of truth, we shall perceive how high above the chances of time and vicissitudes the pedestal of Newton's immortality is founded'.

The lecture closes with a personal application to the hearers to fulfil life's 'great duties,—to suffer with fortitude, or triumph with humility; to expand with science, or warm with philanthropy',—and a tribute of praise to Lord Yarborough for 'co-operating, heart and soul, in that noblest work of British patriotism,—the education, the enlightenment, and the happiness of his fellow-countrymen'.

On its publication the lecture was read with avidity, and it drew upon George Boole the respectful regard of many of the people of Lincoln. A short time afterwards he was employed as an assistant in a large boarding school in Doncaster. Here, where the competition of scholastic establishments is very keen, he gave full satisfaction at once to the head master, his pupils, and their guardians. Here he not only attended to the great law of mental profit, *docendo disce*, 'learn by teaching', but, opportunity offering, he began the study of the languages of Luther and Goethe, of Dante and Manzoni, and of Montaigne and Guizot. Of these languages, at least—German, Italian, and French—he ultimately acquired a singular mastery; and of the varied and extensive literatures of which they are the treasuries he possessed an accurate and catholic knowledge. About this time, too, the poetic mood of his mind found expression in lyric forms, which are credibly asserted to manifest a rare command of elevated thought, suffused with inspiring emotion, and uttered in musical verse. He devoted a large portion of the period capable of being employed studiously, to the acquisition of a well-grounded knowledge of English composition. His success in his newly-adopted profession was unmistakeable; and it soon led to his receiving an invitation from the late Mr. R. Hall, proprietor of a large, popular, and successful educational establishment at Waddington, to take charge of certain of the departments in his seminary. This was an offer which was exceedingly acceptable to George Boole. It was gratifying to the youth's intense domestic affections, by restoring him to the neighbourhood of his family, adding somewhat to his, no doubt, narrow income, and affording him a higher position and more hope.

At Waddington he laboured with assiduity and conscientiousness, giving diligent and earnest heed to all that could promote the advancement and welfare of those who had been placed in his charge, and he was repaid by their rapid progress and their trustful love. His employer soon saw and acknowledged that he had added a man of no common order to his establishment, and reposed in him the utmost confidence—a confidence never withdrawn or abated.

But strong family reasons—the age, ill health, and narrowing circumstances of his parents, and the dependence on him, in part, of some of the other members of the old household—induced George Boole to open, on his own account, a day-school for boys and girls, in Free School Lane, Lincoln. Here he had fair success; his own early years approximating more in feeling to his pupils than usual; and he made many valuable friendships while engaged in the earnest exercise of his vocation. On the death of Mr. Hall, his old employer at Waddington, an advantageous opening for succeeding him occurred, and he accordingly removed there with his parents and other members of his family, and conducted a seminary much used by the agriculturists of Lincolnshire—a wealthy and intelligent class—for several years. An eligible opening having occurred for a boarding (combined with a day) school, he subsequently removed to Minster Close, where he met with great success, and was known and respected as one of the most ingenious and ardent members of the scholastic profession. An eventless life of love, duty, and frequent self-sacrifice, self-reliant yet modest and humble, original yet boastless and blameless, was that which he now passed,—endeavouring to work out the great duty of developing the seeds of thought sown by the Almighty in his soul.

Trying and arduous as were the duties of the day, the energy and enthusiasm of a noble spirit enabled George Boole to give much time and thought to the encouragement and promotion of the objects of all local associations of an ameliorative character. In the Mechanics' Institute he never lost his interest. He had formed its library and arranged its museum, and now he took an active share in the management of its business as a member of committee; and, for the benefit of those whose day of school training had elapsed, but who felt in them yearnings after higher culture, he instituted and taught gratuitously, evening classes for the study of the ancient languages and mathematics, and perseveringly toiled on in his efforts to induce men to enter upon and pursue the paths of improvement and endeavour. Greatly through his influence, advocacy, and resistless business-like attention, a Female Penitents' Home was established, and set a-working on a footing likely to be effective in some measure for the accomplishment of its benign purpose—the reclamation from the paths of woe of the misguided and unfortunate of that sex—

> 'Whose chiefest joy or suffering springs from love'.

Meantime, the most sedulous study on his own part was continued. With tireless persistency he undertook and pursued the most laborious and intent courses of study. He was not contented with acquisition, wide as were the fields from which he reaped; he was anxious to originate. He felt within himself the power of passing beyond the limitations of science as it then was, and of opening to the footway of the studious new and unexplored regions of investigation. He essayed some contributions in the scientific journals, and these met with acceptance and recognition from the men of mark who conducted them. To the 'Transactions' of the Royal Societies, both of London and Edinburgh, he contributed many papers by which he was proved to be worthy of a high place among the first mathematicians of his age. Curiously enough, it is related as a fact that George Boole's first communication to the Royal Society was recommended for rejection by the person entrusted with its perusal, and that only by accident another thinker, whose perception was more acute to the consequences of the views it contained, saw it, commended it, and ultimately, in 1844, it was successful in gaining the gold medal granted for pre-eminent excellence. This paper was entitled, 'On a General Principle in Analysis', and treated of 'the separation of the symbols of operation'; and we are able to state, upon the authority of some of the leading mathematicians of the present time, that it has been the source of almost all the improvements which have of late been introduced into that department of analysis. An additional proof of his high attainments as a mathematician may be found in the fact that on the establishment, in 1846, of the *Cambridge and Dublin Mathematical Journal*, one of the most able and erudite collections of original contributions to the science of quantity and the calculus of operations which have been produced in this country, George Boole was invited to become a contributor. Indeed, so great was the interest excited in many minds by the extraordinary mathematical genius of the self-taught Lincoln schoolmaster, that he was earnestly urged to enter the University of Cambridge, that he might place himself in a position to receive those supreme honours and high rewards which it is in the power of the university alone to bestow. This tempting counsel he had the self-denial to resist, because it would interfere with the fulfilment of those filial obligations to the implementing of which he considered himself morally bound. Even without the advantages of a course of university instruction, without the *éclat* of its honours, of the stimulation of its rewards, by sheer pith of will, and power of pertinacious intellectual application, he succeeded eventually in winning a reputation Europe-wide—echoes of which were heard, too, in the Western hemisphere—as an acute, ingenious, and rarely original mathematical genius; as the first and foremost expounder and extender of the science of operations. It would be neither profitable nor wise to occupy space here with a mere list of the contributions he made to the simplification and extension of the processes arising out of the relations of magnitude and number—contributions which transcend our own power of explanation, and whose value in a scientific point of view we accept as settled upon the faith of those whose specific studies and mental tendencies give them a right to be regarded as able and trustworthy adjudicators.

One of the papers which he contributed to this *Journal of Mathematics* falls within the special scope of this series of articles. It contains the germ of the speculations which he afterwards elaborated, and its main teachings are comprised in the following six canons, which we quote as remarkable at once for clearness of statement and intellectual acuteness; viz.,—

> '1. That the business of logic is with the relations of classes and with the modes in which the mind contemplates those relations. 2. That, antecedently to our recognition of the existence of propositions, there are laws to which the conception of a class is subject—laws which are dependent upon the constitution of the intellect, and which determine the character and form of the reasoning process. 3. That these laws are capable of *mathematical* expression, and that they constitute the basis of an interpretable calculus. 4. That these laws are furthermore such, that all equations that are formed in subjection to them, even though expressed under functional signs, admit of perfect solution, so that every problem in logic can be solved by reference to a *general theorem*. 5. That the forms under which propositions are actually exhibited, in accordance with the principles of this calculus, are analogous to those of a philosophical language. 6. That, although the symbols of the calculus do not depend for their interpretation upon the *idea of quantity*, they nevertheless, in their particular application to syllogisms, conduct us to the quantitative condition of inference'.

The ideas outlined in the foregoing six canons were further elaborated in a treatise entitled 'The Mathematical Analysis of Logic', published in 1847, on the same day as Professor A. de Morgan's 'Formal Logic; or, the Calculus of Inference', appeared. We may remark that these books, notwithstanding the apparent similarity of subject, are little if at all connected with each other. They are so distinct that neither can supply the place of the other. De Morgan's system is all constructed on the ideas of logic already received, but is widened in all directions by those suggestions regarding extension which operate habitually in a mathematician's mind, and is distinguished by the use of a symbolic language, which, though not mathematical, could never have been invented except by a mathematician. Boole's system is distinctly mathematical. It shows that the forms and transformations of algebra can be fitted to meanings of the symbols which will make them express the forms and transformations of thought. 'He brings', as De Morgan somewhere expresses it, 'the laws which work under cover in thought—the *genus*, to show themselves in daylight in algebra—the *species*'. 'He sees thought in its workings *below* language, and tracks its transformations *beneath* the cover of expression'. 'He finds the laws of thought symbolized in algebra'. As, however, this tract has been superseded by a more exhaustive work, which is 'devoted to the same object, though its methods are more general, and its range of applications far wider', it will, we believe, be advisable to adjourn our criticism of these bold generalizations till we can take a single and connected view of the entire scheme of the author whom Sir William Hamilton characterized as 'a very acute philosophical mathematician', while he spoke of this work as 'a very able' one. To Hamilton's opinion we might add that of the late Prof. John Pringle Nichol, who regarded George Boole as 'one of the most accomplished mathematicians of our age'; and of J.D. Morell, who says, in reference to the book under notice, 'These researches promise to be one of the most fruitful of modern contributions to the science of logic; and if fully carried out by the author (as we understand they are to be), to the science of metaphysics as well'.

To the general reputation gained by these works there was added the special renown acquired by his contributions to the 'Transactions' of various societies, the *Mathematical Journal*, and still more, that justly arising from the issue of 'A Treatise on Differential Equations', of which the praise in all mathematical circles was high. It became a class-book even in Cambridge. The poet's axiom, that 'the world knows little of its greatest men', was once again exemplified in Boole's case. While his fame was growing all over the Continent, he was contentedly pursuing his scholastic duties, a well-appreciated citizen, but almost unknown as one who was capable of revolutionizing one of the best-cultivated fields of human knowledge. His hour was not yet come.

The early closing movement secured the sympathy of George Boole; and when the efforts of an 'association for obtaining an abridgment of the hours of business in all trades, with a view to the physical, mental, and moral improvement of those engaged therein', which had been established in Lincoln under the presidency of the Hon. Alexander Leslie Melville, of Branston Hall, succeeded in securing the early shutting of the shops in that town, he, as one of the vice-presidents of the society, endeavoured to turn the minds of its members to the improvement of the time thus placed at their disposal. On 2nd February, 1847, he delivered an address on the 'Right Use of Leisure', which was 'published at the request of the committee'. From this tract, now scarce, but of which, through the kindness of Dr. John Ryall, Vice-Principal and Professor of Greek in Queen's College, Cork, we have been favoured with a perusal, we anticipate that the following excerpts will be found valuable to our readers. After some congratulatory remarks on their attainment of leisure, he proceeds to enforce the truth that opportunity involves responsibility.

'Such a thing as irresponsible right has no existence in this world. Even in the formation of opinion, which is of all things the freest from human control, and for which something like irresponsible right has been claimed, we are deeply answerable for the use we make of our reason, our means of information, and our opportunities of arriving at a correct judgment. It is true, that so long as we observe the established rules of society, we are not to be called upon, before any human court, to answer for the application of our leisure; but so much the more are we bound, by a higher than human law, to redeem to the full opportunities'.

He points thus to the variety of objects from which the human mind may derive innocent gratification:—

'Reflect for a moment on the wealth of ideas which is bequeathed to us in the writings of our poets, on the stores of instruction which are opened in our histories, on the lessons of wisdom contained in the recorded lives of those who have adorned our race. If we desire to acquaint ourselves with the structure of the universe, how many subjects of inquiry and mediation present themselves! What wonders in the heavens! what a glory and beauty in the world around us! what order and intelligence throughout the whole! Dr. Chalmers has remarked that every investigator of nature is led to regard his own particular department of research as the richest in interest and wonder. How replete and gorgeous, he then observes, should we consider the whole to be! If we quit the study of material nature, and apply ourselves to the pursuit of truth in the province of moral and social inquiry, another universe, not less replenished with wonder and interest, presents itself. The rise and fall of extinct nations, the mission they accomplished in the world, and the work which they have left to be done by us; the causes on which national wealth, and virtue, and happiness depend,—these are questions which are second to none within the range of the human intellect in dignity and interest. In short, to whatever province of the kingdom of thought we turn our attention, we find abundant scope and reward for the activity of our inquiry'.

This is his general estimate of 'men's silent monitors', books:—

'There are none of the monuments of past ages in which so much of the accumulated results of human labour is presented to us as in books. Considered merely as a memorial of the industry of man, every well-filled library is more replete with wonder than are the most costly of material structures. The most renowned of ancient cities, could we repair the ruins of time, and restore its lost magnificence, would scarcely represent a larger aggregate of toil than the single library of the British Museum. Such a collection excels every other result of labour, not less in character than in extent. It presents us with the issues of all past time, it unfolds to us the great discoveries of science, it brings us into acquaintanceship with men who were, intellectually and morally, of larger than the common stature of our race. We should esteem it as a high privilege to be admitted to this intimate fellowship with the wise and eminent, not of this place or of the present time alone, but of all times and generations'.

After this he gives advice, good and sound, on the study of history and biography combined with geography. He then proceeds to consider the claims of science,—physical, moral, and intellectual,—indicating, as he goes, the chief easily accessible works which should be read. Here is a passage the spirit of which we highly approve:—

'We ought to seek truth for its own sake; and we cannot set too high an absolute value upon either rectitude of opinion or consistency and accuracy of judgment. But it is our business to act as well as know; and these faculties of our nature, the speculative and the practical, may contribute mutual strength and support. If right judgments are necessary to rectitude of conduct the converse proposition is true also. Rectitude of intention, and an earnest desire to carry into practice the truth to which we have already attained,

> are in some measure necessary to correctness of judgment. For belief is not altogether involuntary; but while it acts is, in turn, acted upon by the habits, the feelings, and the will'.

After advocating the compatibility of scientific study and scriptural sanctity, another passage we relish occurs:—

> 'Consider what truth is. We are not to regard it as the mere creature of the human intellect. The great results of science, and the primal truths of religion and morals, have an existence quite independent of our faculties and of our recognition. We are no more the authors of the one class than we are of the other. It is given to us to discover the truth; we are permitted to comprehend it: but its sole origin is in the will or the character of the Creator; and this is the real connecting link between science and religion. It has seemed to be necessary to state this principle clearly and fully, because the distinction of our knowledge into Divine and human has prejudiced many minds with the belief that there is a mutual hostility between the two—a belief as injurious as it is irrational'.

After speaking of the study of languages, he adverts to the means of education, and gives utterance, on this subject, to an important idea:—

> 'Foremost among the agencies of education we may rank the conscientious discharge of the duties of our lawful business and calling. For the labours of our daily avocation are not merely the equivalent which we pay, in the market of the world, for food and clothing, and other advantages which we may enjoy, but also our sphere of duty—our field of exercise and trial. Now, every faculty we possess is strengthened by exercise; and thus, when the business of our lives is seen by us in the light of duty, it becomes an important means of confirming the power of that principle, and the habit of obedience to its precepts, within us. This appears to me a very just view of the design of life, and it is one which invests with a real dignity and importance the homeliest of our lawful employments'.

He passes next to consider the influences of nature on the mind, the duty of cultivating a love for its beauties, and the blessings to be derived from the exercise of the social faculties in works of beneficence. The expense of printing this lecture was undertaken by Mr. J.T.B. Porter, in admiration of the talent it displayed, the sentiments it contained, and the object it was meant to promote, as well as in token of the personal esteem in which he held the lecturer.

In 1845 the Government of Sir Robert Peel projected the establishment of the Queen's colleges in Ireland, in the hope of overcoming the difficulties in the way of promoting mixed but undenominational education in that country. Parliament consented, and the Catholics at the time accepted the compromise. Buildings for this purpose were erected at Belfast, Galway, and Cork—the latter a fine Tudor-Gothic quadrangular pile. The appointment of professors was a delicate task, for the experiment was fraught not only with peculiar dangers, but with promises of great national advantage. Of the places thus put in the gift of Government, George Boole was selected from among several candidates to fill the position of professor of mathematics in Cork. This promotion he obtained in 1849; it gave him a position he was eminently fitted to adorn, and it supplied him with leisure to promote the favourite studies of his life. Some grumbling at the election of a man unmarked by academic fame or university honours was mooted at first, but the choice was amply justified in the results; for no professor was ever more popular with and beloved by his students than the self-raised thinker of Lincoln, while the success of his students in almost every walk of life proves that he possessed the rare faculty of being 'apt to teach'. The University of Dublin did not long hesitate to confirm with its approval this judicious act of Lord Clarendon's vice-royalty by conferring on him the honorary degree of LL.D.

Professor Boole, on leaving Lincoln, received ample evidence of the esteem in which he was held. Previous to his departure he was entertained by his fellow-citizens at a public supper, presided over by James Snow, Esq., M.D., the mayor. On this occasion he was presented with a handsome silver inkstand and a valuable collection of books. The members of the Mechanics' Institute, remembering his services, arranged for a testimonial presentation too; and the Professor vainly endeavoured to divert them from their purpose by proposing that they should add to their library a copy of Newton's works, or to their apparatus an astronomical telescope. They were determined to recognize his goodness by a gift; and this was finally bestowed on him in the shape of a splendid copy of Johnston's 'Atlas of Physical Geography', a work of great cost and worth. At this point of his fortune, when position and influence are his, and a new future lies before him, we may break off our narrative for the present. We shall resume it in our next, and supply then not only an account of his career at Cork, but such an outline of his thoughts as shall justify us to our readers in giving his name a high place among 'Modern Logicians'.

II

> 'A generation will arise in which the leaders of education will know the value of logic, the value of mathematics, the value of logic in mathematics, and the value of mathematics in logic'.—A. DE MORGAN, F.R.A.S.

The Queen's Colleges of Cork, Galway, and Belfast, though now incorporated into and forming one University, were opened as separate institutions in 1849. In that year the council of the Cuvierian Society (an association founded in 1835) projected a *conversazione*, which was held in the rooms of the Royal Cork Institution and School of Design, under the presidency of A.F. Roche, Esq., then mayor of the city. This reunion was intended as a complimentary reception and welcome to the president and professors of the newly established college, and partook, in some measure, of the nature of a popular demonstration in favour of that institution. The *conversazione* was a great success; much interest was excited in the public mind by the event, and the new staff of officials which Government had added to the equipment of the city found themselves in the midst of friends. At the opening of the succeeding session (1850–51) another *conversazione* was held with equal success, and the college was thereafter considered as one of the home institutions of the Munster seaport.

Of course, the planning, preparation, and arrangement of lectures, class lessons, examination questions, &c., the reception and classification of pupils; and attention to their regular, gradual, and effective mastery of the science of which he had been chosen the teacher, occupied much of the time and thought of the new professor, who was doubly diligent, because he felt that the true test of his own worth would be held to be the success of his students. The ultimate form into which his department settled was two classes—senior and junior—meeting four days in the week; either or both of these classes being subdivided, as might be found convenient for the purposes of special and adapted instruction. All the students of the senior class, as well as the more advanced members of the junior class, received arranged exercises weekly, to which answers in writing were required. The course of study commenced with the arithmetic of fractions, and led through the elements of Euclid and the study of algebra, to the application of these in the sciences and the arts. The senior section was introduced to a knowledge of solid and analytic geometry, the differential and integral calculus, the philosophy of operations, and the adaptation of their principles to the problems arising in mathematical physics and astronomy. In all the work that he did, he was thorough, and his system of teaching, though strictly scientific, was amply relieved from tedium by the ingenuity and facile tact of the proficient instructor. In a very short time he became the idol of his students. They saw in him much more the learned friend than the stern preceptor. He took a personal interest in them, not only in the class-room, but in their own homes or lodgings. He opened his house to them in the most social manner, and not unfrequently gave of his slender substance to help them in temporary straits, and especially in times of sickness he acted not as counsellor only, but consoler—for he carried in his heart the living warmth of practical godliness, and the fadeless light of Christian faith. They knew his sympathy with them, and they repaid him by diligence and with love. Though, in fact, one of the most profound and original mathematicians of this age, he stooped to the humblest capacity, and expounded the merest elements to the veriest tyro with an earnest simplicity and absence of pretension which seldom failed to secure its end—the progress of the student.

But he was not a one-sided soul, of mathematics all compact. He was genial, intelligent, and widely informed, as lovable in the social circle as he was keen-minded in the study. He was held as an acquisition in the city of his adoption, both on account of his frank, candid, and honourable demeanour, and of his enthusiastic readiness to work for the furtherance of any good end. Among his colleagues he rapidly acquired respect and love. They saw his wonderful adaptation to his office; they knew the love he inspired in his students, and they noticed the daily influence he exerted on all classes to harmonize differences and promote peace, and good-will among men. But while he rose in general esteem, he did not fail to prosecute the great life-task laid out for him—his own self-development, and through that the development of the kindred sciences of mathematics and logic.

In 1851, at the opening of the third session of Queen's College, Professor Boole, as Dean of the Faculty of Arts, delivered a 'lecture on the Claims of Science, especially as founded in its relations to human nature'. This is a discourse of remarkable expansiveness of thought. In it he considers the origin of human knowledge, the relations of science to the constitution and design of our own minds, the benefits we owe to it, and the claims which it possesses on our regard. We cull from its pages the following extracts:—

'Science, then, we may regard as the joint result of the teachings of experience, and the desires and faculties of the human mind. Its inlets are the senses; its form and character are the result of comparison, of reflection, of reason, and of whatever powers we possess, whereby to perceive relations, and trace through its successive links the chain of cause and effect. The order of its progress is from particular facts to collective statements, and so on to universal laws. In Nature it exhibits to us a system of law *enforcing* obedience; in the Mind a system of law *claiming* obedience. Over the one presides necessity; over the other the unforced obligations of reason and the moral law. Such I conceive to be the true conception of science'.

The following is an admirable specimen of rhetorical enumeration:—

'The instinctive thirst for knowledge, its disinterested character, its beneficial tendencies, are among the most favoured topics of ancient writers. Cicero dwelt upon them with a peculiar delight, and he has invested them with more than the common charm of his eloquence. Plato made them a chief ground of his speculations concerning the just man and the well-ordered state. Aristotle gave to them the testimony of one of the most laborious of human lives. Virgil devoted the fairest passage of his best poem to the delights of a calm and meditative life, occupied in quest of truth. Lucretius drew from philosophical speculations the matter of what some have regarded as the noblest production of the Latin muse. Sophocles made knowledge, in its aspect of power, the theme of incomparably the finest of his choral odes. Æschylus made knowledge, in its other aspect of patience and martyrdom, the noble burden of his Prometheus. And there is a ground for the conjecture that such influences were not unfelt by those older poets and seers with whom our Milton felt the sympathy of a common fate, and desired to share the glory of a common renown. The early dawn, too, of philosophy, not to speak of its subsequent and higher development in the schools of Athens and Alexandria, is full of suggestive indications. Some records, scattered, indeed, and dim and fragmentary, still exist if the successive attempts which were made in Ionia, in the cities of Southern Italy, and in Greece, to penetrate the mystery of the universe, to declare what it is and whence it came. In those speculations, vague as they are, we discern the irresistible longings of the human mind for some constructive and general scheme of truth, its inability to rest satisfied with the details of a merely empirical knowledge, its desire to escape into some confined sphere of thought, and, if it might be, to hold "converse with absolute perfection". Nor are the efforts to which such feelings gave birth to be regarded as accidental or unmeaning. They had a prospective significance in relation to the science that was yet to appear. They were like the prelusive touches of some great master of harmony, which serve to awaken the feeling of expectancy and preparation. I affirm, and upon deliberate examination, that the peculiar order of the development of human thought which preceded the rise and growth of modern science was not an arbitrary thing, but it is, in its main features, susceptible of explanation. Though for any elucidation of the phenomena of nature it is utterly worthless, upon the human faculties it throws a light of illustration which can scarcely be valued too highly'.—P. 25.

The idea of 'a constructive and general scheme of truth', alluded to in this passage, is one of those which possess a singular fascination for all great minds, and Boole's mind seems in its composition to have flashed back to the thoughts of a former time, when he attempted to pursue Thought into her secret recesses. This appears to have induced him on revision to add note (*a*), p. 31, the most important portion of this tract. From it we gain a brief and plain statement of the germ idea of his entire system of logic:—

'All correct reasoning consists of mental processes conducted by laws which are partly dependent upon the nature of the subject of thought. Of that species of reasoning which is exemplified in algebra, the *subject* is *quantity*, the *laws* are those of the elementary conceptions of quantity, and its implied operations. Of logic the *subject* is our conceptions of *classes* of things, represented by general names; the ultimate facts are those of the above conceptions, and of the operations connected therewith. Let these two systems of thought be placed side by side, expressed, as they admit of being, in the common symbolical language of mathematics, but each with its own interpretations—each with its own laws; and together with much that is obviously common—so much, indeed, as to have fostered the idea that algebra is merely an application of logic—there will be seen to exist real differences and agreements hitherto unnoticed, but not without influence on the course of human thought. The conception of the universe in the one system will occupy the place of that of unity in the other, not through any likeness of nature, as was once supposed, but through subjection to the same formal laws. Moreover, at the root of the logical system there will be found to exist a law, founded in the nature of the conception of "class", to which the conceptions of quantity, as such, are not subject, and which explains the origin, though it does not furnish the justification, of the dualistic tendency above adverted to. I conceive it unnecessary to show that a law of the mind may produce its effect upon thought and speculation without its presence being perceived. Whatever, too, may be the weight of authority to the contrary, it is simply a fact that the ultimate laws of logic—those alone on which it is possible to construct a science of logic—are mathematical in their form and expression, although not belonging to the mathematics of quantity'.

The speculative truth of the principles of logic has often been doubted, sometimes slighted, but if it could be proved that its whole procedure partakes of the perfect character of the sister science mathematics, this would be hereafter impossible. The science of the laws of the action of thought would co-ordinate itself with the science of the necessary matter of thought, and the ultimate calculus of operations employed in each would be seen to be closely allied. Both are sciences capable of exhibition in the rigour of technical forms, and though neither can supersede the other, it is yet a most legitimate exercise of philosophic insight to trace their ground-plan, and to show how nearly their ultimate forms and processes are alike.

To this subject, however, we must recur again; meanwhile we may note, that among other affiliations of himself to the city of Cork, its interests, and institutions, he became a member of the Cuvierian Society; and when leisure came to him, he took his share, not only in the remarks on, but in the production of, papers. Of the latter we may note, and no more, the subjects of the following:—In session 1852–53, under the presidency of R.J. Lecky, he read remarks 'On some astronomical figures from a manuscript of the fourteenth century, representing the earth's motion', and a 'Memoir of Robert Grossetête, Bishop of Lincoln, who died in 1253'; and, in 1853–4, a notice of 'A remarkable echo'. In 1854 he was chosen president of this society; and under his rule a splendid series of *conversazioni*—on the occasion of the inauguration of the Cork Athenæum, which was opened by the *late* able Earl of Carlisle, in May, 1854—were held: of these the former was intended for and attended by the educated classes of the city; and the latter was specially got up for the benefit of the working classes. On opening these Dr. George Boole delivered an address on the 'Union of social enjoyment and intellectual activity', from which we quote the following excerpts:—

'Each generation as it passes away bequeaths to its successor, not only its material works in stone and marble, in brass and iron, but also the truths which it has won, and the ideas which it has learned to conceive; its art, literature, science, and, to some extent, its spirit and morality. This perpetual transmission of the light of knowledge and civilization has been compared to those torch races of antiquity, in which a lighted

> brand was transmitted from one runner to another until it reached the final goal. Thus it has been, as generations succeed each other, borrowing and conveying light, receiving the principles of knowledge, testing their truth, enlarging their application, adding to their number, and transmitting them forward to coming generations,—
>
> "Et quasi cursores vitai lampada tradunt".
>
> Now this connection between intellectual discovery and the progressive history of our race gives to every stage of the former a deep human interest. Each new revelation, whether of the laws of the physical universe, of the principles of art, or of the great truths of morals and of politics, is a step not only in the progress of knowledge, but also in the history of our species. . . . Science, while it is thus a revelation of the laws of the material universe, is also a manifestation of the intellectual nature of man; so, too, all those arts which depend upon the perception of proportion, whether it be in forms, or in sounds, are at least as dependent upon the existence of certain faculties of our nature, which faculties they make known to us, as upon any relation of external things. What a world of sweet and solemn emotions, for instance, does not music awaken within us!—a world of whose existence we should, but for that divine art, be wholly unconscious, and of whose possible limits we are still ignorant. It is not in the instrument, nor in the pulses of the air, nor in the mechanism of the human ear, that the harmony resides, but in ourselves. In the mysterious depths of the human spirit these faculties have their abode, for whose calling forth all these external movements are but a preparation'.

These are only two brief quotations from a speech full of good sense, social urbanity, and popularized thought, and which supplies a lofty idea of the fluency and fertility of the Cork professor of mathematics, when he had seen fit for a season to—

> 'Let Euclid rest, and Archimedes pause';

and how he could enliven social intercourse with 'subjects of thought and conversation, whose interest is not merely of the present time'.

The previous summer (1853) must have been an anxious and busy one for him. He had planned and partly commenced a work on the philosophy of the thinking faculty, of a singularly bold and original character. He had succeeded in thinking together logic and mathematics, and in linking together into one form of thought syllogistic reasoning and algebraic computation. He sedulously and earnestly devoted himself to the completion of this notable production. The book was printed in Dublin, and hither and thither flew the proof sheets, for long months, between author and printer, author and friends, &c., causes of great anxiety and care. On St. Andrew's day, 1853, he saw the work concluded, and wrote the preface, but the business arrangements of publishing kept the work from the hands of readers till the spring of 1854. Considering that Dr. Boole was known to many by his contributions to the *Cambridge and Dublin Mathematical Journal;* for the papers of note, in the scientific world, he had read before the British Association; and for the well-sustained position he had attained, as well as for his former productions, it is surprising that this work attracted so little attention as it did among the cultivators of either of the two great branches of formal science. It has been bitterly remarked of this incident, 'It sought audience fit though few, and found the latter'. Too many of the every-day critics, we fear, gave its pages scant survey, because they would have required to educate themselves up to the power of noticing a work of such consistency and skill, in the searching for and tracing out of ultimate principles and fundamental laws.

The author having given orders for the distribution of copies among those thinkers who were interested in the subject it discussed, and were supposed to be able to comprehend the relations of its speculations to foregone or cognate thought, the writer of this paper had the honour of receiving the book, 'with compliments'. It is only candid to confess that the treatise was felt to be one on which much study must be expended, and that it came to hand when leisure and health were both scanty. From our note-book of readings we quote the following sentences of the opinion, formed after our first perusal of the work:—'The writer displays singular sagacity, stretch and tension of reasoning, an enviable power of lucid exposition, a wonderful capacity for abstract speculation and recondite thinking. The author is profound and erudite; he has brought the far-grasping power of an able mind to the consideration of the subject, and has given in clear, distinct, and expressive language a large development to the philosophy of thought. The subtlety, power, and persistency shown in carrying out the strict and unyielding system of symbolical expression and interpretation for which the author contends are indicative of a mind of superior order. The book opens up views of the relations and forms of thought capable of vast results'.

Our opinions, after a special reperusal, prior to commencing this paper, do not substantially differ from those formed upwards of ten years ago. It is a bold and grandly conceived work. In it, world-old investigations, whose elements were regarded as irrevocably settled, are re-submitted to analysis and found to afford new results; familiar and trite assumptions are subjected to re-questioning, and what before seemed palpable as daylight revealed unnoticed facts in thought and mind. No volume has been, for many years, presented to reflective men containing so much vigorous reasoning, and so successful in gaining so very large an accession of truths relating to apparently exhausted fields of inquiry. It is not only far in advance of previous theories, but in a great measure it takes fresh ground, and proceeds according to new tactics. The work presupposes in its readers a knowledge of the most important terms of logical science, and an acquaintance with the principles of algebra, but we hope in the following outline to lay before our readers an intelligible abstract of the main theory of the treatise,—a treatise fertile in suggestiveness, and not less rich in fulfilment than in promise. The quotations, though sometimes separated by considerable intervals, are believed to be sufficiently knit together to give a concise epitome of the author's main logical ideas.

> The design of this treatise 'is to investigate the fundamental laws of those operations of the mind by which reasoning is performed; to give expression to them in the symbolic language of a calculus, and upon this foundation to establish the science of logic, and construct its method; to make that method itself the basis of a general method for the application of the mathematical doctrine of probabilities; and, finally, to collect from the various elements of truth brought to view in the course of these inquiries, some probable intimations concerning the nature and constitution of the human mind. . . . The operations of the mind are in a real sense subject to laws', and 'a science of the mind is therefore *possible*. . . . Like all other sciences, that of the intellectual operations must primarily rest upon observation. . . . Our knowledge of the laws upon which the science of the intellectual powers rests, whatever may be its extent or its deficiency, is not probable knowledge; for we do not see in the particular example the general truth, but we see it also as a certain truth—a truth our confidence in which will not continue to increase with increasing experience of its practical verifications.

'All sciences consist of general truths, but of those truths some only are primary and fundamental, others are secondary and derived. . . . Shall we err, then, in regarding that, as the true science of logic, which, laying down certain elementary laws, confirmed by the very testimony of the mind, permits us thence to deduce, by uniform processes, the entire chain of its secondary consequences, and furnishes, for its practical applications, methods of perfect generality?

'There is not only a close analogy between the operations of the mind in general reasoning, and its operations in the particular science of algebra, but there is, to a considerable extent, an exact agreement in the laws by which the two classes of operations are conducted. . . . There exist, indeed, certain general principles, founded in the very nature of language, by which the use of symbols, which are but elements of scientific language, is determined. To a certain extent these elements are arbitrary. Their interpretation is purely conventional; we are permitted to employ them in whatever sense we please. But this permission is limited by two indispensable conditions—1st, that from the sense, once conventionally established, we never, in the same process of reasoning, depart; 2nd, that the laws by which the process is conducted be founded exclusively upon the above fixed sense or meaning of the symbols employed'. The tenets of this work 'exhibit logic, in its practical aspect, as a system of processes carried on by the aid of symbols having a definite interpretation, and subject to laws founded upon that interpretation alone. But at the same time they exhibit those laws as identical in form with the laws of the general symbols of algebra, with this single addition, viz., that the symbols of logic are farther subject to a special law ['the rule of interpretation that any expression in which several of these symbols are written together shall represent all the objects or individuals to which their several meanings are together applicable; and to the law that the order in which the symbols succeed each other is indifferent'], to which the symbols of quantity, as such, are not subject. . . . This law constitutes the germ or seminal principle, of which every approximation to a general method in logic is the more or less perfect development.

'Logic is conversant with two kinds of relations—relations among things and relations among facts. But as facts are expressed by propositions, the latter species of relation may, at least for the purposes of logic, be resolved into a relation among propositions. . . . We may then say that the *premises* of any logical argument express *given* relations among certain elements, and that the *conclusion* must express an *implied* relation among those elements, or among a part of them, *i.e.*, a relation implied by or inferentially involved in the premises.

'The requirements of a general method in logic seem to be the following:—1st. As the conclusion must express a relation among the whole, or among a part of the elements involved in the premises, it is requisite that we should possess the means of eliminating those elements which we desire not to appear in the conclusion, and of determining the whole amount of relation implied by the premises among the elements which we wish to retain. Those elements which do not present themselves in the conclusion are, in the language of the common logic, called middle terms; and the species of elimination exemplified in treatises on logic consists in deducing from two propositions, containing a common element or middle term, a conclusion connecting the two remaining terms. But the problem of elimination, as contemplated in this work, possesses a much wider scope. It proposes, not merely the elimination of one middle term from two propositions, but the elimination generally of middle terms from propositions, without regard to the number of either of them, or to the nature of their connection. To this object neither the processes of logic nor those of algebra in their actual state present any strict parallel. 2nd. It should be within the province of a general method in logic to express the final relation among the elements of the conclusion by any admissible *kind* of proposition, or in any selected order of terms': *i.e.*, 'given a set of premises expressing relations among certain elements, whether things or propositions; required explicitly the whole relation consequent among *any* of those elements, under any proposed conditions, and in any proposed form'.

In reply to the assumed objection, 'that the logic of Aristotle, in its rules of syllogism and conversion, sets forth the elementary processes of which all reasoning consists, and that beyond these there is neither scope nor occasion for a general method', he remarks, '1st, that syllogism, conversion, &c., are not the ultimate processes of logic; . . . they are founded upon, and are resolvable into, ulterior and more simple processes, which constitute the real elements of method in logic': and asserts, 'Nor is it true in fact that all interference is reducible to the particular forms of syllogism and conversion, 2nd, if all inference were reducible to these two processes (and it has been maintained that it is reducible to syllogism alone), there would still exist the same necessity for a general method'.

The '*directive* function of Method constitutes its chief office and distinction. In logic, the aid of a directive method, such as a calculus alone can supply, is indispensable. . . . These considerations furnish a sufficient answer to all protests against the exhibition of logic in the form of a calculus. . . . It is the business of science, not to create laws, but to discover them. We do not originate the constitution of our own minds, greatly as it may be in our power to modify their character. And as the laws of the human intellect do not depend upon our will, so the forms of science, of which they constitute the basis, are in all essential regards independent of individual choice'.

So far the statement of the strictly logical part of the author's aim. He, however, insists that 'the subject of probabilities belongs equally to the science of number and to that of logic'. 'It is', he says, 'in recognizing the co-ordinate existence of both these elements' that 'the present treatise differs from all previous ones'. Into this department of his work it is impossible for us, within our limits, to follow the keen-sighted thinker. Indeed, we can now only faintly indicate the chief topics on which in the body of the work he employs his remarkable intellectual powers. The work consists of a system of logical processes carried on by the aid of symbols nearly analogous to those of algebra. Thus the characters x, y, z, &c., are made to represent *things as subjects of our conceptions*; the signs $+$, $-$, $\times$, are employed as marks of *those operations of mind by which our conceptions of things are combined or resolved* so as to form new conceptions involving the same elements. The sign $=$ indicates *identity*, and the laws of thought are expressed by equations. The laws of these symbols of logic are deduced from a consideration of those operations of the mind which are implied in the strict use of language as an instrument of reasoning. This idea may, perhaps, be better given in his own language:—

'That language is an instrument of human reason, and not merely a medium for the expression of thought, is a truth generally admitted. . . . The elements of which all language consists are signs or symbols. Words are signs. . . . A sign is an arbitrary mark, having a fixed interpretation, and susceptible of combination with other signs, in subjection to fixed laws dependent upon their mutual interpretation.

'All the operations of language, as an instrument of reasoning, may be conducted by a system of signs composed of the following elements, viz.:—

'1st. Literal symbols, as x, y, z, &c., representing things as subjects of our conceptions.

'2nd. Signs of operation, as $+$, $-$, $\times$, standing for those operations of the mind by which the conceptions of things are combined or resolved so as to form new conceptions involving the same elements.

'3rd. The sign of identity, $=$.

'And these symbols of logic are in their use subject to definite laws partly agreeing with and partly differing from the laws of the corresponding symbols in the science of Algebra'.

'Let us conceive, then, of an algebra in which the symbols *x, y, z*, &c., admit indifferently of the values 0 and 1, and of these values alone. The laws, the axioms, and the processes of such an algebra will be identical in their whole extent with the laws, the axioms, and the processes of an algebra of logic'. 'The respective interpretations of the symbols 0 and 1 in the system of logic are *Nothing* and *Universe*'; hence 'if *x* represent any class of objects [*e.g.*, men], then will 1 [Universe] $-x$ [men] represent the contrary of supplementary class of objects, that is, the class including all objects which are not comprehended in the class *x*',—in other words, will furnish the *dichotomous* division, 'men' and 'not men'. 'The principle of contradiction which affirms that it is impossible for any being to possess a quality, and at the same time not to possess it, is a consequence of the fundamental law of thought whose expression is $x^2 = x$'.

We are afraid that our zeal to epitomize this system is carrying us too far. One of the notable thinkers of our day strove to dissuade the writer from attempting in a periodical to interest readers in Boole's logic. 'He would be a bold, even a rash man', said he, 'who should venture to invite readers of serials to peruse in abstract the deep issues of Boole's intellect'. Were we addressing common readers we should fear we had overstepped all limits with ours; we hope it is otherwise, and that we have stimulated curiosity, not evoked tedium. If the former, let us commend such of them as possess a little algebraic culture and some mathematical taste to acquire and study this fertile book; if the latter, let us plead the human interest of Dr. Boole's life as a set-off against the weariness which may supervene upon an endeavour to condense into an abstract essence that which to many is abstruse enough when given *in extenso*. Let us farther appease their angry minds, ready to cry out 'Impractical and impossible! Who could use a system so intricate and cumbrous'? by assuring them that Dr. Boole has no desire to introduce this *x-y-z-ity* into speech or writing. 'The perfection of the method of logic', he says, 'may be chiefly valuable as an evidence of the speculative truth of its principles. To supersede the employment of common reasoning, or to subject it to the rigour of technical forms, would be the last desire of one who knows the value of that intellectual toil and warfare which imparts to the mind an athletic vigour, and teaches it to contend with difficulties and to rely upon itself in emergencies'. It is as an educational agent he proposes his scheme, as a gymnastic training that we press its study. We commonly reason by the aid of words, and in the forms of a well-constructed language. This is both an actual condition of and an important safeguard in the reasonings and discourses of common life; for thus not only every step of the connected train of thought, and every mediate result established in the course of thought, is thus brought forward to show its correspondency with experience and intelligible expression; but in doing so we seldom attend to the ulterior grounds upon which the forms alike of language and of thought have been established; and we often remain quite unconscious that any (or at least what) law underlies the processes by which premiss and conclusion are knit together. It is to force our contemplations below the mere surface of reasoning that symbolic logic is employed, and it is to prove that these symbols may be not only representative of what *is* thought, but of what *is in* thought, that George Boole has elaborated his interpretation of formal thinking by mathematical laws yielding general axioms involved in all the operations of the mind. 'The object of science, properly so called, is the knowledge of laws and relations. To be able to distinguish what is essential to this end from what is only accidentally associated with it is one of the most important conditions of scientific progress'. This is precisely what is aimed at in this treatise on the Laws of Thought.

The author, however, sees in these indications of law, which he traced out with such pertinacity, glimpses of higher truths than those with which mere logic is concerned, as may be gleaned from the following quotation:—

> 'Among those conclusions relating to the intellectual constitution which may be considered as belonging to the realm of positive knowledge we may reckon the scientific laws of thought and reasoning, which have formed the basis of the general method of this treatise, together with the principles (Chap. V.) by which their application has been determined. The resolution of the domain of thought into two spheres, distinct but co-existent (IV., XI.); the subjection of the intellectual operations within those spheres to a common system of laws (XI.); the general mathematical character of those laws and their actual expression (II., III.); the extent of their affinity with the laws of thought in the domain of number, and the point of their divergence therefrom; the dominant character of the two limiting conceptions of universe and eternity among all the subjects of thought with which logic is concerned; the relation of those conceptions to the fundamental conception of unity in the science of number,—these, with many similar results, are not to be ranked as merely probable or analogical conclusions, but are entitled to be regarded as truths of science. Whether they be termed metaphysical or not is a matter of indifference. The nature of the evidence upon which they rest, though in kind distinct, is not inferior in value to any which can be adduced in support of the general truths of physical science. Again, it is agreed that there is a certain order observable in the progress of all the exacter forms of knowledge. The study of every department of physical science begins with observation; it advances by the collation of facts to a presumptive acquaintance with their connecting laws; the validity of such presumption it tests by new experiments, so devised as to augment, if the presumption be well founded, its probability indefinitely; and finally, the law of the phenomenon having been with sufficient confidence determined, the investigation of causes, conducted by the due mixture of hypothesis and deduction, crowns the inquiry. In this advancing order of knowledge the particular faculties and laws whose nature has been considered in this work bear their part. It is evident, therefore, that if we would impartially investigate either the nature of science, or the intellectual constitution in its relation to science, no part of the two series above presented ought to be regarded as isolated. More especially ought those truths which stand in any kind of *supplemental* relation to each other to be considered in their mutual bearing and connection'.

It can scarcely be denied by any impartial inquirer that Dr. Boole has shown that there is a very singular supplemental relation between logic and mathematics—so great, indeed, as amply to justify Professor De Morgan's statement,—

> 'The connection of the formulæ of probability with those of logic in general has been most strikingly illustrated by Professor Boole, in his "Mathematical Analysis of Logic", Cambridge, 1847, 8vo.; and subsequently in his "Investigation of the Laws of Thought", London, 1854, 8vo. In these works the author has made manifest that the symbolic language of algebra, framed wholly on notions of number and quantity, is adequate, by what is certainly not an accident, to the representation of all the laws of thought'. [See Prof. A. De Morgan's 'Syllabus of Logic'.]

But 'the mathematical laws of reasoning are, strictly speaking, the laws of *right* reasoning only, and their actual transgression is a perpetually recurring phenomenon. The moral duty, therefore, of the study of logic is undeniable, while the *design* exhibited in the possibility of reaching the selfsame formal laws of thinking from inductive experience, or from mathematical excogitation, proves that right reasoning, because easily possible to, is justly demanded of, man'. Originality is the test of a great mind, but it must be the originality of good sense joined to keen thought and freshness of mind.

Interesting, however, as are the far-reaching idea suggested by the perusal of Dr. Boole's 'exceedingly subtle and able book', the most interesting point in connection with it, to our mind, is, that it is the product of a mind chiefly self-trained, and that, spurred on mainly by the in-felt energies of his own soul, he has been able to explore the whole region of logical science, and to discover the reign therein of formal laws, precisely similar in essence to those which hold sway in the mathematical sciences, and thus lay open to mankind a new and independent theory of the order and system of trustworthy reasoning.

In 1855, Dr. Boole was married to Miss M. Everest, daughter of the late Rev. T.R. Everest, rector of Wickwar, in Gloucester. This lady possesses high intellectual faculties, and has made notable scientific acquisitions. She is the niece of Dr. Ryall, Vice-President of Cork College, and of that Colonel Everest after whom the loftiest peak of the Himalayas has been named. Five daughters have been the fruit of this union, which was one of great mutual happiness. In his new family relations he found fresh spurs to exertion, and he was helpfully assisted by Mrs. Boole in all his efforts. Of his later labours we may note an important memoir, following up the subject-matter of the twenty-first chapter of 'The Laws of Thought', and treating of 'The Combination of Testimonies and of Judgment', read before the Royal Society of Edinburgh. Shortly afterwards the Society honoured itself by electing him Fellow. This honour was succeeded by the greater one of his being selected by the University of Oxford as a fit and proper person upon whom to confer the honorary degree of D.C.L.

In 1857 he had projected and in part prepared a work, which was to bear the title of 'The Philosophy of Logic', which Messrs. MacMillan and Co., of Cambridge and London, were authorized to announce as 'nearly ready'. He was, however, so ill to satisfy with himself that he shrunk from laying the work before the public at that time; and we understand that he had laboured long and diligently upon the matter it contains, and that he had renewed arrangements for putting it in press. When the Queen's Colleges were united, so as to form the Queen's University of Ireland, he was appointed one of the public examiners for degrees, thus giving a sort of national recognition of his merit and worth. In 1860 he published another very admirable mathematical treatise, on 'The Calculus of Finite Differences', just issued under the able editorship of I. Todhunter, himself an admirable instructor in mathematics.

In 1862 he added to the 'Transactions of the Royal Society of Edinburgh' a paper which he had read on the 19th June, 'On the Theory of Probabilities'. His work on 'Differential Equations' had acquired such authority on the subject that a re-issue was called for. To accomplish this he set himself gladly, but the singular originality of his mind was such that this treatise grew into quite a new book, with developments stretching into farther reaches of the science. He gave himself thoroughly to whatever he had in hand, and the latest vacation season he enjoyed he occupied a portion of it in researches in London tending to extend or complete his work. He was looked upon in all scientific circles with hope, and his repute had widened and greatened so much that the honours of the French Academy were about to be proposed for him, when the unresting hand of death struck him to the heart and called him hence. The work on equations is announced now as republished.

The lesson of Boole's life for us is of high value. It proves, by another important example, that the right method of action is that which is determined by duty; that the true manliness is that of conscientious conviction; that self-sacrifice is the real heroism of humanity; and that self-culture is fitted to attain the greatest heights, if steadily persevered in, not for mere worldly ambition, but from a determination to work out the divine purpose implied in the possession of the powers of thought. We hold it to be of high importance in the midst of material aspirations and secularizing influences to be convinced that persistence in duty, however apparently adverse it may seem to self-enjoyment, will ultimately and unmistakeably lead to a happiness and an elevation of character which self indulgence never reaches. Dr. Boole's life never rose above the need—and he never fell below the love—of work. Mind-sweat earned him all that he had, and effort made him all that he was. Let us be faithful in our several life-paths to the duties which the Giver of life imposes on us. Be it ours to work His will; let us leave with Him the results, fearlessly, 'either in this life or that which is to come'.

In his anxiety to compass his college-work, and yet supply the much-wanted work on which he was employed, Boole overtaxed his strength. When, therefore, fever attacked him, he had but a scanty stock of strength to resist the effects of the derangement of functions of which his disease was the result. Hence his illness ran a rapid course; and after only a short intermission of labour this hard-wrought toiler was taken to his rest, and died Dec. 8, 1864, having but shortly before entered the fiftieth year of his age. Immediately on his demise being made known a general sorrow was expressed. He was carried to his resting-place amid the regrets of men of all parties and sects, the sadness of devotedly loving students, and the ardent grief of friends. His fine Christian enthusiasm imparted to all he did a sort of radiance from heaven. Without obtruding his religious views upon any one, he had the happy tact of showing that a spirit suffused his entire nature which was not of this world. A rare tolerance and magnanimity, and an almost rarer humility, combined with steadfastness of principle and conscientious seriousness in regard to every matter of duty, were always manifest in him. His sense of responsibility to the great Father of Lights, his lowly reliance on the merits of Jesus, his sanctified use of the gifts of intellect, might be noted, but cannot be enlarged upon. These affairs concern the inner sanctuary of home life and the individual conscience, and are not on slight occasion to be brought out into the garish light of common day. But it seems right in our times to mention that in him there co-existed the hardiest independence and the bravest originality of scientific thought with the most composed and serene Christian faith and the lost elevated practical piety. From the word of the living God he drew much of the wisdom and worth which gave a genuine nobility to his life and a solemn grandeur to his death. It is seldom, so far as we know, that there passes from the jostled pathways of professional life a man so justly admired, so deservedly loved, as Dr. George Boole.

He has left behind him a large quantity of manuscripts on various subjects of scientific inquiry, which had occupied his mind during his later years. These, we believe, are in the hands of literary friends, and some of them may yet—indeed, we trust, will—be issued as a memorial of their self-raised and justly celebrated author.

Other memorials, too, are about to be instituted, so that the life, works, and genius of George Boole may not pass from the recollections of men. In Cork, the scene of his later labours, whose cemetery is enriched by his dust, a monument over his early grave is likely to be erected; a mortuary tablet will preserve the remembrance of the earliest mathematical professor in his old class-room in Cork college; and, that the influences of his life and worth may not be lost, a scholarship bearing his name is about to be established in Cork, which shall combine the two useful purposes of preserving the renown of a deservedly notable man and of stimulating and rewarding succeeding students of the formal sciences. His own city and county, Lincoln, will honour themselves by placing in their splendid cathedral a memorial window worthy of themselves, the building, and the dead illustrator of science, thought, and life, who has added a glory to the land we live in. We are happy to note that Government, rightly appreciating its duty in this instance, has conferred on Mrs. Boole a civil list pension of £100 per annum, 'in consideration of her late husband's distinguished attainments as an original mathematician, and of his remarkable labours towards the extension of the boundaries of science'. This is as it should be. The memories of her great men are a nation's best treasures. They prove her glory in the past, they encourage the children of the future to strive, too, to be worthy of their land, of themselves, and of the Mighty Life-Giver.

EDITOR'S NOTE

This article first appeared in *The British Controversialist and Literary Magazine*, no 80 n.s. pp. 81–94 (August 1865) and no 81 n.s. pp. 161–174 (September 1865), in a series entitled 'Modern Logicians'. Professor Ivor Grattan-Guinness has kindly provided the following background information: Samuel Neil (1825–1901) was a Scottish Dissenter who was involved with the *British Controversialist* from its start in 1850 and edited it for most of the time till it closed in 1872 after 6 + 6 + 28 volumes. The title of the journal referred to its practise of posing some question, maybe religious, philosophical, social, even military, and publishing 'affirmative' and 'negative answers' to it together. Very keen on logic, he wrote a series on 'the art of reasoning' in 1850–1852, which was expanded into a book with historical preface in 1853 (London: Walton & Maberley). A decade later he started an extensive series of biographical and expository articles on 'Modern Logicians', often rather gushing but doubtless useful to his public. The one on Boole, reprinted here, is by far the most significant; other subjects included Mill, Kant, Whewell, Thomson, Whately, Hegel and Hamilton. (Another series on 'Toiling Upward' included the American astronomer and mathematician Nathaniel Bowditch in its roster.)

GEORGE PAXTON YOUNG

REMARKS ON PROFESSOR BOOLE'S MATHEMATICAL THEORY OF THE LAWS OF THOUGHT

In a recent issue we announced the death of Professor George Boole, of Queen's College, Cork, a man of varied and profound acquirements, and of singular originality of mind. The work on which his fame will mainly rest is undoubtedly his 'Investigation of the Laws of Thought, on which are founded the Mathematical Theories of Logic and Probabilities'. We have long purposed to call attention to this remarkable production, though various circumstances have hitherto prevented us from doing so. The present seems a suitable occasion for testifying our admiration of the genius of the deceased philosopher, and, at the same time, endeavouring to give a brief account, inadequate as it must necessarily be, of what may be termed his Mathematico-logical speculations.

The primary, though not the exclusive, design of the 'Investigation', is to express in the symbolical language of a Calculus, the fundamental Laws of Thought, and upon this foundation to establish the science of Logic and construct its method.

The elementary symbols of Professor Boole's Calculus are of three kinds: 1st. Literal symbols, as x, y, &c., representing the objects of our conceptions; 2nd. Signs of operation, as $+$, $-$, $\times$; and 3rd, the sign of identity, $=$. The sign $+$ is used to express the mental operation by which parts (of extensive quantity) are collected into a whole. For instance, if x represent *animals,* and y *vegetables,* $x + y$ will represent the class made up of *animals and vegetables together.* On the other hand, the sign $-$ is used to express the mental operation of separating a whole (of extensive quantity) into its parts. Thus, x representing *human beings,* and y representing *negroes,* $x - y$ will represent *all human beings except negroes.* With regard to the sign $\times$, $x \times y$ or xy (as it may be written) is used to denote those objects which belong at once to the class x and to the class y; just as, in common language, the expression *dark waters* denotes those objects which are at once *dark* and *waters.* Hence we obtain a method of representing a concept taken particularly. For, if x denote *men*, then, since *some men* may be viewed as those who besides belonging to the class x belong also to some other class v, *some men* will be denoted by vx. In general,

$$vx = \text{some } x \tag{1}$$

It can easily be shown, that, as in Algebra, so in the logical system which we are describing, the literal symbols, x, y, &c., are commutative; that is,

$$xy = yx; \tag{2}$$

J. Gasser (ed.), A Boole Anthology, 27–43.

and that they are also distributive; that is,

$$z(x \pm y) = zx \pm zy \tag{3}$$

Another relation between Algebra and the Logical System under consideration is, that, in the latter as well as in the former, a literal symbol may be transposed from one side of an equation to the other by changing the sign of operation, + or –. But there is an important relation which subsists in the science of Thought, and not generally in Algebra, namely,

$$x^2 = x \tag{4}$$

That this is true in the Logical system, is plain; for x^2, which is another form of xx, denotes (by definition) those things which belong at once to the class x and to the class x; that is, it denotes simply those things which belong to the class x; and it is therefore identical with x. But though the equation (4) does not generally subsist in Algebra, it subsists when x is unity or zero. If, therefore, we take the science of Algebra with the limitation that its unknown quantities can receive no values distinct from unity and zero, the analogy between the two sciences will still be preserved.

It is necessary to observe that unity and zero (1 and 0) are virtually included by Professor Boole among his literal symbols. Of course we can give 1 and 0 any meaning we please, provided the meaning once imposed on them be rigidly adhered to. By 0, then, Professor Boole understands Nothing—a class (if the expression may be permitted) in which no object whatever is found. On the other hand, by 1 he understands the universe of conceivable objects. Thus 1 and 0 are at two opposite poles; the former including every thing in its extension; the latter, nothing. The meaning which has been affixed to 1 and 0 preserves, in the Logical system as in Algebra, the equations,

$$\left.\begin{aligned} 1 \times x &= x \\ \text{and, } 0 \times x &= 0; \end{aligned}\right\} \tag{5}$$

for, the meaning of the former is, that objects which are common to the universe and to the class x are identical with those which constitute the class x; and the latter means, that there are no objects which are common to a class in which nothing is found and to a class x: both of which propositions are self-evident. From the meaning affixed to 1, we see what the meaning of $1 - x$ must be. In fact, x and $1 - x$ are logical contradictories; the latter denoting all conceivable objects except those which belong to the former; so that

$$1 - x = \text{not } x \tag{6}$$

This value of the symbol 1 being admitted, we can, by the principles of transposition and distribution [see (3)] reduce equation (4) to the form,

$$x(1 - x) = 0 \tag{7}$$

The law here expressed, which is termed the Law of Duality, plays a most important part in the development of logical functions, and in the elimination of symbols. In fact, it may be described as the germ out of which Professor Boole's whole system is made to unfold itself.

Having shown how concepts, whether taken universally or particularly, are represented, and also how the contradictory of a concept is represented, we have next to notice the manner of expressing judgments. All judgments are regarded by our author as affirmative; the negation, in those which are commonly called negative, being attached by him to the predicate. But an affirmative judgment is nothing else than an assertion, through immediate comparison, of the identity of concepts. Suppose, therefore, that we are required to express the judgment, 'Some stones are precious'. Let x denote *stones*; and y, *precious*. The proposition means, that some stones are identical with some precious things. Consequently, its symbolical expression [see (1)] is,

$$vx = vy.$$

If the judgment to be represented had been, 'Some stones are not precious', its expression would [see (6)] have been

$$vx = v(1 - y).$$

These examples in the meantime may suffice. More complicated forms will present themselves afterwards.

With the few simple preliminary explanations which have been given, and which were necessary to render intelligible some of the criticisms presently to be offered, we are now prepared to state the view which our author takes of the science of Logic. Logic he regards as the science of Inference; and the problem which it seeks to solve is this: Given certain relations among any number of concepts (x, y, z, &c.), it is required to find what inferences can be drawn regarding any one of these or regarding a given function of any one of them. A properly constructed science of Logic would require to solve this problem adequately, and by a definite and invariable method. Now, Professor Boole claims that the view which he presents of the problem which Logic has to solve, is both deeper and broader than that commonly taken; and he claims at the same time that he has devised an adequate method, different from all existing methods, for solving this problem, and that his method is one of definite and invariable application.

The objections brought against the logic of the schools, that it is neither sufficiently deep nor sufficiently broad, will probably take our readers by surprise. It is not difficult to understand how a question might be raised as to the practical utility of the scholastic logic; but most persons who have examined the subject will be ready to admit, both that the scholastic logic is well founded, and that, when properly developed from its first principles, it forms a complete and perfect system. In the opinion of our author, however, it is so defective in its foundation, and so incomplete in its superstructure, as not to be entitled to the name of a science. 'To what final conclusions', he says, 'are we then led respecting the nature and extent of the scholastic logic? I think to the following: that it is not a science, but a collection of scientific truths, too incomplete to form a system of themselves, and not sufficiently fundamental to serve as the foundation upon which a perfect system may rest'.

In order that it may be understood in what sense it is held that *the foundation of the scholastic logic is defective,* we make two other quotations. 'That which may be regarded as essential in the spirit and procedure of the Aristotelian, and of all cognate systems of logic, is the attempted classification of the allowable forms of inference, and the distinct reference of those forms, collectively or individually, to some general principle of an axiomatic nature, such as the Dictum of Aristotle'. Again: 'Aristotle's Dictum de omni et nullo is a self-evident principle, but it is not found among those ultimate laws of the reasoning faculty to which all other laws, however plain and self-evident, admit of being traced, and from which they may in strictest order of scientific evolution be deduced. For though of every science the fundamental truths are usually the most simple of apprehension, yet is not that simplicity the criterion by which their title to be regarded as fundamental must be judged. This must be sought for in the nature and extent of the structure which they are capable of supporting. Taking this view, Leibnitz appears to me to have judged correctly when he assigned to the principle of contradiction a fundamental place in logic; for we have seen the consequences of that law of thought of which it is the axiomatic expression'. The sum of what is contained in these passages, in so far as they bear on the point before us, is, 1st, that the foundation of the Aristotelian, and of all cognate systems of logic, is some such canon as the Dictum; 2nd, That that canon, and other maxims of a like description, though self-evident, are not deep enough to serve as a basis for a science of logic in which all the forms of thought are to be exhibited; and, 3rd, That the only principle sufficiently fundamental to form the basis of a complete science of logic is the principle of contradiction. Now what is the real state of the case? Nothing is more certain than that the Dictum was not considered by Aristotle as either the exclusive or the ultimate foundation of his logical system. Not the exclusive foundation; for, as a matter of fact, many of the forms of thought embraced in the Aristotelian logic receive no direct warrant from the Dictum, but can be derived from it only by the aid of the principle of contradiction. Not the ultimate foundation; for what is the Dictum, but a particular case of a more comprehensive, and (in this sense) more fundamental, law? Aristotle saw this, and has expressed it as clearly as any man that ever lived. 'It is manifest', he says, 'that no one can conceive to himself that the same thing can at once be and not be, for thus he would hold repugnant opinions, and subvert the reality of truth. Wherefore, all who attempt to demonstrate, reduce everything to this as the ultimate doctrine; for this is by nature the principle of all other axioms'.

Professor Boole's acceptance of the Leibnitzian maxim (though it was much older than Leibnitz) that the true foundation of the science of logic is the principle of contradiction, has the appearance of being at variance with some extraordinary statements which he elsewhere makes, to the effect that the principle of contradiction is a consequence of the law of duality. We may remind our readers that the law of duality [see (4) and (7)] is substantially the principle out of which all the details of Professor Boole's own doctrine are evolved. Now, under the influence of what was, perhaps, not an unnatural desire to vindicate for his system a peculiar depth of foundation, Professor Boole has been betrayed into observations by which his fame as a philosophic thinker must be seriously affected. For instance: 'that axiom of metaphysicians which is termed the principle of contradiction, and which affirms that it is impossible for any being to possess a quality and at the same time not to possess it, is a consequence of the fundamental law of thought, whose expression is $x^2 = x$'. And again: 'the above interpretation has been introduced, not on account of its immediate value in the present system, but as an illustration of a significant fact in the philosophy of the intellectual powers, viz., that what has commonly been regarded as the fundamental axiom of metaphysics is but the consequence of a law of thought, mathematical in its form'. In thus speaking of the principle of contradiction as a consequence of the law of duality, Professor Boole seems to take away the fundamental character of the principle of contradiction; for, if that principle be, in the proper sense of the term, a consequence of something else, it cannot be itself truly fundamental. Yet, as we have seen, Professor Boole admits that it is the real and deepest foundation of the science of logic. What, then, does he mean? On the one hand, he certainly does not intend to deny that the principle of contradiction is self-evident. On the other hand, it is plain that he does hold that the principle of contradiction can be deduced from the law of duality. But (we ask) how? Can the principle of contradiction be deduced from the law of duality, without our assuming the principle of contradiction itself as the basis of the deduction? This would be absurd; for a conclusion can be established in no other way than by pointing out that the supposition of its being false involves a contradiction. In the particular case before us, the equation $x(1 - x) = 0$, which is that expression of the law of duality in which the principle of contradiction is regarded as being brought to light, is only reached by a process of reasoning, every step of which takes the principle of contradiction for granted. The only interpretation, therefore, which Professor Boole's words can bear, unless we give them a meaning palpably absurd, is, that a formula, which we are enabled to state by assuming the law of contradiction, contains a symbolic representation of that law. This hardly seems to us a very significant fact in the philosophy of the intellectual powers. If indeed the formula in question could be shown to represent some law of thought of wider application than the law of contradiction, that would be a very significant fact. But such is not the case. The equation $x(1 - x) = 0$ is just the law of contradiction symbolically expressed: neither more nor less.

The Aristotelian logic is charged with being *incomplete,* as well as with being not sufficiently fundamental. By this our author does not mean that Aristotle and his followers have casually omitted some forms of thought which their system ought to have embraced: had they done so, the fault would have been chargeable—not upon the system, but upon its expounders; but he means, that, from the very nature of the system, there is an indefinite variety of problems belonging to the science of inference, which their system is incapable of solving, or for the solution of which at all events it furnishes no definite and certain method.

It will be observed that there are two questions here, which, as radically distinct from one another, require to be considered separately: the one being, whether the Aristotelian logic is capable of solving all the problems belonging to the science of inference; and the other, whether it furnishes a definite and certain method for the solution of these.

The former of these questions may, with perfect confidence, be answered in the affirmative. It admits of absolute demonstration, that there is no chain of valid inference which the ordinary logic is incompetent to express, or, in other words, which is not reducible to conversion or syllogism. Some logicians have been of opinion that conversion is nothing else than syllogism at bottom; but, for what we have at present in view, it is unnecessary to discuss this question. Suffice it to say, that, whether conversion and syllogism be substantially identical or not, all immediate inference is of the nature of conversion, and all mediate inference (or reasoning proper) of the nature of syllogism. Does Professor Boole deny this? Formally, and in plain terms. 'Possibly', he writes, 'it may here be said that the logic of Aristotle, in its rules of syllogism and conversion, sets forth the elementary processes of which all reasoning consists, and that beyond these there is neither scope nor occasion for a general method. I have no desire to point out the defects of the common logic, nor do I wish to refer to it any further than is necessary, in order to place in its true light the nature of the present treatise. With this end alone in view, I would remark: 1st. That syllogism, conversion, &c., are not the ultimate processes of logic. It will be shown in this treatise that they are founded upon, and are resolvable into, ulterior and more simple processes which constitute the real elements of method in logic. Nor is it true that all inference is reducible to the particular forms of syllogism and conversion. 2nd. If all inference were reducible to these processes alone (and it has been maintained that it is reducible to syllogism alone), there would still exist, &c.'. In illustration of the statement, that some inference is not reducible to the forms of syllogism and conversion, Professor Boole examines the case of conversion, and arrives at the result that 'conversion is a particular application of a much more general process in logic, of which', he adds, 'many examples have been given in this work'. In like manner he examines the case of syllogism; and his conclusion is as follows: 'Here, then, we have the means of definitely resolving the question, whether syllogism is indeed the fundamental type of reasoning,—whether the study of its laws is co-extensive with the study of deductive logic. For if it be so, some indication of the fact must be given in the system of equations upon the analysis of which we have been engaged. No sign, however, appears that the discussion of all systems of equations expressing propositions is involved in that of the particular system examined in this chapter. And yet writers on logic have been all but unanimous in their assertion, not merely of the supremacy, but of the universal sufficiency of syllogistic inference in deductive reasoning'. These statements, that conversion and syllogism are branches of a much more general process, have of course no meaning except on the supposition that the 'much more general process' is not reducible to conversion and syllogism. If reducible to these, it would not be a more general process. Now we take our stand firmly on the position, that a chain of valid reasoning, which cannot be broken into parts, every one of which shall be an instance either of conversion or of syllogism, is not possible. We are prepared to show this in the case of every one of the examples of his 'more general process' which Professor Boole gives in his work. Nay, we go farther, and as was intimated above, hold it to be absolutely demonstrable, that, from the nature of the case, inference cannot be of any other description than conversion or syllogism.

To make this out, let it be remarked that the conclusion of an argument exhibits a relation between two terms, say *X* and *Y*. It is an important assumption in Professor Boole's doctrine, that a proposition may exhibit a relation between many terms. This is not exactly true. A proposition may involve a relation between a variety of terms implicitly; but explicitly exhibits a relation only between two. Take, for instance, the proposition—'Men who do not possess courage and practise self-denial are not heroes'. Here, on Professor Boole's method, a variety of concepts are supposed to be before the mind, as, *men, those who practise self-denial, those who possess courage,* and *heroes*. But in reality, when we form the judgment expressed in the proposition given, the separate concepts, *men, those who practise self-denial, those who possess courage*, are not before the mind; but simply the two concepts, *men who do not possess courage and practise self-denial,* and *heroes.* What is a judgment but an act of comparison? And the comparison is essentially a comparison of two concepts, each of which may no doubt involve in its expression a plurality of concepts, but these necessarily bound together by the comparing mind into a unity. Now, if the conclusion of an argument exhibits a relation between two terms X and Y, this conclusion must be drawn (what other way is possible?) either through an immediate comparison of X and Y with one another, or by a mediate comparison of them through something else. If it be drawn by an immediate comparison of X and Y, then no concepts enter into the argument except X and Y, and the argument is reduced to conversion. But if the conclusion be drawn mediately, it must be by the comparison of X and Y with some third thing: not with a plurality of other things, but with some single thing. Here we have the mind drawing its inference in a syllogism. What the various admissible forms of conversion and syllogism may be, or whether these forms have been correctly specified by particular eminent logicians, are minor questions. The essential thing in a philosophical respect is, that the mind, in the inferences which it draws, does and can work in no other moulds that those described. All this seems to us so plain that we confess ourselves utterly puzzled to comprehend how men of profound and original genius have been beguiled into an assertion of the contrary.

Professor Boole himself, in summing up his assault on the Aristotelian Logic, comes very near admitting what we contend for. 'As Syllogism', he says, 'is a species of elimination, the question before us manifestly resolves itself into the two following ones: 1st. Whether all elimination is reducible to Syllogism; 2nd. Whether deductive reasoning can, with propriety, be regarded as consisting only of elimination. I believe, upon careful examination, the true answer to the former question to be, that it is always theoretically possible so to resolve and combine propositions that elimination may subsequently be effected by the syllogistic canons, but that the process of reduction would in many instances be constrained und unnatural, and would involve operations which are not syllogistic. To the second question I reply, that reasoning cannot, except by an arbitrary restriction of its meaning, be confined to the process of elimination'. With regard to this second question, we merely note in passing, that we have proved in the preceding paragraph that inference, where not immediate or of the nature of conversion, can be nothing else than elimination. It is, however, with the first question, whether elimination is reducible to syllogism, that we have now more particularly to do; and we accept with satisfaction the admission, guarded and (to some extent) neutralised as it is, that every line of argument may be thrown into a form in which the eliminations that take place are effected by the syllogistic canons. It is quite irrelevant to notice, as Professor Boole does, that the process of reduction would, in many instances, be constrained and unnatural; for we are not here in the province of Rhetoric. Much more to the purpose is the charge, that the process of reduction would involve operations which are not syllogistic. The operations referred to are those embraced in the 'much more general process' in which, as we have seen, our Author holds conversion and syllogism to be contained. Of course, the ground which we take in reply is, on the one hand, to challenge the production of an instance of valid inference, which cannot be reduced to either conversion or syllogism; and on the other hand, to fall back upon the demonstration which we have given of the absolute impossibility of valid inference being anything else than conversion or syllogism.

In stating the charge of incompleteness brought by our Author against the Aristotelian system, we explained his meaning to be, that, from the very nature of the system, there is an indefinite variety of problems belonging to the science of inference, which the system is incapable of solving, or for the solution of which, at all events, it furnishes no definite and certain method. We have, we trust, fully refuted the opinion that there are problems in the science of inference which the Aristotelian logic is incapable of solving. But Professor Boole urges, that, even if all inference were reducible to conversion and syllogism, 'there would still exist the same necessity for a general method. For it would still be requisite to determine in what order the processes should succeed each other, as well as their particular nature, in order that the desired relation should be obtained. By the desired relation I mean that full relation which, in virtue of the premises, connects any elements selected out of the premises at will, and which, moreover, expresses that relation in any desired form and order. If we may judge from the mathematical sciences, which are the most perfect examples of method known, this directive function of method constitutes its chief office and distinction. The fundamental processes of arithmetic, for instance, are in themselves but the elements of a possible science. To assign their nature is the first business of its method, but to arrange their succession is its subsequent and higher function. In the more complex examples of logical deduction, and especially in those which form a basis for the solution of difficult questions in the theory of probabilities, the aid of a directive method, such as a Calculus alone can supply, is indispensable'.

Now, we at once admit that the Aristotelian logic neither has, nor professes to have, any such method as that here described. But can it justly, on that account, be charged with incompleteness? A science must not, because it does not teach everything, be therefore reckoned incomplete: enough, if it teaches the whole of its own proper circle of truths. The special question which the scholastic logic proposes to itself is: what are the ultimate abstract forms according to which all the exercises of the discursive faculty proceed? The science is complete, because it furnishes a perfect answer to this question.

But, it may be said, is it not desirable to have a method enabling us certainly to determine, in every case, the relation which any of the concepts explicitly or implicitly entering into a group of premises bear to the others? Most desirable. And herein consists the real value of Professor Boole's labours. He has devised a brilliantly original Calculus by which he can, through processes as definite as those which the Algebraist applies to a system of equations, solve the most complicated problems in the science of inference—problems which, without the aid of some such Calculus, persons most thoroughly versed in the ordinary logic might have no idea how to treat. In expressing our dissent, as we have been obliged very strongly to do, from much that is contained in Professor Boole's treatise, we have no desire to rob that eminent writer of the credit justly belonging to him. Our wish has been simply to separate the chaff from the wheat, and to point out accurately what constitutes, as far as the 'Investigation' is concerned, Professor Boole's claim to renown.

Our readers will, however, be now anxious to obtain some fuller information regarding the method about which so much has been said, and which is the same with 'the more general process' under which the processes of the scholastic logic are held by Professor Boole to be comprehended. This part of our article must necessarily be altogether technical; and we shall require to ask our readers to take a few things on trust; but we hope to be able to present the subject in such a manner as to give at least some idea of the system we are to endeavour to describe. Those who desire to become thoroughly acquainted with it will of course study the 'Investigation' for themselves.

We begin by referring to the development of logical functions. An expression which in any manner involves the concept x, is called a function of the concept, and is written $f(x)$. Now there is one standard form to which functions of every kind may be reduced. This form is not an arbitrary one, but is determined by the circumstance that every conceivable object must rank under one or other of the two contradictory classes x and $1 - x$. Hence every conceivable object is included in the expression,

$$ux + v(1 - x); \tag{8}$$

proper values being given to u and v. For, if a given concept belong to the class x, then, by making $v = 0$, the expression (8) becomes ux, which, by (1), means *some* x; and if the given concept belong to the class $1 - x$, then, by making $v = 0$, the expression (8) becomes $v(1 - x)$, which, by (1) and (6), means *some not* x. Therefore $f(x)$ being any concept depending on x, we may put

$$f(x) = f(1) + f(0)(1 - x) \tag{9}$$

It has been shown that one of the coefficients, u, v, must always be zero; but the forms of these coefficients may be determined more definitely. For, by making $x = 0$ in (9), the result is $v = f(0)$; and by making $x = 1$, there results $u = f(1)$; by substituting which values of u and v in (9), we get

$$f(x) = f(1)x + f(0)(1 - x) \tag{10}$$

This is the expansion or development of the function x. The expressions x, $1 - x$, are called the constituents of the expansion; and $f(1)$ and $f(0)$ are termed the coefficients. The same phraseology is employed when a function of two or more symbols is developed.

Any one in the least degree acquainted with mathematical processes will understand how the development of functions of two or more symbols can be derived from equation (10). In fact, by (10), we have

$$f(x, y) = f(1, y)x + f(0, y)(1-x).$$

But again, by (10),

$$f(1, y) = f(1, 1)y + f(1, 0)(1 - y),$$

and

$$f(0, y) = f(0, 1)y + f(0, 0)(1 - y).$$

$$\therefore f(x, y) = f(1, 1)xy + f(1, 0)x(1 - y)$$

$$+ f(0, 1)y(1 - x) + f(0, 0)(1 - x)(1 - y) \qquad (11)$$

The development of a function of three symbols may be written down, as we shall have occasion in the sequel to refer to it:

$$f(x, y, z) = f(1, 1, 1)xyz + f(1, 1, 0)xy(1 - z)$$

$$+ f(1, 0, 1)xz(1 - y) + f(1, 0, 0)x(1 - y)(1 - z)$$

$$+ f(0, 1, 1)yz(1 - x) + f(0, 1, 0)y(1 - x)(1 - z)$$

$$+ f(0, 0, 1)z(1 - x)(1 - y)$$

$$+ f(0, 0, 0)(1 - x)(1 - y)(1 - z) \qquad (12)$$

As the object of the expansion of logical symbols may not be evident at first sight, and as the process may consequently be regarded by some as barbarous, we may observe that not only is there a definite aim in the developmeut, but the thing aimed at, has, in our opinion, been most felicitously accomplished. Of this our readers will probably be satisfied when they are introduced to some specimens of the use which is made of the formulæ obtained; in the meantime it may throw some light on the character of these formulæ if we notice that the constituents of an expansion represent the several exclusive divisions of what our author terms the universe of discourse, formed by the predication and denial in every possible way of the qualities denoted by the literal symbols. In the simplest case, that in which the function is one of a single concept, it will be seen by a glance at (10) that there are only two such possible ways, x and $1 - x$. In the case of a function of two symbols, there are [see (11)] four such ways, xy, $x(1 - y)$, $y(1 - x)$, $(1 - x)(1 - y)$. In a function of three symbols there are eight such ways; and so on. A development in which the constituents are of this kind prepares the way for ascertaining all the possible conclusions, in the way either of affirmation or denial, that can be deduced, regarding any concept, from any given relations between it and the other concepts.

If S be the sum of the constituents of an expansion, and P the product of any two of them, then

$$S = 1, \qquad (13)$$

$$\text{and } P = 0. \qquad (14)$$

The truth of these beautiful and important propositions will easily be gathered by an intelligent reader from an inspection of the formulæ, (10), (11), (12). Another important proposition is involved in (14), namely, that, if $f(x) = 0$, either the constituent or the coefficient in every term of the expansion of $f(x)$ must be zero. For, let

$$f(x) = Q + AX + A_1X_1 + \dots\dots\dots + A_nX_n;$$

where A, A_1, &c., are the coefficients which are not zero, their corresponding constituents being X, X_1, &c.; while Q represents the sum of those terms in which the coefficients are zero. Then we say that

$$X = 0 \qquad (15)$$

For, since $Q = 0$, and $f(x)$ is supposed to vanish,

$$AX + A_1X_1 + \&c. = 0$$

$$\therefore AX^2 + A_1XX_1 + \&c. = 0$$

But, by (14), $XX_1 = XX_2 = \ldots\ldots\ldots = XX_n = 0$. Therefore

$$AX^2 = 0.$$

But A is not zero. Therefore X must be zero.

These principles having been laid down, our best course will probably now be to take a few examples, and to offer in connection with them such explanations as may seem necessary of the mode of procedure which they are intended to illustrate.

Our first example shall be one in which but a single proposition is given: 'clean beasts are those which both divide the hoof and chew the cud'. Let

x = clean beasts,

y = beasts dividing the hoof,

z = beasts chewing the cud.

Then, the given proposition, symbolically expressed, is,

$$x = yz,$$

or, by transposition,

$$x - yz = 0 \qquad (16).$$

This premiss contains a relation between three concepts; and, according to Professor Boole, a properly constructed science of inference should enable us, by some defined process, to show what consequence, as respects any one of these, follows from the premiss. Now, the definite and invariable process which Professor Boole applies, with the design which has been indicated, to an equation such as (16), is to develop the first member of the equation. Writing, then,

$$f(x, y, z) = x - yz,$$

we have, $f(1, 1, 1) = 0,$

$$f(0, 0, 0) = 0,$$

and so on. Hence [see (12)] the development required is

$$\begin{aligned} x - yz = {} & xy(1-z) + xz(1-y) \\ & + x(1-y)(1-z) - yz(1-x) \\ & + 0xyz + 0y(1-x)(1-z) \\ & + 0z(1-x)(1-y) \\ & + 0(1-x)(1-y)(1-z). \end{aligned}$$

Therefore, by (16),

$$xy(1-z) + xz(1-y) + x(1-y)(1-z) - yz(1-x) = 0:$$

and therefore, by (15),

$$\left.\begin{aligned} xy(1-z) &= 0, \\ xz(1-y) &= 0, \\ x(1-y)(1-z) &= 0, \\ yz(1-x) &= 0. \end{aligned}\right\} \tag{17}$$

Still farther, since, by (13), the sum of the constituents of an expansion is unity; and since four of the constituents in the expansion of $x - yz$ have been shewn to be zero; it follows that the sum of the remaining constituents in the expansion of $x - yz$ is unity.
That is,

$$xyz + y(1-x)(1-z) + z(1-x)(1-y) + (1-x)(1-y)(1-z) = 1. \tag{18}$$

It is obvious that this method can be applied in every case. To what then does it lead? First of all, in the group of equations (17), we have brought before us all the different classes (if the expression may be permitted) to which the given proposition warrants us in saying that nothing can belong; and next, in equation (18) we have brought before us those different classes to one or other of which the given proposition warrants us in asserting that everything must belong. For instance, the first of equations (17) denies the existence of beasts which are clean (x) and divide the hoof (y) but do not chew the cud ($1 - z$); the second denies the existence of beasts which are clean (x) and chew the cud (z) but do not divide the hoof ($1 - y$); and so on. Equation (18), again, informs us that the universe, which is represented by 1, is made up of four classes, in one or other of which therefore every thing must rank; the first denoted by xyz, the second by $y(1 - x)(1 - z)$; and so on. As an example of the interpretation of the expression by which these classes are denoted, we may take the last, $(1 - x)(1 - y)(1 - z)$. This represents things which are neither clean beasts, nor beasts chewing the cud, nor beasts dividing the hoof.

By the method employed, we have been able to indicate certain classes which do not exist, and also to indicate certain classes in one or other of which every thing existing is found. But this, it may be said, is not a solution of the most general problem of inference. The most general problem is: to express (speaking mathematically) any one of the symbols entering into the given premiss, or any function thereof, as an explicit function of the others. To the problem as put even thus in its widest generality, Professor Boole's processes extend. It would make our article too lengthened were we to go into minute details; but we must endeavour to give some idea of the course here followed, as it both is extremely interesting as a matter of pure speculation, and forms an important part of the system under consideration.

Take the equation in (16), $x - yz = 0$; and, as a simple instance will serve the purpose of illustration as well as a complicated one, let the inquiry be: how can z be expressed in terms of x and y? In ordinary Algebra we should have

$$z = \frac{x}{y} \tag{19}$$

But though both sides of an equation may, in Logic as in Algebra, be multiplied (so to speak) by the same quantity, they cannot, in Logic, be legitimately divided by the same quantity. For instance, let the objects common to the class X and to the class U be identical with those common to the class Y and to the class U; in other words, let

$$UX = UY;$$

it does not follow that X is identical with Y, or symbolically, that

$$X = Y.$$

Hence equation (19) could not, in Logic, be legitimately deduced from (16), even if y were an explicit factor of x. But still further, when x has not y as one of its factors, the expression $\frac{x}{y}$ is not, in the logical system, interpretable. Nevertheless, Professor Boole shows that conclusions both interpretable and correct will *ultimately* be arrived at, if the value of z be deduced Algebraically, as in (19), and the expression $\frac{x}{y}$ be then, as a logical function, subjected to development. Now, if $\frac{x}{y}$ be developed by (11), and the expansion equated to z, we get

$$z = xy + \frac{1}{0}x(1-y) + 0(1-x)y + \frac{0}{0}(1-x)(1-y) \tag{20}$$

Here we have two symbols, $\frac{0}{0}$ and $\frac{1}{0}$, the meaning of which has not yet been determined. Our author shows that the former, which in Algebra denotes an indefinite numerical quantity, denotes in the logical system an indefinite class. In Algebra $\frac{1}{0}$ denotes infinity; and, as is well known, when it occurs as the co-efficient in a term in an equation all of whose other terms are finite, this indicates that the quantity of which it is the co-efficient is zero. So, in the logical system, if, in any term of an equation obtained in the manner in which equation (20) has been obtained, the co-efficient be $\frac{1}{0}$, the corresponding constituent must be 0. These are certainly very remarkable analogies. But let us see what follows. We have first, from (20),

$$x(1-y)=0.$$

Hence as the equation (20) describes the separate classes of which z consists, and as there is no such class as $x(1-y)$ in existence, the second term on the right hand side of equation (20) may be rejected. The third term also may be omitted, its co-efficient being zero. This reduces the equation to the form,

$$z = xy + \frac{0}{0}(1-x)(1-y):$$

which means, that beasts which chew the cud consist of the class xy, together with an indefinite remainder of beasts common to the classes $1-x$ and $1-y$.

Before leaving the subject of inference from a single premiss, we must say a few words regarding elimination; for though, in Algebra, elimination is possible only when two or more equations are given, Professor Boole shows that, in Logic, a class symbol may be eliminated from a single equation. In fact, elimination from two or more premises is ultimately reduced by our author to elimination from a single premiss. And yet, as if to preserve the analogy between Algebra and Logic, even where the two sciences seem to differ most widely from one another, the possibility of eliminating x from a single premiss in the latter science, arises from the circumstance, that, in that science the equation previously referred to as expressing the Law of Duality always subsists; and it is by the combination of that equation with the given proposition that the elimination of x from the given proposition is effected. For let the given proposition be

$$f(x) = 0 \tag{21}$$

Then, by (10),

$$f(1)x + f(0)(1-x) = 0.$$

$$\therefore\ x\{f(0) - f(1)\} = f(0),$$

$$\text{and, } (1-x)\{f(0) - f(1)\} = -f(1).$$

$$\therefore\ x(1-x)\{f(0) - f(1)\}^2 = -f(0)f(1).$$

But, by the Law of Duality, $x(1-x)=0$. Therefore

$$f(0)f(1)=0 \tag{22}$$

which is the result of the elimination of x from equation (21). We cannot pause to give examples of the use of the formula (22); but we must quote an interpretation of it, viewed as the result of the elimination of x from (21), which strikes us as extremely elegant. The formula implies that either $f(0)=0$, or $f(1)=0$. Now the latter equation $f(1)=0$ expresses what the given proposition $f(x)=0$ would become if x made up the universe; and the former $f(0)=0$ expresses what the given proposition would become if x had no existence. Hence, (22) being derived from (21), it follows that *what is equally true whether a given class of objects embraces the whole universe or disappears from existence, is independent of that class altogether.*

The principle of elimination is extended by our author to groups of equations, by the following process. Let

$$\left.\begin{array}{l} T=0, \\ V=0, \\ U=0, \\ \ldots\ldots\ldots \end{array}\right\} \tag{23}$$

be a series of equations, in which T, V, U, &c., are functions of the concept x. Then

$$T^2+V^2+U^2+\text{\&c.}=0 \tag{24}$$

It is shown by Professor Boole that the combined interpretation of the system of equations (23) is involved in the single equation (24). Indeed, had all the terms in the developments of T, V, U, &c., been such as to satisfy the Law of Duality, it would have been sufficient to have written

$$T+V+U+\text{\&c.}=0.$$

In order now to eliminate x from the group (23), it is sufficient to eliminate it, by the method described in the preceding paragraph, from the single equation (24); and, if the result be

$$W=0,$$

this equation will involve all the conclusions that can legitimately be derived from the series of equations (23) with regard to the mutual relation of the concepts, exclusive of x, which enter into these equations.

We do not see how it is possible for any one not blinded by prejudice against every thing like an alliance of Logic with formulæ and processes of a mathematical aspect to deny that these are very remarkable principles. By way of instance, we select from the work under review the following problem, in which two premises are given. Let it be granted, first, that the annelida are soft-bodied, and either naked or enclosed in a tube; and, next, that they consist of all invertebrate animals having red blood in a double system of circulating vessels. Put

A = annelida, s = soft-bodied animals,

n = naked, t = enclosed in a tube,

i = invertebrate, r = having red blood in &c.

Then the given premises are

$$A = vs\{n(1-t) + t(1-n)\}, \tag{25}$$

$$A = ir \tag{26}$$

Suppose the problem then to be: to find the relation in which soft bodied animals enclosed in tubes stand to the following elements, viz., the possession of red blood, of an external covering, and of a vertebral column. Professor Boole would doubtless have granted that this problem admits of being solved by what he calls the ordinary logic; but he would probably have contended that the ordinary logic does not possess any definite and invariable method of solution. A skilful thinker may be able to find out how syllogisms may be formed so as ultimately to give him the relation which soft bodied animals enclosed in tubes bear to the elements specified; but what of thinkers who are not very skilful? How are they to proceed? In Professor Boole's system, the process is as determinate, and as certain of leading to the desired result, as the rules for solving a group of simple equations in Algebra. Eliminate v, the symbol of indefinite quantity, from (25). Reduce (25), thus modified, and (26), to a single equation, by the method described in a previous paragraph. The equation is

$$A\{1 - sn(1-t) - st(1-n)\} + A(1-ir) + ir(1-A) + nt = 0.$$

Then, since the annelida are not to appear in the conclusion, we must eliminate A, by (22), from this equation. This will be found to give us

$$ir\{1 - sn(1-t) - st(1-n)\} + nt = 0.$$

And ultimately we get

$$st = ir(1-n) + \frac{0}{0}i(1-r)(1-n) + \frac{0}{0}(1-i)(1-n);$$

the interpretation of which is: *Soft bodied animals enclosed in tubes (st) consist of all invertebrate animals having red blood (ir) and not naked (1 – n), and an indefinite remainder* ($\frac{0}{0}$) *of invertebrate animals (i) not having red blood (i – r) and not naked (1 – n) and of vertebrate animals (1 – i) which are not naked (1 – n).*

We have entered so fully into the explanation of Professor Boole's system in its bearing on what he terms Primary (virtually equivalent to Categorical) Propositions, that we cannot follow him into the field of Secondary (virtually equivalent to Conditional, that is, Disjunctive and Hypothetical) Propositions. Nor is it necessary that we should do so; for our object is not to give a synopsis of the 'Investigation', but simply to make the nature of the work understood; and, for that purpose, what has been said is sufficient. The application of the Calculus to Secondary Propositions is exceedingly similar, in respect not only of the general method followed, but even of the particular formulæ obtained, to its application to Primary. All that is peculiar in the treatment of Secondary Propositions arises from the introduction of the idea of Time. For instance, the proposition, 'If X is Y, A is B', is held to be not substantially different in meaning from this: 'the time in which X is Y, is time in which A is B'. Such being the fundamental view taken, symbols like x and y are used to represent the portions of time in which certain propositions (e.g., X is Y, A is B) are true. Then, the symbol 1 denoting the universe of Time, or Eternity, the expressions, $1 - x$, $1 - y$, will denote those portions of time respectively in which the propositions, X is Y, A is B, are not true; and so on.

The extension of his method, by Professor Boole, to the theory of Probabilities, is a splendid effort of genius on the part of the author, and furnishes a most convincing illustration of the capabilities of the method. The part of the 'Investigation' which is devoted to this subject, is much too abstruse to admit of being here more particularly considered; but, to show what the method can accomplish—though the bow of Ulysses perhaps needs the arm of Ulysses to bend it—we may simply state one of the problems of which Professor Boole gives the solution. 'If an event can only happen as a consequence of one or more causes, A_1, A_2, ... , A_n, and if generally C_1 represents the probability of the cause A_1, and p_1 the probability that, if the cause A_1 exists, the event E will occur, then the series of C_1 and p_1 being given, required the probability of the event E'.

To those who have followed us thus far, it will be evident what final judgment we are to pass on the work under review. On the one hand, as a contribution to philosophy, in the strict sense of that term, it does not possess any value. Professor Boole distinctly, though modestly enough, avows the opinion, that, in his 'Investigation', he has gone deeper than any previous inquirers into the principles of discursive thinking, and that he has thus thrown new light on the constitution of the human mind. We are sorry to be unable to accept this view. But, on the other thand, Professor Boole is entitled to the praise of having devised a Method, according to which, through definite processes, it can be ascertained what conclusions, regarding any of the concepts entering into a system of premises, admit of being drawn from these premises. This Method depends on a Calculus, original, ingenious, singularly beautiful both in itself and in its relations to the science of Algebra, and capable (in hands like those of its inventor) of striking and important applications. In a word, the merit of the Treatise lies in that part of it which has nothing to do with the Laws of Thought, but which is devoted to showing how inferences, from data however numerous and complicated, and whatever be the matter of the discourse, can be reached through definite mathematical processes.

LUIS M. LAITA

THE INFLUENCE OF BOOLE'S SEARCH FOR A UNIVERSAL METHOD IN ANALYSIS ON THE CREATION OF HIS LOGIC

Summary. This paper deals with the influence exerted by Boole's own work on differential equations on his creation of algebraic logic. The main traits of Boole's methodology of logic, and the particular algorithms which he used in his 1847 *The Mathematical Analysis of Logic*, are first pointed out. An examination of the mathematical papers which Boole wrote before the publication of the mentioned logical treatise shows that both the methodology leading to the production of his logic and the algorithms used in its development were repeatedly used by him in his earlier work in analysis.

1. INTRODUCTION: METHODOLOGY AND ALGORITHMS USED BY BOOLE IN THE CONSTRUCTION OF HIS LOGIC

The purpose of this paper is to show that George Boole's work on differential equations significantly influenced his creation of algebraic logic. There were other external and internal influences of utmost importance in this creation, but they will not be dealt with here. Let us simply point out that among the external influences which led Boole to build his system of logic were the controversy about the quantification of the predicate between Augustus De Morgan and William Hamilton,[1] and that among the internal influences an interesting metaphysical idea entertained by Boole must have had special relevance.[2]

In order to carry out our project we begin by giving a short account of both the general methodology and the main particular algorithms used by Boole in his first logical treatise, *The Mathematical Analysis of Logic* of 1847.[3] We shall complete our argument by showing that this methodology and these algorithms were developed by Boole in his mathematical work prior to the publication of this book.

The general methodology in *The Mathematical Analysis of Logic* may be briefly described as follows. Boole presented first a known mathematical theory, the theory of developments of functions by series.[4] His second step was to center his attention on the symbols of the expressions of the mentioned theory, leaving aside their accepted meaning. Next, he operated with those symbolical expressions according to algebraic laws and to operating devices taken from the theory of linear differential equations. Finally, the results obtained were submitted by him to a careful and judicious process of interpretation in terms of known logical concepts. This general methodology was based on Boole's own conviction that the validity of the process of mathematics does not depend upon the interpretation of the symbols but solely upon the laws of their combination.[5] For him, logic was a branch of a more universal mathematical calculus which embodied not only analysis but also other existing or

J. Gasser (ed.), A Boole Anthology, 45–59.

possible calculi; for this reason he did not hesitate to apply mathematical processes in order to obtain logical results.

In regard to the particular algorithms or expressions which appear in *The Mathematical Analysis of Logic*, the following features should be pointed out for later comment:

(1) In the formulation of the laws of operation of the human mind, Boole translated the acts of mental selections of objects of a given universe into written symbols which he considered to be symbolical operators. These had the property of acting on the objects of the universe either directly or through other operators.[6] As a matter of interest, Boole used the word 'subject' in the formulation of the laws of mental selection with a meaning which cannot be understood unless one considers logic as an extension of a larger calculus in which 'subject' means 'subject of a symbolical operator'.

(2) The symbolical expressions of the above mentioned laws of mental selection happened to coincide with the mathematical expressions of two well-known mathematical laws (the distributivity and commutativity laws) and a third, apparently 'unorthodox' one; the law $x^n = x$.

(3) Techniques for analyzing or solving differential equations and systems of these equations, such as the introduction of auxiliary functionality symbols, developments of functions stopping at the second term, Lagrange's multipliers, and so on, were thoroughly used by Boole in his study of the elective functions of logic.[7]

An important remark ought to be made at this point: when we claim that the methodology and algorithms used by Boole in his work in logic had been developed in his earlier work in analysis, we are not claiming that Boole's logic was mere applied mathematics, if by applied mathematics one understands the set of expected applications of the known results of 'established' mathematics. What we suggest is that Boole's logic was just one of the possible materializations of a universal calculus of symbols which embodied logic as well as analysis and other calculi: only in this way Boole's logic may be considered as applied mathematics.[8] Nevertheless, as the universal calculus was not known, the only way to produce its particular materializations was to work by analogy with the already known ones; the particular materialization called 'logic' was built by analogy with the calculus of differential and finite difference operators. The break which marked the distinction between the calculus of logic and the calculus of these two latter operators was the introduction of the law $x^n = x$; but as we shall see, even this law, in which conception Boole seems to have been guided by an extrascientific idea,[9] has a counterpart in the calculus of differential operators.[10]

In the remainder of the paper we intend to show, by means of a detailed examination of Boole's mathematical works published before or during 1847, the existence of striking similarities, both in method and in particular traits, between his logic and his work in analysis. The latter appeared in papers which were published, with a few exceptions, in *The Cambridge Mathematical Journal* (from 1846 *The Cambridge and Dublin Mathematical Journal*), founded by the mathematician Duncan F. Gregory. Gregory had an important influence on Boole, as will be indicated in sections 2 and 3.

2. THE SEARCH FOR A GENERAL METHODOLOGY OF ANALYSIS: THE PAPERS WRITTEN BY BOOLE IN 1841

Boole's first paper, which was published in *The Cambridge Mathematical Journal* in 1841,[11] already showed the recourse to symbolical operators typical of Boole's later logical developments. These operators in this case were merely substitution operators: P, which substituted for a function of two variables x and y; and P', which expressed the same function in which x and y had been respectively substituted by the two functions $f(x', y')$ and $f'(x', y')$, x' and y' being new variables (the primes here do not represent derivatives).[12] The resulting symbolical equation, $P = P'$, which did not have any particular meaning, was subjected to operations following the ordinary rules of 'algebra' (that is, analysis) in order to obtain a system of equations which were universally valid regardless of particular interpretations.[13] This trend towards generalization from symbolization was the main characteristic of Boole's early methodology; and it was to be at the very basis of his methodology of logic.

Two more aspects of Boole's first published paper deserve consideration because of their significance for the topic of our study: his ideas on interpretation and his care with regard to the order of procedure. For Boole, if not every step of a mathematical description was to have its own interpretation, at least some interpretation was necessary from time to time, both to direct the successive steps of the description and to avoid getting lost midst a blind play of symbolism. In this particular paper Boole interpreted his general formula $P = P'$ as a transformation of an equation representing a surface, produced by a transformation of axes.[14] Interpretation was later to play an important role in Boolean logic, as was noted above. In logic, as in the whole of Boole's mathematical work, the most important idea was that once the symbols and rules have been defined, one should proceed to obtain results without interpreting every step of the procedure. But even though interpretation was in this way playing a secondary role, this role was very important. It was secondary inasmuch as the theory required only a coherent use of the rules applied to the defined symbols, but it was essential as an external tool which enabled him to pick out among all the results those which were relevant to known facts or ideas.

In regard to Boole's extreme care in the order of procedure one reads in his paper: 'In making use of the differential equations of the first or higher orders . . . we must be careful to introduce new conditions only after performing the differentiations'.[15] In some parts of *The Mathematical Analysis of Logic* his care in not omitting a single case is absolutely amazing; for example, in chapter 2, where all the possible ways of expressing propositions were determined in the most minute detail. As a matter of biographical interest, Boole's care in carrying mathematical operations out seems to have been a consequence of a pronounced trait of his personality—an almost exaggerated love of detail.[16]

The second of Boole's published papers (but the first written),[17] begins with an introduction which explains its purpose. In this introduction Boole wrote, referring to Lagrange:

> Every Mathematician is aware of the important uses which the illustrious author has there made of the Calculus of Variations. The mode in which it is employed as an instrument of demonstration consists in almost every instance in a comparison of the

> terms of developed expressions, a mode of investigation by which the abstract existence of truths is shown, much more clearly than the nature of their mutual dependence. Now as the results obtained presuppose, in a majority of instances, the fulfilment of a certain primitive condition (the integrability of $Xdx + Ydy + Zdz$), it appeared to the writer of this paper, that a mode of demonstration, which should establish in a more direct way the connection between the above condition, and the great secondary principle of dynamics thence deducible, would in many respects possess the advantage. From these considerations the following attempts originated.[18]

Boole was here, as in most of his papers, aware that he was developing a new method or a new viewpoint.[19] Moreover, the quotation shows that he held the opinion that mathematical descriptions should show the connections and mutual interdependence between a set of principles and the set of their consequences rather than the abstract existence of mathematical entities. This opinion was to influence decisively his ideas about the construction of logic.

In the paper under examination, the set of principles was given by the conditions of integrability of $Xdx + Ydy + Zdz$, and the consequences were the principles of 'vis viva'[20] and of least action.[21] The symbols of the language were the operators d, δ, $\int$, $\frac{d}{dx}$, endowed with formal properties justified on interpretation; the properties were the transposability of the symbols and the relation $dU\delta V = dV\delta U$.[22] The last relation was universal, but the propriety of its introduction was justified in terms of interpretable and well-known cases which were capable of meaningful interpretation in mechanics.[23] In the second place the exposition was very close to the modern ideal of a mathematical description: the symbols of the mathematical language were introduced first, the rules were stated immediately afterwards, the principles were written down, and finally the consequences were proved. Nevertheless, after this promising start for a modern reader, in which the rules of the mathematical language are given, one realizes that such a language is not complete, and that addenda are necessary to proceed with the proofs. In the paper which we are examining, this difficulty appears when trying to apply the calculus of variations in its known state to an integral formula $\int Udx$, in which U is a function of the two variables x (independent) and y and of the successive derivatives of y with respect to x. At this point Boole introduced a seemingly arbitrary supposition; that there can be considered two kinds of variations: 'The reason of this [difficulty] appears to be, that in the former case [when U is a function of x and y and the derivatives of y; the case when U is a function of x, y, and the successive differential of x and y presented no problem] y and its differentials are susceptible of two different kinds of variation, . . . '.[24] With this addendum Boole arrives in a relatively simple way to a lengthy theorem proved by Lagrange, and from it to a new demonstration of the principles of 'vis viva' and of least action.[25]

These facts have been mentioned because Boole kept the same pattern (not to give a completely self-contained theory, but one which needed successive presuppositions) when writing *The Mathematical Analysis of Logic*. Nevertheless, these presuppositions were not introduced without a reason: they were aimed at some known fact inside the theory, but once Boole had reached this aim (usually in a different and simpler way than the established one), he was able to find new applications of the results obtained (in the paper on variations, the applications were the two principles of mechanics mentioned above). In the second chapter of his first treatise on logic, Boole introduced both the language and the rules of his system; but later in the book he needed to add implicit or explicit suppositions (as the supposition that the development of elective functions followed the same rules as other kinds of functions, or the supposition that the sum of the symbolical expressions of all the mutually exclusive propositions was the Unity[26]), aimed to obtain more or less expected results. As in his work in analysis, these presuppositions were not arbitrary: they were based either on the conception that logic was a calculus analogous to the calculus of differential operators, or on the metaphysical idea referred to above that the Unity can be reached from the contemplation of opposites.[27]

Finally, Boole described an operator which acted on y through x.[28] The mechanics of this procedure happened to be the same as the mechanics of the elective operators, as it appears in the first chapter of *The Mathematical Analysis of Logic*. The resemblance should not be carried too far, because Boole himself probably did not think of this particular operator when, later, he introduced his elective operators. Here it need only be pointed out that his ease in manipulating the operators of logic has as background his previous detailed manipulation of differential operators.

In his third paper, entitled 'On the integration of linear differential equations with constant coefficients',[29] Boole gave a simplification of the so-called 'method of separation of symbols'.[30] The philosophy underlying this method was to 'separate' symbols from their meanings and to operate with the meaningless symbols according to given algebraic rules. Before making any contact with other mathematicians, Boole had been using this method implicitly in his early work in analysis, but his actual acquaintance with it probably took place during one of his first contacts with Gregory. In this paper Boole translated a differential equation into a rational expression which had the same form, so that the ordinary rules of algebra could be applied to it in order to resolve it into a sum of fractions. After writing down this rational expression, he continued:

> Now the method of the resolution of this into a sum of partial fractions, is independent of any properties of the variable, except the three which have been shown by Mr. Gregory . . . to be common to the symbol $\frac{d}{dx}$, and to the algebraical symbols generally supposed to represent numbers. Consequently the same means which enable us to determine the form of the partial fractions in ordinary algebra, may be applied to the circumstances of the case now under consideration. This, it will be seen is nothing more than a farther extension of the application of the principles on which the whole method of the separation of symbols is founded.[31]

This was the first explicit statement made by Boole of what he had already done in other papers and was to continue doing without stating it: that when dealing with particular problems of mathematics, one should proceed to a substitution of the interpretable symbols by general ones to which the general rules of algebra, with some modifications, could be applied. Here, the symbols 'z' and 'd/dx' followed the same rules, so that a differential symbolical function $(f(d/dx))^{-1}$ could be developed in exactly the same way as $(f(z))^{-1}$. When he later introduced his elective operators of logic he proceeded almost exactly in the same way: he made the symbols representing these operators follow the same rules (with the exception of the special rule $x^n = x$) as the rules followed by differential operators.

A clear statement of Boole's mathematical methodology appears at the end of the paper which we are considering:

> But the process which I have here exhibited shows, that the form of the solution depends solely on the method of decomposing the original operating factor; and this decomposition is effected by means of processes which are common to the two operations under consideration, being founded only on the common laws of the combination of the symbols.
>
> It is thus seen that every step of the solution of Differential equations and Equations of Finite Differences is reduced to the known theorems of ordinary Algebra, with the exception of the two theorems . . . which are necessary for passing to the interpretation of the expressions at which we arrive. This seems to be as great a simplification of the problem as the present state of mathematics admits of, for any further improvement must involve the invention of new processes for the treatment of ordinary algebraical expressions. With such we are not presently concerned; our object being to reduce the more complicated processes of the higher analysis to the simpler results which have been already obtained, and which may be looked on in the military phrase as bases for further operations.[32]

This interesting quotation speaks for itself. The main traits of Boolean methodology: generalization by symbolization, 'blind' applications of the laws of algebra to the symbolical expressions obtained, and correct introduction of interpretative devices, appear clearly stated.

Another paper by Boole, 'Analytical Geometry',[33] also appeared in 1841. The paper dealt with the application of a theorem noticed by a writer in the first volume of *The Cambridge Mathematical Journal* to problems related to distances between lines, planes and loci in general. From the consideration of these particular problems, Boole reached again the solution of a general system of equations which was broadly applicable to different aspects of analytical geometry. As in the previous papers, Boole showed in this one a rare ability to produce generalizations. The pattern followed was approximately the same; he began with some already known mathematical statements, these were then considered from the most general viewpoint possible, and finally the application of the obtained generalization was extended to several seemingly unrelated fields of mathematics or even physics. A skilful use of symbols and of mathematical operators was also displayed in this paper.

The paper began by presenting a given theorem, after which Boole proceeded step by step to generalization: 'We shall next proceed to investigate a general expression for the minimum distance of two loci of the first degree; to accomplish which it will be sufficient, as might easily be shown, to discuss generally the case of two straight lines'.[34] The consideration of this case led to a symbolical formula

$$D = D' \cos x,$$

which expressed the minimum distance between two loci of the first degree. But he observed that because of the generality given by symbolization, such a formula was also valid for loci whose analytical equations were not of the first degree. Further generalizations were possible by an intelligent use of interpretation. If straight lines were considered as equations of intersecting planes, a general value of the symbol *D* could be found.[35] This going back and forth from generalizations to symbolizations to produce new generalizations was typical of Boole. The paper even provided a way to find the paths of rays of light.

No paper of Boole appeared in 1842, but in 1843 he published a long memoir on linear transformations. Even though it was published in 1843, it had been worked out in 1841, a short time after he had written the papers which we have already examined. As in other papers, this one presented first an already known calculus (in this case it was the determination of the coefficients of a differential equation resulting from a linear transformation). The solution of the problem was not known for degrees (in the transformation equation) larger than two. Again Boole intended to attack the problem by a symbolization presenting a general formula relative to which the problem in question was merely a particular case. His method was based in the theory of developments, which was to be the topic of one of his later papers, and, as will be seen, a tool for the study of the elective functions of logic.

In the paper we are studying Boole presented an elimination method;[36] this is pointed out because he used elimination in several forms of his logic. A functionality symbol was also used in this paper;[37] a similar functionality symbol and a similar process was used in *The Mathematical Analysis of Logic*.[38] These illustrations support our claim that the whole theory of elective functions of Boolean logic was based on the analogy with Boole's own ideas on the theory of linear differential equations. Boole himself explicitly said referring to his elective equations: 'It is to be observed of such equations, that from the very nature of elective symbols, they are necessarily linear, and that their solutions have a very close analogy with those of linear differential equations, arbitrary elective symbols in the one, occupying the place of arbitrary constants in the other'.[39] The development of elective functions, which together with the symbolical expressions of the laws of thought were the foundations of Boole's system of logic, was possible because of not only his ingenious idea of assigning symbols to the operations of the mind, but also his skilfulness in the treatment of linear differential equations.

3. COMPLETION OF BOOLE'S GENERAL METHOD IN ANALYSIS

The work which represents the peak of all Boole's publications on analysis was undoubtedly the paper 'On a General Method in Analysis'.[40] The work was too weighty for *The Cambridge Mathematical Journal*, and Boole considered printing it at his own expense or applying for its publication in the *Philosophical Transactions of the Royal Society*. After some delay the paper was finally printed in the *Transactions* and was awarded the Royal Medal for mathematical science for 1844. The study of this paper, which is a masterpiece of mathematical exposition, should convince the reader that the creation of algebraic logic in the hands of Boole was not a happy coincidence. When he was extending the ordinary laws of analysis to symbols representing mathematical entities different from quantities, the theoretical foundations of an extension to logical symbols were laid down.

'On a General Method in Analysis' begins with an introduction in which Boole acknowledged some of the mathematicians who had worked before on related topics. After this introduction, he wrote:

> Mr. Gregory lays down the fundamental principle of the method in these words: 'there are a number of theorems in ordinary algebra, which, though apparently proved to be true only for symbols representing numbers, admit of a much more extended application. Such theorems depend only on the laws of combination to which the symbols are subject, and are therefore true for all symbols, whatever their nature may be, which are subject to the same laws of combination'. The laws of combination which have been hitherto recognized are the folowing, π and ρ being symbols of operation, u and v subjects.
>
> 1. The commutative law, whose expression is $\pi\rho u = \rho\pi u$.
> 2. The distributive law, $\pi(u + v) = \pi u + \pi v$.
> 3. The index law, $\pi^m\pi^n u = \pi^{m+n}u$.
>
> Perhaps it may be worth while to consider whether the third law does not rather express a necessity of notation, arising from the use of general indices, than any property of the symbol.
>
> The above laws are obviously satisfied when π and ρ represent such symbols as $\frac{d}{dx}$, δ, &c., or with constant quantities.[41]

The reference to Gregory was taken from his recent book *Examples of the Processes of the Differential and Integral Calculus*.[42] Gregory had written there:

> In this chapter [chapter XV, 'On General Theorems in the Differential Calculus'] I shall collect those theorems in the differential calculus which, depending only on the laws of combination of the symbols of differentiation, and not on the functions which are operated by these symbols, may be proved by the method of separation of symbols: but as the principles of this method have not as yet found a place in the elementary works of calculus, I shall first state briefly the theory on which it is founded.[43]

At this point Gregory stated the fundamental principle quoted by Boole in the quotation above. Gregory continued as follows:

> The laws with which we have here concern are few in number, and may be stated in the following manner. Let a, b, represent two operations, u, v, two *subjects* on which they operate, then the laws are:
>
> 1. $ab(u) = ba(u)$
> 2. $a(u + v) = a(u) + a(v)$
> 3. $a^m \cdot a^n \cdot u = a^{m+n} \cdot u$
>
> ... The third law is not so much a law of combination of the operation denoted by a but rather of the operation performed on a which is indicated by the index affixed to a ... these are the laws employed in the demonstration of the principal theorems in Algebra..., but they are not confined to symbols of numbers; they apply also to the symbol used to denote differentiation.[44]

Thus Gregory was the inspiration for the main paper of Boole, as the latter acknowledged by his reference both to the book *Examples* and to the articles written by Gregory on the same topic. Gregory on his side might have received inspiration from John Herschel,[45] but Herschel as well as Gregory and Boole undoubtedly were independently inspired by the work on differential operators achieved by Lagrange,

Laplace, Arbogast, Brisson, Poisson, Fourier, Servois, Cauchy and others.[46] Boole also recognized the influence of Murphy and De Morgan.[47] Nevertheless, it has to be remarked again that Boole arrived by himself, if not to the precise formulation of the method, then at least to its underlying philosophy of obtaining universal results through symbolization. Moreover, while Gregory was able to use the method in only a few instances (Taylor's theorem, a special theorem for differentials, decomposition into rational fractions, and a few more), Boole extended it to differential equations with variable coefficients, series, generating functions and, which is more important, to finite differences, suggesting in this way that the method of separation of symbols might be extended to 'new' calculi.

But it is important to consider now that the first principles of Boolean logic were almost a *direct application* of the Gregory-Boole theory of the calculus of operators to the elective operators of logic. For Boole wrote in *The Mathematical Analysis of Logic*:

> Further, let us conceive of a class of symbols x, y, z, possessed of the following character.
>
> The symbol x operating upon any subject comprehending individuals or classes, shall be supposed to select from it all individuals of the class *X*s which it contains . . .
>
> When no subject is expressed, we shall suppose 1 (the Universe) to be the subject understood, so that we shall have
>
> $$x = x(1)$$
>
> . . . The result of an act of classificating is independent of the grouping or classification of the subject . . . We may express this law mathematically by the equation
>
> $$x(u + v) = xu + xv$$
>
> . . . It is indifferent in what order two successive acts of election are performed.
>
> . . . The symbolical expression of this law is
>
> $$xy = yx$$
>
> . . . The result of an act of election performed twice or any number of times in succession, is the result of the same act performed once . . . supposing the same operation to be n times performed, we have
>
> $$x^n = x$$
>
> . . . [These] laws are sufficient for the basis of a calculus . . . . The third law [$x^n = x$] we shall denominate the index law. It is peculiar to elective symbols and will be found of great importance in enabling us to reduce our results to terms meet for interpretation.[48]

The quotation is revealing. Logic happened to be a calculus with the same distributive and commutative laws as in the calculus of operators plus a third law, *required for interpretation*, which also reflected a law of the calculus of operators, which 'probably did not express any property of the symbol, but rather a necessity of notation', these last words reflecting exactly what Boole had stated in his 'On a General Method in Analysis'. The parallel between Boolean analysis and Boolean logic is, as the reader can see, remarkable. Some of the critics of Boole's logic have necessarily judged partially Boole's results, because they have not taken in account that Boolean logic does not stand as an isolated construct but rather as a completion of

Boolean analysis, *if* by analysis one considers a universal calculus of symbols. As has been said above, Boole's logic was not the result of a direct application of the known analysis, but rather a system analogous to this analysis, logic and analysis both being particular branches of a universal calculus. Both the calculus of symbols representing differential or difference operators, and the calculus of symbols representing the elective operators of logic, followed the commutativity and distributivity laws; the break was marked by the law of logic $x^n = x$, which was different from the law $a^m \cdot a^n = a^{m+n}$. Nevertheless, *methodologically* the break was not complete: both laws were exponential laws, and both were requisites for interpretation, each one in its field. In the differential calculus $a \cdot a \cdot a \ \ldots (u)$ was $a^n(u)$, in logic interpretation required $x \cdot x \cdot x \ \ldots (u)$ to be $x(u)$. Once more it has to be recalled that the law $x^n = x$ might have been expressed as the result of Boole's metaphysical conception. If this was the case, Boole's analysis and Boole's logic, even though analogous in form, had indeed different origins and aims.

Boole's 'On a General Method in Analysis' ended with the following words, which clearly show him grasping the existence of a universal calculus of symbols, capable of being materialized in several different calculi:

> Fearful of extending this paper beyond its due limits, I have abstained from introducing any researches not essential to the development of that general method in analysis which it was proposed to exhibit. It may however be remarked that the principles on which the method is founded have a much wider range. They may be applied to the solution of functional equations, to the theory of expansions, and, to a certain extent, to the integration of non-linear differential equations. The position which I am most anxious to establish is that any great advance in the higher analysis must be sought for in an increased attention to the laws of the combination of symbols. The value of this principle can scarcely be over-rated, and I only regret that in the absence of books, and under circumstances unfavorable for mathematical investigation, I have not been able to do that justice to it in this essay which its importance demands.[49]

When Boole was here suggesting the possibility of construction of new calculi, his suggestion was made with the anxiety of one who knew that a new road was being opened for mathematics and wanted others to become aware of its breadth, with the anxiety of one who grasped how wonderful a new field had been suddenly revealed but regretted not having had enough material means to explore it. Boole was indeed prepared to 'discover' the calculus of logic.

4. BOOLE'S PAPERS PUBLISHED BETWEEN 1844 AND 1847, THE YEAR OF THE PUBLICATION OF *THE MATHEMATICAL ANALYSIS OF LOGIC*

From 1844, the year of the publication of 'On a General Method in Analysis', to the year 1847, when *The Mathematical Analysis of Logic* appeared, ten more papers by Boole were published, eight in *The Cambridge Mathematical Journal* (after 1846 *The Cambridge and Dublin Mathematical Journal*), one in *The British Association for the Advancement of Science Reports*, and one in the *Philosophical Magazine*. Most of these papers dealt either with applications of the method presented in his 1844 paper or new generalizations of the method. Occasionally he referred to the main purpose of his enquiries, which was to suggest methodologies rather than to solve particular problems:

> I apprehend that the same principles may be extended . . . to the inverse calculus of definite multiple integrals. At present, however, I have not leisure to pursue this enquiry, nor am I so much concerned to multiply results as to establish principles. . . .[50]

> I have entered with more particularity into the details of the above solution, than to some might have appeared necessary; but it was my object in this paper not only to integrate the equations of Laplace, but also to illustrate, and in so doing if it may be, to recommend a method in Analysis. . . .[51]

> This equation is a symbolical one, and it admits of two conjugate solutions . . . their validity does not depend on the significance of the symbols which they involve, but only on the truth of the laws of their combination. One interpretation of these symbols gives us Laplace equations, but a more general interpretation than this is possible in the integral calculus, and there is perhaps, for I have not examined the question, an interpretation in the Calculus of Finite Differences.[52]

If these remarks are illuminating in regard to methodology, the consideration of two papers entitled 'On the Equation of Laplace Functions' and 'On the Theory of Developments' throws light in other directions. In the first of these papers, Boole referred to Hargreave's expression of the complete integral of the equation of 'Laplace's functions':

$$u = \ldots \int \cos^{-2}\frac{y-x}{2}\left(\int \cos^{2n}\frac{x-y}{2}\chi(y)dy + \psi(x)\right)dydy\ldots [n \text{ times}].^{53}$$

Such an expression included the two arbitrary functions χ and ψ. Boole, using the method of separation of symbols, found another solution simpler than Hargreave's in being free of integrals. After that, he suggested a proper determination of the two auxiliary functions in order to obtain the actual form of the coefficients in the Laplace functions.[54] But in the study of the solutions of the elective equations in *The Mathematical Analysis of Logic* he referred to the possibility of finding these solutions also by the introduction of auxiliary functions, and he gave an example.[55] The method used by him was not an exact copy of what he had done in 'On a Certain Symbolical Equation', but the basic similarity which exists between the two processes shows, in a very illustrative way, the strong influence of Boole the mathematician on Boole the logician.

The same consequence can be drawn from the examination of 'On the Theory of Developments'. One finds in this paper an instance of a development which stops at the second term.[56] When Boole expanded an elective function in *The Mathematical Analysis of Logic* by means of a MacLaurin expansion he obtained the expression

$$\phi(x) = \phi(1)x + \phi(0)(1 - x),^{57}$$

that is, an expansion which also stops at the second term. Boole himself noted the similarity, saying: 'The mathematician may be interested in the remark that this is not the only case in which an expansion stops at the second term',[58] referring the reader to his paper on developments.[59]

5. CONCLUSION

All our considerations above support the claim made at the beginning of the paper that both the methodology on which Boole based the construction of his logic, and several of the algorithms used in its development were the same as those used by him in his prior work on analysis. Moreover, and this is more important, they also support our suggestion that the discovery of algebraic logic at the hands of Boole was not a happy coincidence: as he grasped the idea of the existence of a general calculus of symbols embodying the known differential calculus as a particular branch, he became immediately aware of the possibility of construction of other branches. One was the calculus of finite differences, another one happened to be logic. All branches had common laws, and all had distinguishing laws, based on the interpretations which each field required. In the actual statement of the distinguishing law $x^n = x$ of logic, a curious metaphysical idea entertained by Boole[60] might have had a crucial influence. If this was the case, the origin and aim of Boole's logic were different from those of other calculi, but all these calculi and logic were in his methodology and particular traits *basically analogous*.

ACKNOWLEDGMENTS

This paper arose from discussions with Professor Michael J. Crowe, of the Graduate Program in the History and Philosophy of Science at Notre Dame University. Professor Crowe, to whom I am wholly indebted, directed my studies in the history of mathematics. I also wish to extend my thanks to Dr. Ivor Grattan-Guinness, editor of *Annals of Science*, for the valuable information on particular points he has given me, and especially for his question: may Boole's logic, under the perspective given in this paper, be considered *applied mathematics?* The study of this question has helped me to clarify my opinion that Boole's logic was a consequence of Boole's work on analysis, precisely because he considered analysis to be a universal calculus of symbols; Boole's logic was a consequence of this conception, not a direct consequence of the 'known' analysis. Naturally neither Professor Crowe nor Doctor Grattan-Guinness should be assumed necessarily to be in agreement with this opinion or with my statements in the paper.

NOTES

[1] At the very beginning of *The Mathematical Analysis of Logic* (the reference is given in footnote 3 below) Boole stated that he was moved to resume his enquiries about an algebra of logic by the controversy between Hamilton and De Morgan. Besides having been the external stimulus which moved Boole to pursue his enquiries, the controversy may have had another kind of influence on Boole; it seems that several of the ideas interchanged in the controversy confirmed and clarified Boole's own ideas, and that probably the latter stated his philosophy of mathematics in the introduction to *The Mathematical Analysis of Logic* as a response to Hamilton's ideas about the same subject.

[2] Mary Everest Boole, Boole's wife, referred in many places in her writings (which are cited in footnote

16 below) to the existence of a mystico-pedagogico-psychological background and aim of Boole's logic. According to her, the equation $x^2 = x$ was the expression of a dualistic philosophy which claimed that the Eternal Unity of God could be reached from the contemplation of all opposite opinions and facts; $(1 - x)$ was the opposite of x, and $(1 - x) + x = 1$. Such conception involved a psychological theory of the mind and had pedagogical implications. Several historians of logic have despised Mary Everest's claim, but in fact, it seems that she was basically right. It would take too long to prove it here; I have examined Mary Everest's claims in 'A Study of the Genesis of Boolean Logic', a doctoral dissertation submitted to the Graduate School of the University of Notre Dame. This dissertation was directed by Professor Michael J. Crowe.

Among the psychological implications of Boole's—probably existing—metaphysical conception was that God works directly through some hidden resort of the human mind, so that the latter does not need to be continuously conscious of its operations. In pedagogy, this idea materializes in the stress on the importance of using symbols which only at determinate steps of the argument need to have meaning, in the processes of learning and teaching. One interesting problem is to know the possible influence on Mary Everest (and Boole too?) of a symbolical school which developed in the early and mid-nineteenth century (for references to secondary literature on this theme, see footnote 46 below).

[3] G. Boole, *The Mathematical Analysis of Logic, Being an Essay towards a Calculus of Deductive Reasoning* (1847, Cambridge and London: repr. 1958, New York). Also in G. Boole, *Studies in Logic and Probability* (ed. R. Rhees: 1952, London), 49–124. This work is cited hereafter as '*MAL*'.

[4] Neither the application of McLaurin expansions to obtain Boole's elective functions, nor the calculus of solutions of the corresponding equations by means of the theory of linear differential equations, appears in *MAL* until the last two chapters. Nevertheless, the most important results of the first part of the book are based on the results obtained in those two last chapters. This fact shows that, if not in order of presentation, the theory of elective functions was first in order of procedure. But this theory was, on its hand, based in the theory of developments of functions. (See footnote 7 below for an explanation of the elective operators and functions.)

[5] See the Introduction to *MAL*.

[6] *MAL*, 17–18.

[7] Boole stated the laws of logic from the contemplation of the operations performed by the mind when electing or classifying elements from a given universe. For instance, x is a symbol which represents a selection of groups of individuals, namely u and v. The result of an act of election is independent of the grouping of the individuals, so that we have the distributivity law $x(u + v) = xu + xv$. In the same way, Boole inferred the commutativity law $xy(u) = yx(u)$ and the special law $x^n = x$. The elective functions are functions the variables of which are precisely the elective operators, $x, y, \ldots$. He supposed that all the operative devices provided by ordinary algebra and analysis could be applied to these functions, with the only care of taking in account the irregularities provided by the condition $x^n = x$.

[8] This view that Boole's logic was *applied mathematics* of a universal calculus of symbols *supports* Russell's view that 'Pure mathematics was discovered by Boole' (*Mysticism and Logic* (1917, London; repr. 1963), 59), precisely because of the universal meaning attributed by Boole to the word 'analysis'. Nevertheless, Russell's assertion was too exclusive: even though it seems that Boole grasped by himself the existence of a universal calculus capable of embodying several known and possible fields, this idea was already at least implicitly contained in the works of the French mathematicians of the immediately prior period, and in those of Peacock, Herschel, Hamilton, Gregory, De Morgan and others. In mathematics it is often misleading to assign a particular discovery to a particular person. Russell himself retracted somewhat his comment on Boole discovering pure mathematics. (I owe the last information to Dr. Ivor Grattan-Guinness, who cites Russell's letter in *Proc. Roy. Irish Acad.*, 57 (A) (1954–56), 28.)

[9] See footnote 2 above.

[10] See the last part of section 3 of this paper.

[11] G. Boole, 'Researches on the Theory of Analytical Transformations, with a Special Application to the Reduction of the General Equation of the Second Order', *Cambridge Mathematical Journal*, 2 (1841), 64–73.

[12] *Ibid.*, 64.

[13] *Ibid.*, 65.

[14] *Ibid.*

[15] *Ibid.*

[16] M. Everest Boole. *Collected Works* (4 vols., ed. E.M. Cobham and E.S. Dummer: London, 1931), vol. 1, 4, 5, 35.
[17] G. Boole, 'On Certain Theorems in the Calculus of Variations', *Cambridge Mathematical Journal*, 2 (1841), 97–102.
[18] *Ibid.*, 97.
[19] See, for instance, *Cambridge Mathematical Journal*, 2 (1841), 64–73 (p. 64); 2 (1841), 114–119 (p. 115); 3 (1843), 1–20 (p. 2); and *Cambridge and Dublin Mathematical Journal*, 2 (1847), 7–12 (p. 7).
[20] G. Boole (footnote 17), 99.
[21] *Ibid.*, 101.
[22] *Ibid.*, 97.
[23] *Ibid.*, 98.
[24] *Ibid.*
[25] *Ibid.*, 101–102.
[26] *MAL*, 49–50, 52, 64, 77.
[27] It may be argued that what was arbitrary was precisely Boole's methodological or philosophical assumptions, but this has been the way many (not to say most) scientific discoveries have taken place: by recourse to methodological and philosophical ideas extraneous to the field of direct investigation.
[28] G. Boole (footnote 17), 98.
[29] G. Boole, 'On the Integration of Linear Differential Equations with Constant Coefficients', *Cambridge Mathematical Journal*, 2 (1841), 114–119.
[30] See footnotes 45–47 and text.
[31] G. Boole (footnote 29), 115.
[32] *Ibid.*, 119.
[33] G. Boole, 'Analytical Geometry', *Cambridge Mathematical Journal*, 2 (1841), 179–188.
[34] *Ibid.*, 180.
[35] *Ibid.*, 182.
[36] G. Boole, 'Exposition of a General Theory of Linear Transformations', *Cambridge Mathematical Journal*, 3 (1843), 1–20, 106–119 (pp. 6–7).
[37] *Ibid.*, 8.
[38] *MAL*, 73.
[39] *MAL*, 70.
[40] G. Boole, 'On a General Method in Analysis', *Philosophical Transactions of the Royal Society of London*, 134 (1844), 225–282.
[41] *Ibid.*, 225.
[42] D.F. Gregory, *Examples of the Processes of the Differential and Integral Calculus* (1841, Cambridge and London).
[43] *Ibid.*, 235.
[44] *Ibid.*
[45] J. Herschel, 'Considerations on Various Points in Analysis', *Philosophical Transactions of the Royal Society of London*, 104 (1814), 440–468.
[46] These important developments still await the historical study that they deserve, but interim information can be recovered from: S. Pincherle, 'Pour la bibliographie de la théorie des opérations distributives', *Bibl. math.*, (2) 13 (1899), 13–18; his 'Funktionale Operationen und Gleichungen', *Enc. math. Wiss.*, Bd. 2, Teil A (1903–21, Leipzig), 761–824; H. Burkhardt, 'Entwicklungen nach oscillierenden Funktionen...', *Jber. Dtsch. Math.-Ver.*, 10, pt. 2 (1901–08), esp. ch. 13; E. Koppelman, 'The Calculus of Operations and the Rise of Abstract Algebra', *Arch. Hist. Exact Sci.*, 8 (1972), 155–242; and I. Grattan-Guinness and J.R. Ravetz, *Joseph Fourier* 1768–1830 . . . (1972), Cambridge, Mass.), esp. pp. 464–466.
[47] G. Boole (footnote 40), 225.
[48] *MAL*, 15–18.
[49] G. Boole (footnote 40), 282.
[50] G. Boole, 'On the Inverse Calculus of Definite Integrals', *Cambridge Mathematical Journal*, 4 (1845), 82–87 (p. 87).
[51] G. Boole, 'On the Equations of Laplace's Functions', *Cambridge and Dublin Mathematical Journal*, 1 (1846), 10–22 (p. 22).
[52] G. Boole, 'On a Certain Symbolical Equation', *Cambridge and Dublin Mathematical Journal*, 2 (1847), 7–12 (p. 7).
[53] G. Boole (footnote 51), 11.

[54] *Ibid.*, 11, 13, 20.
[55] *MAL*, 70–72.
[56] G. Boole, 'On the Theory of Developments', *Cambridge Mathematical Journal*, 4 (1845), 214–223 (p. 219).
[57] *MAL*, 60.
[58] *Ibid.*
[59] *Ibid.*, 60–61.
[60] See especially footnotes 1 and 2 above.

Departamento de Algebra y Fundamentos, Sección de Matemáticas,
Universidad de Sevilla, Sevilla, Spain

THEODORE HAILPERIN

BOOLE'S ALGEBRA ISN'T BOOLEAN ALGEBRA

A Description, Using Modern Algebra, of What Boole Really Did Create

To Boole and his mid-nineteenth century contemporaries, the title of this article would have been very puzzling. For Boole's first work in logic, *The Mathematical Analysis of Logic,* appeared in 1847 and, although the beginnings of modern abstract algebra can be traced back to the early part of the nineteenth century, the subject had not fully emerged until towards the end of the century. Only then could one clearly distinguish and compare algebras. (We use the term **algebra** here as standing for a formal system, not a structure which realizes, or is a model for, it—for instance, the algebra of integral domains as codified by a set of axioms *versus* a particular structure, e.g., the integers, which satisfies these axioms.) Granted, however, that this later full degree of understanding has been attained, and that one can conceptually distinguish algebras, is it not true that Boole's "algebra of logic" is Boolean algebra?

Conventional opinion without exception is on the affirmative side of this question. Briefly put, this opinion has it that Boole was the first to give a successful algebraic setting for doing logic (of class terms, or properties) and, as is well known, the abstract form of the calculus of classes (classes = sets), is Boolean algebra. But if we look carefully at what Boole actually did—his major work is *The Laws of Thought* of 1854—we find him carrying out operations, procedures, and processes of an algebraic character, often inadequately justified by present-day standards and, at times, making no sense to a modern mathematician. Not only that, but even though his starting algebraic equations (expressing premises of an argument) were meaningful when interpreted in logic, the allowed transformations often led to equations without meaning in logic. Boole considered this acceptable so long as the end result could be given a meaning, which he did give by a specifically introduced procedure. Writing in 1870 the logician (and economist) W.S. Jevons complained:

> ...he (Boole) shrouded the simplest logical processes in the mysterious operations of a mathematical calculus. The intricate trains of symbolic transformations, by which many of the examples in the Laws of Thought are solved, can be followed only by highly accomplished mathematical minds; and even a mathematician would fail to find any demonstrative force in a calculus which fearlessly employs unmeaning and incomprehensible symbols, and attributes a signification to them by a subsequent process of interpretation [4, p. 143].

As this quote indicates, it appears that to determine what sort of algebra Boole did use to do logic will require a fairly substantial exegetical effort, and what the outcome will be, whether Boolean algebra or something else, is not immediately clear. Before entering into a presentation of Boole's ideas, we depict the foundations-of-algebra *milieu* in which he worked.

J. Gasser (ed.), A Boole Anthology, 61–77.

SYMBOLICAL ALGEBRA (CIRCA 1830–1840)

Although in 1837 W.R. Hamilton had shown how to define complex numbers as ordered pairs of ordinary numbers, there was, as yet, no real understanding of the various number systems as being successive, more inclusive, extensions of natural numbers; nor was there any conception of such number structures as models of appropriate formal algebraic systems. Such fundamental algebraic properties as commutativity and distributivity had only recently been so named (by F.-J. Servois in 1814/15), and associativity not until 1844 (by Hamilton). And, despite the fact that over the course of several centuries the techniques and application of numerical algebra had become highly developed and refined, there was no adequate body of justifying principles. Symbolical Algebra, as it was called, claimed to provide these absent principles. Its chief promulgator was George Peacock, Fellow and Tutor of Trinity College, Cambridge. Also prominent in its advocacy were Duncan F. Gregory, editor of the journal which published Boole's earliest mathematical research, and Augustus De Morgan, an early champion of Boole's work in logic, who also wrote extensively on formal logic.

The doctrine espoused by Peacock was that there were two sciences of algebra, *arithmetical* and *symbolical.* In the former the general symbols and operations refer to the numbers and operations of "common arithmetic" whose meaning required in many cases restriction on the performability of the operations. For example, in arithmetical algebra one couldn't subtract a larger from a smaller number and hence the form $a - b$ involving the general symbols a and b could be meaningless if so interpreted. In symbolical algebra, however, all restrictions on operations are removed—what meaning expressions then had was to be subsequently determined from the assumed laws of symbols. The laws are obtained from the following principles (stated in Peacock's language to retain historicity):

(i) Whatever forms in general symbols are equivalent in arithmetical algebra, are also equivalent in symbolical algebra.

(ii) Whatever forms are equivalent in arithmetical algebra when the symbols are general in their form, though specific in their value, will continue to be equivalent when the symbols are general in their nature as well as in their form.

Thus although $a(b - c) = ab - ac$ is true in arithmetical algebra only if b is not less than c, this restriction is removed in symbolical algebra and the equation is considered to be true. Symbolical algebra is to be an extension of the arithmetical in the sense that when its symbols are numbers and its operations arithmetical, the results must be identical with that of arithmetic. However, "...inasmuch as in many cases, the operations required to be performed are impossible, and their results inexplicable, in their ordinary sense, it follows that the meaning of the operations performed, as well as the results obtained under such circumstance, must be derived from the assumed rules, and not from their definitions or assumed meanings, as in Arithmetical Algebra" [9, p. 7]. In contrast to this view, contemporary mathematics does not consider formal rules (or axioms) as being able to create meanings for operations, only to set limits for them—e.g., adoption of $a + b = b + a$ excludes the symbol + from being interpreted as a noncommutative operation. It should be emphasized that nowhere in Peacock's *Treatise* do we find a full listing of algebraic properties of the operations analogous to a present day set of axioms.

Despite such inadequacy, and much more that one could find fault with, symbolical algebra did provide an environment in which algebra could be freed from its exclusive use with number. For example, it provided a measure of justification for the calculus of operations or separation of symbols, as it was also known. In this calculus one separates off the symbols of operation and performs algebraic operations on them. A simple example is the well-known symbolic formula $(D + a)^2 = D^2 + 2aD + a^2$ of elementary differential equations in which D stands for the operation of differentiation and the square on D indicates a two-fold application. This type of procedure, of applying algebra to "symbols" and using special interpretation, was suspect in the view of some mathematicians—Cauchy for one—but was widely used in England. Boole was a leading practitioner of the art.

Though a confirmed adherent of the symbolical algebra school of thought, Boole was no rigid follower. He significantly modified Peacock's principles by relaxing the requirement that when the general symbols and the operations have their common arithmetical meaning, the result should be an arithmetical truth. For example, in a paper of 1844 for which he received a Royal Society medal, Boole introduced non-commuting types of operators. Turning to the immediate subject at hand, the characteristic law $x^2 = x$ which he added to algebra so as to do logic is not true arithmetically save—as Boole points out—if one restricts x to being 0 or 1. Did Boole thus create Boolean algebra by adding $x^2 = x$? Before addressing ourselves to the question we turn to a brief exposition of Boolean algebra.

BOOLEAN ALGEBRA

Nowadays we describe a formal algebraic system by way of axioms. For Boolean algebra especially, such axiom systems come in a large variety of shapes (basic operations, relations) and sizes (number of axioms). The one we choose to present here focuses attention on analogies with numerical algebra. It uses two binary operation symbols, + and ×, one unary operation symbol ′, and two constant symbols 0 and 1.

Axioms for Boolean Algebra

(i)	$ab = ba,$	$a + b = b + a,$
(ii)	$(ab)c = a(bc),$	$(a + b) + c = a + (b + c),$
(iii)	$1a = a,$	$0 + a = a,$
(iv)	$aa' = 0,$	$a + a' = 1,$
(v)	$a(b + c) = ab + ac,$	$a + (bc) = (a + b)(a + c),$
(vi)	$aa = a,$	$a + a = a.$

The evident analogy with ordinary algebra which is displayed here can be further enhanced by the introduction of a subtraction symbol in terms of which ′ can be expressed: we let $x - y$ stand for xy' so that, with (iii), we have $1 - x = 1x' = x'$. Hence in place of (iv) we can write

$$\text{(iv*)} \quad a(1 - a) = 0, \qquad a + (1 - a) = 1.$$

We need to distinguish **a Boolean algebra** from the *general concept of,* or the *formal theory of,* Boolean algebras. When using the indefinite article, we are referring to a particular mathematical structure which, via an appropriate interpretation, satisfies the above axioms. The best known example of a Boolean algebra is the set of subsets of a fixed set, with the set operations of union, intersection and complementation interpreting, respectively, the symbols +, ×, ′, and with the empty set and the fixed set interpreting 0 and 1.

One can come to Boolean algebra from another direction which adds a little more to the analogy with numerical algebra. The Boolean addition in the above axioms lacks the property of linear solvability or, equivalently, existence of an additive inverse; that is, $a + x = b$ need not have a solution for x nor, if it did, a unique one. If, however, one introduces by definition an operation $+_\Delta$ ("symmetric difference") by putting

$$x +_\Delta y = x - y + y - x = xy' + yx',$$

then it is true that in any Boolean algebra the equation $a +_\Delta x = b$ has a unique solution for x, for any a and b. One can show that, under the operations $+_\Delta$ and ×, with 0 and 1 as the zero and unit, a Boolean algebra is a commutative ring with unit. A commutative ring having the idempotency property $a^2 = a$ is called a **Boolean ring**. It is straightforward to show that the theories of Boolean algebra and Boolean rings (with unit) are equivalent. The particular axiom system given above with the redundant $a + bc = (a + b)(a + c)$ deleted for Boolean algebras has the property that it becomes an axiom system for Boolean rings just by replacing the axiom $a + a = a$ by $a + a = 0$ (with all the +'s then taken to be symmetric difference). A well-known simple example of a mathematical structure which is a Boolean ring is the field of integers modulo 2.

The close affinity of any Boolean algebra with two-valued arithmetic is expressed, in sophisticated language, by the following representation theorem [10, p. 50]: *every (non-degenerate) Boolean algebra is isomorphic to a subalgebra of a direct union of two-element Boolean algebras.* In unsophisticated language this says that any Boolean algebra "looks like" an algebra (Boolean, of course) on a subset of all n-tuples $U = \{\langle v_1, v_2, \ldots\rangle\}$ with each v_i either 0_i or 1_i (the zero and unit of a two-element Boolean algebra), and with operations on the n-tuples defined componentwise. The importance of the two-valued arithmetic in connection with Boolean algebras is also brought out in the following useful result: any Boolean polynomial equation in n variables is true for all Boolean algebras (i.e., is a Boolean identity) if true for all 2^n possible assignments of 0's and 1's for the variables. We also note that by virtue of idempotency ($x^2 = x$) a Boolean polynomial is linear in each of its variables.

Since a Boolean algebra can be viewed as a ring (with symmetric difference as the addition) the notions of ideals and of residue classes can be introduced in customary fashion.

BOOLE'S LOGIC OF CLASS TERMS

Boole believed he was the first to use a mathematical approach to logic. But much earlier Leibniz had conceived of the idea of a formal mathematical system which could be used to conduct logical inference. In manuscript notes and papers—which did not become generally known until some 200 years later—he had made some remarkable starts but did not bring the project to fruition. One of the minor hampering items may have been his regarding general terms (e.g., "human", "sheep") primarily as designating attributes rather than classes (i.e., thinking *intensionally* rather than *extensionally*). In this view the compound term "human sheep" has a wider, more inclusive, intension than either "human" or "sheep", whereas as a class it has a less inclusive extension than either of the component terms. Thinking extensionally is far simpler; for example, the terms "human" and "featherless biped" designate distinct attributes but as classes are identical (plucked chickens being ruled out). Leibniz could, and did, think either way, but Boole's thinking was exclusively extensional.

A more substantial difference, however, was the circumstance that Leibniz, faced with *infiniti modi calculandi,* had to grope his way towards a calculus of logic, whereas Boole was sure he had one ready-made, namely symbolical algebra, furnished with a suitable interpretation for its general symbols and operations. Having as a guide a theoretical system, even if only approximately correct, can be decidedly advantageous. As we now know, neither of them had at their disposal a clear and sufficiently extensive body of formal logical usage—what was available, to Boole as well as to Leibniz, wasn't much more than the Aristotelian syllogistic, a body of rules not lending itself well to formulation as an algebra of the equational type, and, moreover, burdened with the distracting problem of existential import (Does "All S are P" imply "There are S"?).

In his initial writings on logic Boole associated the general symbols x, y, z,... of symbolical algebra with *operators* which selected classes out of a universal class. (Subsequently, this aspect became less pronounced and the symbols came to stand for the classes themselves.) Selecting out the x's and then the y's produces the same class as selecting out the y's and then the x's. Thus Boole wrote

$$xy = yx.$$

For Boole to think of using multiplication for logical intersection was quite natural since this was what was used in the calculus of operators to indicate successive application of operations. In keeping with this "selecting out" idea the symbolic laws

$$x \cdot 1 = x, \quad x \cdot 0 = 0$$

inevitably suggest the logical interpretation for 1 as the universal class and 0 as the empty class. Selecting out the x's and then from this the x's yields the x's; hence the algebraic property,

$$xx = x^2 = x,$$

a departure from ordinary algebra which we have already mentioned.

To symbolize the aggregate of two classes (e.g., "mountains and vales") Boole used the symbol of addition but with the proviso that the classes be disjoint and hence the operation is only partially defined. Boole had no objection to partially defined operations—compare, for instance, $a - b$ in arithmetical algebra. Similarly, subtraction required that the "subtrahend" class be contained in the "diminunend" class. Under these conditions he found

$$x(y + z) = xy + xz$$

and

$$x(y - z) = xy - xz$$

are true for classes. With hardly much more than these examples Boole boldly proposed [1, p. 37]:

> Let us conceive, then, of an Algebra in which the symbols, x, y, z, etc., admit indifferently of the values 0 and 1, and of these values alone. The laws, the axioms, and the processes, of such an Algebra will be identical in their whole extent with the laws, the axioms, and the processes of an Algebra of Logic. Difference of interpretation will alone divide them. Upon this principle the method of the following work is established.

In these sentences we heve, for the period, some remarkably innovative ideas: that there can be an algebra of 0 and 1, that such an algebra has at least two interpretations, and that one of these is logic. These items give rise to the following questions.

(1) What is Boole's algebra of 0 and 1?
(2) Is arithmetic modulo 2 one of the interpretations?
(3) How does one interpret the algebra to get logic?

It will be convenient for us to address these questions in reverse order.

Re (3). As already indicated, the symbols x, y, z, etc., are to stand for classes, with 1 being the universe and 0 the empty class. The product xy represents the intersection of the classes represented by x and y and is meaningful in all circumstances. The aggregate class $x + y$ has meaning only if x and y are mutually exclusive; likewise $x - y$ only if y is contained in x (and thus is *not* the same as the xy' of Boolean algebra). The expression $1 - x$ is meaningful in any case and represents the complement of x in the universe 1. Assertions are always represented by equations—e.g., "No x's are y's" is "$xy = 0$" and "All x's are y's" is "$x = vy$" where v is an indefinite class symbol. This use of the special symbol v was adversely criticized by other logicians but can in part be justified by considering it as an understood existentially quantified variable [2, pp. 97–98]. Logical inference—about which we will say very little in this paper—is performed by algebraic transformations on equations, allowing all the operations of "common" algebra. This creates a problem (to be discussed in detail in our next section) in that expressions arise, e.g., -1, $x + x$, $x - 1$, x/y, to which no logical meaning has been assigned. Boole considered this perfectly acceptable in intermediate steps of a deduction so long as the starting and ending equations were interpretable in logic. He likened this to "the employment of the uninterpretable symbol $\sqrt{-1}$ in the intermediate processes of trigonometry". Aside from this dubious feature, the direct translating of logical statements involving class terms *into* algebraic form goes tolerably well.

Re (2). It is well known that an interpretation for Boolean algebra (i.e., the algebra of section 2) can be given as an algebra over the two-element set $\{0, 1\}$ whose arithmetic differs from ordinary arithmetic in that one has either (i) $1 + 1 = 1$ or (ii) $1 + 1 = 0$, the latter case yielding arithmetic modulo 2. However, neither (i) nor (ii) can be the interpretation Boole has in mind, for he always writes "1 + 1" as "2", considering 2 as a "numerical" factor which does not obey the general law of thought $x^2 = x$. Correspondingly, for any algebraic expression A, the sum "$A + A$" is written "$2A$" and never simplified. In a letter to Jevons written in 1863 in response to an inquiry, Boole flatly asserted that $x + x = x$ was not true (i.e., a law) in logic, so that (by implication) $1 + 1 = 1$ would not be true in his algebra of 0 and 1. Clearly then Boole would consider neither arithmetic modulo 2 nor the alternative with $1 + 1 = 1$ as an interpretation for his algebra.

Re (1). It is not easy to determine what Boole's algebra is since he gives no explicit list of its laws (i.e., axioms). He talks about $x^2 = x$ being the general law of thought, but $A^2 = A$ for arbitrary A is not one of his algebraic laws—it holds for *class symbols* x, y, etc., and for expressions such as

$$xy,\ 1 - x,\ xy' + x'y$$

which he refers to as "independently interpretable", but not for expressions such as

$$2x,\ x + y,\ x - y$$

which nevertheless do occur in his procedures. An expression A not satisfying the idempotency condition $A^2 = A$ is dubbed "uninterpretable". A careful examination of those algebraic properties of + and × which Boole actually makes use of, for interpretables as well as uninterpretables, show that they are those of a

commutative ring with unit having no additive or multiplicative nilpotents. **(SM)**

(No additive or multiplicative nilpotents means that $nA = 0 \Rightarrow A = 0$ and that $A^n = 0 \Rightarrow A = 0$.) To use this algebra for logic (of class terms) one needs entities which satisfy Boole's law of thought $x^2 = x$. Let us single out and call such entities the Boolean (or idempotent) elements and designate the set of such elements by $\mathbf{B}$. Clearly the 0 and 1 of the ring are in $\mathbf{B}$. Also, for any $x, y \in \mathbf{B}$, the commutative ring properties imply that the elements

$$xy,\ 1 - x,\ xy' + x'y, \quad \text{and} \quad xy + x'y + xy'$$

(here $x' = 1 - x$) are also in $\mathbf{B}$. It is easy to show that the elements of $\mathbf{B}$ will constitute a Boolean algebra (or equivalently a Boolean ring) with the operation

$$x \cup y = xy + x'y + xy'$$

as the Boolean sum, or with

$$x +_{\Delta} y = x'y + xy'$$

as the Boolean ring addition. Thus by restricting himself to the idempotents of his algebra, Boole would have had Boolean algebra—and we rightfully honor him by attaching his name to this algebra. Only *Boole* didn't know it. He steadfastly refused to acknowledge any operation but his +. When Jevons claimed that $x + x = x$, and Boole emphatically denied this, they were really talking about different operations. Jevons' + is indeed the present day $\cup$; however Boole's + is not, as is generally believed, the $+_\Delta$ of the Boolean ring but the + of the ring **SM**. This algebra merits a brief discussion.

What kinds of models (realizations) does Boole's algebra **SM** have? An answer to this question comes from a structure theorem of N.H. McCoy [8, p. 123] which, adapted to our circumstance reads: any model of **SM** is isomorphic to a subdirect sum of integral domains which are without additive nilpotents. Intuitively such a model looks like n-tuples of elements in which each component ranges over an integral domain without additive nilpotents, and with the operations of + and × for these n-tuples defined componentwise.

Not all models or interpretations of a set of axioms are of equal general interest. In the case of Boolean algebra there is a sense in which one can say that the principal interpretation is that of an algebra of sets. For Boole's algebra **SM** we consider the principal interpretation to be that of an **algebra of signed multisets**: a multiset is like an ordinary set except that multiple occurrences of elements are allowed, and by a signed multiset we mean one in which negative multiplicities are allowed. For example, during a poker game your pile of chips contains various positive or zero multiples of red, white or blue chips, all of each color indistinguishable as far as the game is concerned; and if you borrow from the pot, you are adding negative multiples to your pile. When the multiplicities are restricted to being either 0 or 1, then multisets become ordinary sets. As for the operations, Boole's + corresponds to dumping the contents of the two multisets together, and × corresponds to multiplying the respective multiplicities. (For more details see [2, pp. 91, 92–96]; in that work we used the term "heap", not realizing that "multiset" was already in use.)

If all one wanted to use Boole's algebra for was to do class (i.e., term) logic, then there is no need to go beyond the idempotents—and this is the path history chose. Nevertheless, the richer structure of Boole's algebra can be of interest in its own right, for example, in the logic of multisets. As a simple instance, the equivalence

$$(x + x)y = 0 \Leftrightarrow xy = 0$$

when interpreted for multisets tells us that duplicating the elements of a set does not affect its being exclusive from another set. It is possible that Boole's algebra can be useful in the subject of pseudo-Boolean functions [3]. Pseudo-Boolean functions, a notion used in operations research, are functions from a two-element set $\{0, 1\}$ into the integers, and they have an obvious representation by means of a polynomial in Boolean variables with integer coefficients.

BOOLE'S INTERPRETATION PROCEDURE

As we have mentioned earlier, Boole considered his algebraic methods for doing logic to be sound so long as the starting and end formulas were interpretable in logic.

But many of the algebraic processes—especially division, which plays an important role in Boole's method—led to uninterpretable forms. For example, in order to determine what the Boolean equation $xw = y$ implies about the class w, Boole solves this equation for w, obtaining

$$w = \frac{y}{x}.$$

What does y/x mean in logic? What is the interpretation for this equation? Boole ignores the first of these questions, but to handle the second he introduces the idea of the **expansion** (or **development**) **of a fonction** and along with it a general method for interpreting an equation between a class symbol and such an expansion. By this technique—widely decried as "mysterious"—Boole had a general and uniform method for obtaining the logical import of any equation in class symbols (i.e., Boolean variables). We give a brief description of this procedure.

If $f(x)$ is an algebraic expression (Boole only considered linear fractional forms in Boolean variables), its expansion is given by

$$f(x) = f(1)x + f(0)(1-x).$$

Boole "establishes" this identity by assuming $f(x) = ax + b(1-x)$ and determining a and b by setting $x = 1$ and $x = 0$. Similarly (using x' as an abbreviation for $1 - x$) for an expression in two variables he gives

$$f(x,y) = f(1,1)xy + f(1,0)xy' + f(0,1)x'y + f(0,0)x'y'.$$

Thus if $f(x,y) = \frac{y}{x}$, then

$$\frac{y}{x} = 1 \cdot xy + 0 \cdot xy' + \frac{0}{0}x'y' + \frac{1}{0}x'y.$$

(We follow Boole's custom of putting the 1/0 term last.) Boole's argument for equating a function with its expansion is faulty in that it assumes (without justification) that any such function is linear in its Boolean variables. To conclude, Boole's rule for interpreting an equation $w = f(x,y,...)$ is to equate w to the sum of those terms in the expansion of $f(x,y,...)$ which have 1 as their coefficient plus an indefinite multiple of the sum of those terms having 0/0 as their coefficient and, as an independent condition, the sum of those terms having 1/0 as coefficient is equated to 0. The reasons Boole gives for his rule are weak and unconvincing and we shall not reproduce them. As an example of its use, the interpretation for the equation $w = y/x$ is that the class w consists of all of xy, none of xy', an indefinite amount (some, none, or all) of $x'y'$ and, independently, $x'y$ is set equal to 0. The correctness of this as an equivalent to $xw = y$ is easily seen in FIGURE 1.

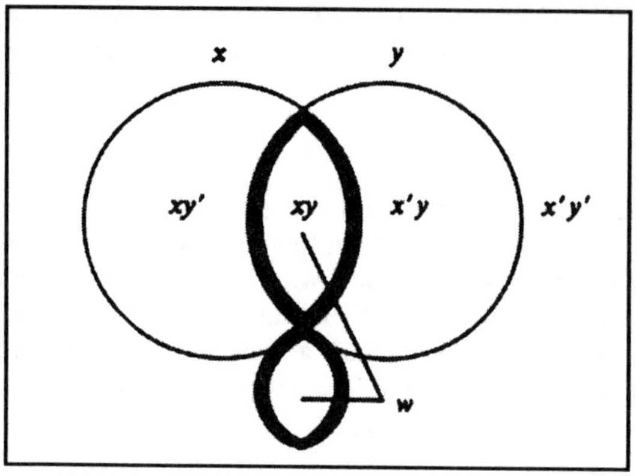

FIGURE 1.

$xw = y$
equivalent to

$$\begin{cases} x'y = 0 \\ w = xy + vx'y', v \text{ indefinite.} \end{cases}$$

In the *Laws of Thought* there are lots of worked-out examples, all with correct conclusions. One is intrigued by this fact and wonders how such a process involving "division" of class expressions, and using entities such as 0/0 and 1/0, can be meaningful. We venture an explanation for this in our next section.

DIVISION IN BOOLEAN ALGEBRAS

In the arithmetic of integers the quotient of m by n exists if $n \neq 0$ and m is a multiple of n. Thus for integers the operation of division is a restricted one and is not always possible. The process of extending the integers to the more extensive structure of the field of rationals is well known: one defines the rationals as equivalence classes of ordered pairs of integers (here ordered pairs = fractions) and with appropriate definitions of addition, multiplication, zero and unit, one has a new structure in which division is possible without exception (save for 0). This structure, moreover, contains a substructure isomorphic to the integers, so that one can still do integer arithmetic within the rationals.

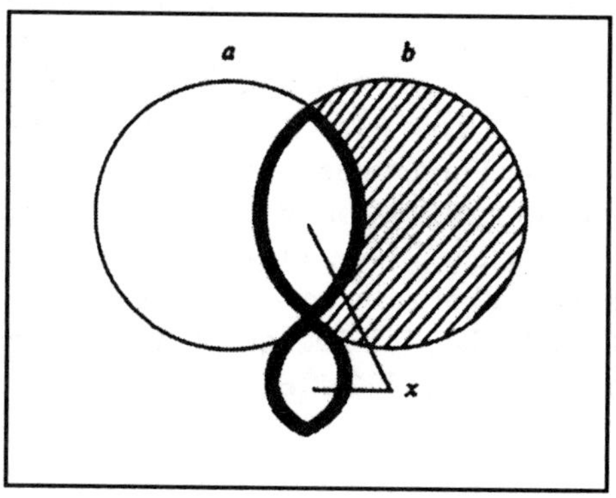

FIGURE 2

Consider now in a Boolean algebra **B** the equation

$$ax = b, \tag{1}$$

where a and b are given and x is unknown. It is a simple exercise in Boolean algebra to show that this equation has a solution for x if and only if b is included in a, i.e., if and only if $a'b = 0$. Thus, as with the integers, quotients do exist for some pairs of elements of **B**. However, unlike the integers, such a quotient need not be unique. As we have already seen in illustrating Boole's interpretive procedure, a solution x for $ax = b$ will be equal to ab plus any amount of $a'b'$ (see FIGURE 2); or, to say it in other words, any two members of the solution set for (1) will differ by a Boolean multiple of (i.e., a part of) $a'b'$. We now show how to "factor out" the differences so as to have a unique quotient (if any exist).

To say that any two solutions of (1) differ by a multiple of $a'b'$ is to say that the solution set is a residue class in the factor ring $\mathbf{B}/(a'b')$, where $(a'b')$ is the principal ideal generated by $a'b'$ (= set of classes included in $a'b'$). If in the factor ring $\mathbf{B}/(a'b')$ we denote by

$$[r] = r + (a'b')$$

the residue class determined by an element r of **B**, then the mapping $r \to [r]$ is a homomorphism of rings and hence (1) implies that

$$[a][x] = [b] \tag{2}$$

holds in the ring $\mathbf{B}/(a'b')$. Conversely, (2) implies (1). For if, for $v_1, v_2, v_3 \subset \mathbf{B}$, we have

$$(a + v_1a'b')(x + v_2a'b') = b + v_3a'b',$$

then by multiplying this equation through by $(a'b')'$ $(= a \cup b)$ and using simple Boolean identities, one readily obtains (1). Thus the problem of solving for x in (1) is equivalent to finding $[x]$ satisfying (2) in the reduced Boolean algebra $\mathbf{B}/(a'b')$. If there is a solution, say $[x] = [p]$, then, on using the fuller notation for a residue class, this solution is

$$[x] = p + (a'b'). \tag{3}$$

In order to better bring out the relation between what we are doing with Boole's procedure, we now alter the customary notation for a principal ideal in a ring and write $\frac{0}{0}a'b'$ in place of $(a'b')$ so that (3) becomes

$$[x] = p + \frac{0}{0}a'b' \tag{4}$$

which, in terms of elements of **B**, says

$$x = p + va'b' \tag{5}$$

where v is an element of **B**. By going over to $\mathbf{B}/\frac{0}{0}a'b'$ we are now at a stage similar to solving an equation $mx = n$ with m and n integers, where only under limited circumstances does a (unique) quotient exist. We would like the greater algorithmic freedom that one has when working with rationals where division is unrestricted. This is essential if we are to reach our goal of justifying Boole's unusual technique, which involves solving equations such as

$$Ew = F$$

for the unknown w with E and F polynomials in variables $x, y, z, \ldots$. To this end we turn to the problem of extending a Boolean algebra by the introduction of quotients.

EXTENDING A BOOLEAN ALGEBRA WITH QUOTIENTS

In contemplating the introduction of quotients and the operation of division into a Boolean algebra, one is confronted with the difficulty that the usual definitions of addition and multiplication of two fractions, namely

$$\frac{r_1}{s_1} + \frac{r_2}{s_2} = \frac{r_1 s_2 + s_1 r_2}{s_1 s_2}$$

$$\frac{r_1}{s_1} \cdot \frac{r_2}{s_2} = \frac{r_1 r_2}{s_1 s_2},$$

involves the product of the two denominators. In the case of the integral domain of integers, where there are no divisors of zero, if s_1 and s_2 are not 0, then neither is the product $s_1 s_2$. However in the case of Boolean algebra, where $aa' = 0$ for any a, *every* nonzero element is a divisor of zero. Thus no element of a Boolean algebra can function with complete freedom as a denominator.

Resolution of the difficulty (though not without a price) comes by suitably restricting the denominators to being in a multiplicative set: we say S is a **multiplicative set** if

(i) $0 \notin S$ and
(ii) $s_1, s_2 \in S \Rightarrow s_1 s_2 \in S$ for any s_1, s_2.

For any commutative ring R with unit, by restricting the formation of fractions to those having denominators coming from a multiplicative subset S of R, one can alleviate the aforementioned difficulty. But now the customary definition of the equivalence of two fractions $\left(\frac{r_1}{s_1} \sim \frac{r_2}{s_2} \textit{ iff } r_1 s_2 - s_1 r_2 = 0\right)$ has to be modified to

$$\frac{r_1}{s_1} \sim \frac{r_2}{s_2} \text{ iff for some } s \in S,\ s(r_1 s_2 - s_1 r_2) = 0. \tag{6}$$

Denote by $\left[\frac{r}{s}\right]$ the equivalence class determined by a pair of elements $r, s \in R$, $s \in S$, and by RS^{-1} the set of all such equivalence classes. Then with operators + and × defined by

$$\left[\frac{r_1}{s_1}\right] + \left[\frac{r_2}{s_2}\right] = \left[\frac{r_1 s_2 + s_1 r_2}{s_1 s_2}\right],$$

$$\left[\frac{r_1}{s_1}\right]\left[\frac{r_2}{s_2}\right] = \left[\frac{r_1 r_2}{s_1 s_2}\right],$$

and with $\left[\frac{0}{1}\right]$ and $\left[\frac{1}{1}\right]$ as the zero and unit, one can show that RS^{-1} is a ring—the ring of quotients of R by S [12]. And now here is the price we have to pay in order to have quotients in a general ring: in the resulting ring of quotients not every member of R can be a denominator, nor need the ring of quotients be a field; moreover, although the mapping $h: R \to RS^{-1}$, $h(r) = \left[\frac{r}{1}\right]$, is a homomorphism of rings, it need not be injective—as would be the case for the rationals.

To apply this theory to a Boolean algebra (which we take in the form of a ring with + standing for symmetric difference) one chooses as a denominator set any filter. If e is a nonzero element of a Boolean algebra **B**, then the filter determined by e consists of all elements which include e. We may conveniently designate this set by $e \cup \mathbf{B}$. Clearly $e \cup \mathbf{B}$ is a multiplicative set and thus $\mathbf{B}(e \cup \mathbf{B})^{-1}$ is a ring of quotients of **B** by $e \cup \mathbf{B}$. As the usual notation for quotients is here insufficient in that it contains no indication of the denominator set, let us use $\left(\frac{r}{s}\right)_e$ where the e suffices to indicate the denominator set $e \cup \mathbf{B}$.

There are a few results about Boolean quotients to which we will refer:

THEOREM 1. *For any element a in the denominator set* $e \cup \mathbf{B}$

$$\text{(i)} \quad \left(\frac{ab}{a}\right)_e = \left(\frac{b}{1}\right)_e,$$

$$\text{(ii)} \quad \left(\frac{b}{a}\right)_e = \left(\frac{ab}{1}\right)_e.$$

Proof. Both results are immediate consequences of definition (6); (i) by virtue of $e(1 \cdot ab - a \cdot b) = 0$, and (ii) by $a(b \cdot 1 - a \cdot ab) = 0$.

THEOREM 2. *The factor ring* $\mathbf{B}/(e')$ *and the ring of quotients* $\mathbf{B}(e \cup \mathbf{B})^{-1}$ *are, under the mapping* $b + (e') \to \left(\frac{b}{1}\right)_e$, *isomorphic structures.*

Proof. [2, p. 34].

We now show how to reproduce, with due mathematical rigor, Boole's solution of a Boolean equation by division and expansion, and to justify his interpretation for the algebraic solution. Since the solution of a Boolean equation is directly obtainable in Boolean algebra, our procedure, which uses a number of new structures, will appear complicated and circuitous. We emphasize, however, that it is the reproducing of Boole's method, not the solution of the equation, which is of interest.

Consider an equation

$$ax = b \tag{7}$$

where a ($\neq 0$) and b are elements of a Boolean algebra $\mathbf{B}$. As we have seen above, this equation has a solution for x in $\mathbf{B}$ if and only if the equation

$$[a][x] = [b] \tag{8}$$

has a solution for $[x]$ in $\mathbf{B}^* = \mathbf{B}/\frac{0}{0}a'b'$, where $\frac{0}{0}a'b'$ is the principal ideal generated by $a'b'$ and $[r]$ designates the residue class modulo $\frac{0}{0}a'b'$ determined by r; moreover, if $[x] = [p]$ is the solution of (8), then the solution set for (7) is given by

$$x = p + va'b', \qquad v \text{ ranging over all elements of } \mathbf{B}.$$

Suppose now that (8) has a solution. In the ring of quotients $\mathbf{Q} = \mathbf{B}^*([a] \cup \mathbf{B}^*)^{-1}$ we have, from (8),

$$\left(\frac{[a][x]}{[a]}\right)_{[a]} = \left(\frac{[b]}{[a]}\right)_{[a]}, \tag{9}$$

from which, by cancellation (Theorem 1(i)),

$$\left(\frac{[x]}{[1]}\right)_{[a]} = \left(\frac{[b]}{[a]}\right)_{[a]}. \tag{10}$$

This is what for us corresponds to Boole's $x = b/a$. Using Theorem 1(ii) and the fact that the mapping $r \to [r]$ is a homomorphism to replace $[a][b]$ by $[ab]$, we obtain from (10) that

$$\left(\frac{[x]}{[1]}\right)_{[a]} = \left(\frac{[ab]}{[1]}\right)_{[a]} \tag{11}$$

in the ring **Q**. We now go over to the ring $\mathbf{B}^*/([a]')$ which, by Theorem 2, is isomorphic to **Q**. But first we note that $[a]' = [a']$, and that $[a'] = [a'b]$ since a' and $a'b$ differ by $a'b'$ and thus determine the same residue class modulo $a'b'$. Hence the ideal $([a]')$ is $([a'b])$. We change the usual notation and represent this ideal by $\frac{1}{0}a'b$. (Note that $\frac{1}{0}a'b$ is an ideal of $\mathbf{B}^*$, whereas $\frac{0}{0}a'b'$ is an ideal of **B**.) Using "$\approx$" to indicate the relation of isomorphism between elements of $\mathbf{B}^*([a] \cup \mathbf{B}^*)^{-1}$ and $\mathbf{B}^*/([a]')$ we have by Theorem 2

$$\left(\frac{[b]}{[a]}\right)_{[a]} \approx [ab] + \frac{1}{0}a'b$$

or, in fuller notation,

$$\left(\frac{[b]}{[a]}\right)_{[a]} \approx 1 \cdot ab + 0 \cdot ab' + \frac{0}{0}a'b' + \frac{1}{0}a'b \tag{12}$$

which is what for us corresponds to Boole's

$$\frac{b}{a} = 1 \cdot ab + 0 \cdot ab' + \frac{0}{0}a'b' + \frac{1}{0}a'b.$$

Since he also has $x = b/a$, he can then write

$$x = 1 \cdot ab + 0 \cdot ab' + \frac{0}{0}a'b' + \frac{1}{0}a'b,$$

whereas we can only equate the isomorphic images of the members of (11) to obtain

$$[x] + \frac{1}{0}a'b = [ab] + \frac{1}{0}a'b$$

$$= 1 \cdot ab + 0 \cdot ab' + \frac{0}{0}a'b' + \frac{1}{0}a'b. \tag{13}$$

Now for the Boolean interpretation (i.e., in terms of **B**) of (13). If $ax = b$ then $a'b = a'ax = 0$, so that

$$\frac{1}{0}a'b = ([a'b]) = ([0])$$

is the zero ideal of **B***. Thus from (13)

$$[x] = [ab] = 1 \cdot ab + 0 \cdot ab' + \frac{0}{0}a'b'$$

and so by (5),

$$x = ab + va'b' \qquad v \in \mathbf{B}.$$

Consequently (7) implies, through (13), that

$$\begin{cases} x = ab + va'b' & v \in \mathbf{B} \\ a'b = 0 \end{cases} \tag{14}$$

which is precisely what Boole gives as his interpretation for

$$x = 1 \cdot ab + 0 \cdot ab' + \frac{0}{0}a'b' + \frac{1}{0}a'b.$$

Boole didn't consider a converse argument (i.e., from (14) to (7)) since it was unnecessary from his point of view. For him the processes of "common" algebra were applicable to logical equations and hence b/a would obviously be the correct solution to $ax = b$. The only problem for him was to obtain the logical meaning of $x = b/a$.

SUMMARY

We have briefly described the nascent abstract algebra ideas within which Boole originated his algebra of logic. While he never made his algebra fully explicit, we inferred that what he did use was, if clarified, a commutative ring with unit, without nilpotents, and having idempotents which stood for classes. By thus hewing closely to "common" algebra Boole could use familiar procedures and techniques. He did not realize that class calculus needed only the idempotents and operations closed with respect to them (i.e., Boolean algebra). Instead he used the ring operations and, in particular, its addition, which is not closed with respect to idempotency. Boole also freely used division to solve equations, introducing then a special, not clearly explained or justified, method for extracting the logical content of the resulting quotient. To explain this we found it necessary to introduce additional structures—factor rings, and rings of quotients for such rings.

We have alluded to the possibility of applications for Boole's algebra—to multisets and to pseudo-Boolean functions. As an additional possibility we would also like to mention that in his *Laws of Thought* Boole uses his peculiar expansions with 1/0 to present an original approach to conditional probability [2, p. 195].

REFERENCES AND READINGS

For matters relating to the history of algebra relevant to this paper, we recommend [5] and [7], and for the history of logic, [6]. For an alternative to our justification of Boole's methods, one may consult [11, pp. 191–197].

[1] G. Boole, An Investigation of the Laws of Thought, on which are founded the Mathematical Theories of Logic and Probabilities, Walton and Maberley, 1854. Reprinted by Open Court Publ. Co., 1916, 1940, and by Dover, New York, 1951.
[2] T. Hailperin, Boole's Logic and Probability, North-Holland, New York, 1976.
[3] P.L. Hammer (Ivanescu) and S. Rudeanu, Boolean Methods in Operations Research and Related Areas, Springer-Verlag, New York, 1968.
[4] W.S. Jevons, Pure Logic and Other Minor Works, Macmillan, New York, 1890.
[5] M. Kline, Mathematical Thought from Ancient to Modern Times, Oxford Univ. Press, New York, 1972, Chapter 32.
[6] W. Kneale and M. Kneale, The Development of Logic, Oxford Univ. Press, New York, 1962.
[7] E. Koppelman, The calculus of operations and the rise of abstract algebra, Arch. Hist. Exact Sci., 8 (1971) 155–242.
[8] N.H. McCoy, Rings and Ideals, Carus Mathematical Monograph No. 18, The Mathematical Association of America, Washington, D.C., 1948.
[9] G. Peacock, A treatise on algebra, 1842; reprinted, Scripta Math., Yeshiva College, 1940.
[10] R. Sikorski, Boolean Algebras, 3rd ed., Springer, New York, 1969.
[11] N.I. Styazhkin, History of Mathematical Logic from Leibniz to Peano, The M.I.T. Press, 1969 (Originally published as Stanovleniye idey mathematischeskoy logiki, Nauka, 1964).
[12] A.I. Uzkov, On rings of quotients of commutative rings (Russian), Mathematicheskii sbornik (N.S.), 22(64) (1948) 439–441. (Translated as American Mathematical Society Transl. no. 3, 1–5, American Mathematical Society, 1949.)

Lehigh University, Bethlehem, PA 18015 USA

MICHAEL DUMMETT

REVIEW OF BOOLE*

The volume edited by Rhees contains reprints of Boole's original booklet *The Mathematical Analysis of Logic* (1847) (I), and of his essays 'The Calculus of Logic' (1848) (II), 'Of Propositions Numerically Definite' (c. 1850, pub. 1868) (IV), and 'The Claims of Science' (1851) (V), and some notes on I and two essays on logic (VI and VII) from Boole's manuscripts published for the first time. There is another essay (III) from the manuscripts which is partly concerned with probability theory. I, II, and VII are printed with omissions. The long 'Note on Editing' by Rhees quotes some passages omitted from VII and another unpublished essay. There are also ten reprints of essays by Boole on probability theory, and reprints of essays on the same subject by Wilbraham and Bishop Terrot; these fall outside the scope of this Journal.

There can be no doubt that Boole deserves great credit for what he achieved, in the sense that in those historical circumstances what he did must have been very difficult to do. If, however, we ask after the historical significance, and in particular, the present interest of Boole's writings, a different answer must be given. Boole cannot correctly be called 'the father of modern logic'. *The* discoveries which separate modern logic from its precursors are of course the use of quantifiers (or, more generally, of operators which bind variables and can be nested) and the concept of a formal system, both due to Frege and neither present even in embryo in the work of Boole. Boole has indeed a great historical importance both for abstract algebra and for logic. As had Leibniz two centuries earlier, he devised a general theory of classes under Boolean operations, a theory which of course contained the traditional theory of the syllogism. This move gained its importance for logic rather from the novelty of *any* extension of logical theory than from the magnitude of the extension itself; and anyone unacquainted with Boole's works will receive an unpleasant surprise when he discovers how ill-constructed his theory actually was and how confused his explanations of it. He thought that the word 'nothing' denotes the null class, that a proposition like 'Some men are vegetarians' is a conventional ellipsis for 'Some men are some vegetarians', and that the latter proposition contains a reference to *four* classes, denoted respectively by 'men', 'vegetarians', and by each of the two occurrences of 'some' (pp. 128–129). His method of dealing with particular propositions, by means of his symbol 'v', is irretrievably confused; 'v' sometimes denotes an arbi-

*Originally published as a review of George Boole, *Studies in Logic and Probability*, ed. R. Rhees, London and La Salle, Illinois, 1952, and of 'Celebration of the Centenary of *The Laws of Thought* by George Boole', *Proceedings of the Royal Irish Academy*, vol. 57, section A, n° 6, 1955, in *Journal of Symbolic Logic*, vol. 24, n° 3, September 1959, pp. 203–209.

J. Gasser (ed.), A Boole Anthology, 79–85.

trary class, sometimes an arbitrary *non-empty* class—and this is not all that is at fault with his use of the symbol. He introduced a division sign for the operation inverse to intersection, and never succeeded in unravelling the complicated tangles which resulted from this. He stated the fundamental theorem that $\phi(x) = \phi(1)\cdot x + \phi(0)\cdot(1 - x)$ for any Boolean function ϕ without ever giving a reasonable proof of it; indeed, in I he professes to derive it from Taylor's theorem.

The presentation of Boole's logical theory contained in these essays is markedly inferior to that of *Laws of Thought* and it seems to the reviewer that there is in this respect little of interest to be derived from them that cannot be derived from that book. In particular, although Boole's logic was intended to be extensional, he throughout these essays explains '1' as denoting the class of all *conceivable* objects 'whether actually existing or not'—a mistake he does not make in *Laws of Thought* and which is presumably attributable to his supposing that logic has to do with what is *mental*.

Boole uses as primitive two constants 0 and 1 (the empty and universe classes), and three binary operators $\cdot$, $+$, and $-$. He explains generally that $x \cdot y$ is the intersection of x and y, but gives explicitly no quite general interpretation of '+' and '–'; instead he lays it down only that when x and y are disjoint, $x + y$ is the union of x and y, and when y is included in x, $x - y$ is the class of those members of x which are not members of y. However, there is not in his system any restriction on the formation of terms of the forms '$x + y$' and '$x - y$' to such combinations as may be interpreted by direct appeal to the rules of interpretation which he explicitly states; and in formalising logical inferences he regularly transforms equations in which no such 'uninterpretable' terms occur into equations in which they do occur and these in turn into equations in which they do not. This procedure obviously stands in considerable need of justification, a fact of which Boole's commentators seem aware without their giving a very clear account of the matter. Thus it is often correctly remarked that Boole's successors used '+' to mean union for any two classes; but this observation is seldom accompanied by noting that this interpretation is inconsistent with Boole's system as it stands, since it makes illegitimate the cancellation law, *If* $x + y = z + y$, *then* $x = z$, which Boole uses; and it has the consequence that there is no operation, $-$, which satisfies the condition, $x - z = y$ *if and only if* $x = y + z$, which Boole also assumes.

Boole states various axioms and in particular the law of idempotence, $x \cdot x = x$. Though he does not lay down any explicit restriction on the use of this axiom, it appears probable from several passages (e.g. p. 21) that he intended it to be understood rather differently from the others. In the law $x + y = y + x$, for example, we are permitted in accordance with the usual practice in mathematics to substitute any terms whatever for 'x' and 'y'. Boole appears, however, to have intended the law $x \cdot x = x$ to apply only to 'interpretable' expressions, i.e., we are permitted to substitute only single variables and other 'interpretable' terms for 'x' (it will in fact be sufficient if we permit the substitution of single variables only). If we make explicit this tacit restriction, we arrive at a possible interpretation of the resulting system by taking the variables to range over the integers 0 and 1, and the operations as representing ordinary arithmetical addition, multiplication and subtraction. The 'uninterpretable' terms will then denote integers other than 0 and 1. The possibility of this arithmetical interpretation proves the consistency of the system, but does not justify it with respect to its intended logical interpretation. Under the arithmetical interpretation, there is, relative to a given assignment to the variables, a denotation for every term including the 'uninterpretable' ones, although the range of denotation of the 'uninterpretable' terms is wider than the range of the variables. On the logical interpretation, however, there is really no general interpretation for the operators '+' and '–', and hence no denotation for the 'uninterpretable' terms. Hence a proof is needed that if we start with interpretable equations which are true on the logical interpretation under a certain assignment to the variables, transform these equations in accordance with Boole's (tacit) transformation rules into uninterpretable ones and these in turn into interpretable ones, the equations thus arrived at will also be true on the logical interpretation under the original assignment. Boole never gave a satisfactory proof of this, and his commentators seem quite content to leave the matter where he left it. It is true that he claimed (pp. 131–132) that every equation is interpretable; but the grounds on which this claim is made are extremely dubious.

It is in fact easy to supply the missing proof by considering a modification of Boole's system reached by dropping the restriction on substitution in the law of idempotence. We then have quite generally: $y + y = (y + y)^2 = 4y$, and hence $y + y = 0$. Therefore $x + y + y = x$, whence $x + y = x - y$. Further, $x - y = x - x{\cdot}y - y + x{\cdot}y = (x - x{\cdot}y) - (y - x{\cdot}y) = (x - x{\cdot}y) + (y - x{\cdot}y)$. Thus in this system we prove the general equivalence of '$x + y$' and '$x - y$' to each other, and of both to an expression which is always interpretable; '+' and '–' both denote symmetric difference (this corresponds to a well-known fact about rings). In the extended system, therefore, the partial interpretation of '+' and '–', taken together with laws of the system, determines a certain general interpretation of '+' and '–'. The interpretation thus obtained is of course a possible general interpretation for Boole's (restricted) system, which agrees with his partial interpretation, and hence ensures the correctness of his transformation rules under his interpretation. The arithmetical interpretation of the extended system would of course be the arithmetic of the integers mod 2. The extended system also shows a way in which Boole's system could be greatly simplified; if we drop the restriction on substitution, we need only two operators instead of three, and we can dispense with numerical coefficients as well as with numerical exponents.

Rhees does not see the matter in this light, and quotes Boole with approval as saying that '$x + y$' is interpretable only when x and y are disjoint, without apparently considering that such a procedure stands in need of justification. Boole indeed points out that if '$x + y$' is interpretable, we have: $x + x{\cdot}y + y{\cdot}x + y = x + y$, and hence: $x{\cdot}y + y{\cdot}x = 0$. But he goes on to say that we can have $x{\cdot}y + y{\cdot}x = 0$ only when $x{\cdot}y = 0$, 'since in Logic the only class, . . . which being added to itself produces Nothing, is Nothing'. This argument, if correct, would establish the impossibility of extending the partial logical interpretation to a general interpretation. Boole is tacitly appealing to the rule, *If* $z + z = 0$, *then* $z = 0$. This is a rule of which he makes no use in practice, and which gains plausibility only from the implicit assumption that there *is* a general interpretation for '+', and that it means union under this interpretation;

whereas in fact '+' cannot be taken to mean union (for arbitrary classes) consistently with the laws of Boole's system.

Rhees may here be following a very questionable remark of Boole's. In I and II Boole states that if we have an equation $\phi(x, y) = 0$, and we expand $\phi(x, y)$ in accordance with the fundamental theorem, then we may set any term whose coefficient is not 0 equal to 0. Thus (Boole's example) the coefficient of $y \cdot (1 - x)$ in the expansion of $x + 2y - 3x \cdot y$ is 2, and hence from $x + 2y - 3x \cdot y = 0$ we may derive $y \cdot (1 - x) = 0$. It might be possible to prove that whenever $x + 2y - 3x \cdot y = 0$ was derivable from a set of interpretable equations, so was $y \cdot (1 - x) = 0$, but Boole has given no reason whatever for believing this. It is certainly possible to give an assignment to the variables such that under the general interpretation suggested above the former equation comes out true while $y \cdot (1 - x) = 0$ comes out false, since in the extended system without restriction on substitution, the former equation is equivalent to $x \cdot (1 - y) = 0$. Hence the one equation is certainly not derivable from the other by the ordinary rules of Boole's system; the statement should be amended to read, 'any term whose coefficient is *odd*'. Boole does not repeat this mistake in *Laws of Thought*. Much of the rest of Rhees's 'Note' is similarly unhelpful, owing to his following Boole too closely. In particular, he follows him in admitting division into the system, and trying in consequence to 'interpret' such expressions as '$\frac{1}{0}$' and '$\frac{0}{0}$'; his explanations of these have no greater clarity or rigour than those of Boole. Boole's procedure for 'solving' Boolean equations is indeed a useful method, and can be stated without 'division' as the generalisation for functions of several arguments of the following theorem:

Let ϕ, ψ be Boolean functions such that $x \cdot \phi(y) = \psi(y)$. Write '$x_0$' for '$1 - x$' and '$x_1$' for '$x$', and similarly for '$y$'. Then for $i = 0$ or 1:

(i) if $\phi(i) = \psi(i) = 1$, then $x_{1-i} \cdot y_i = 0$;

(ii) if $\phi(i) = 0$, $\psi(i) = 1$, then $y_i = 0$;

(iii) if $\phi(i) = 1$, $\psi(i) = 0$, then $x_i \cdot y_i = 0$.

'Logic and Reasoning' (VI) is an interesting attempt on Boole's part to expound a theory of logical inference directly based on his symbolic calculus but with no overt reference to it. Boole was always exceedingly impressed by the fact that his calculus could be interpreted as the arithmetic of the integers 0 and 1. In VI he attempts to state the logical significance of this fact by means of the barely intelligible statement that 'it is the essence of the terms or names between which propositions express relations to represent something which we must regard as existent or non-existent'. He also came to regard his method for 'solving' Boolean equations as constituting the essence of his theory, and hence the four expressions '$\frac{1}{1}$', '$\frac{1}{0}$', '$\frac{0}{1}$', and '$\frac{0}{0}$' which occur in such 'solutions' as the fundamental constants. These appear in VI as four fundamental categories—the universal, the impossible, the non-existent, and the indefinite. The 'four operations' of Boole's calculus—multiplication, addition, subtraction, and 'division'—likewise appear as the 'mental operations' of composition, addition, subtraction, and abstraction. This essay serves to convince the reviewer, at least, only of the unfruitfulness for logic of the way in which Boole developed his symbolic theory.

IV is an essay concerned, in effect, with the introduction into monadic predicate calculus of operators of the form 'For exactly *n* *x*'s' and 'For at least *n* *x*'s'. This is an interesting topic, which, as far as is known to the reviewer, has seldom been systematically treated, perhaps because it has not been found to lead to very exciting results. There is, however, a far more elegant treatment of the topic in *Laws of Thought*, chap. 19: for example, Boole there avoids the muddle at the beginning of IV due to his handling of the symbols '+' and '–', and gives in Prop. I a much neater proof of Prop. V of IV. The editor does not say whether IV is the manuscript referred to in *Laws of Thought*, p. 310 fn., as having been lost.

'On the Mathematical Theory of Logic' repeats VI but with explicit mention of Boole's symbolism. II is a summary of I. V contains highly general remarks of little interest. That part of 'Sketch of a Theory' (III) which concerns logic rather than probability theory is transitional between I and *Laws of Thought*, but contains nothing not found in one or other. The only misprint I have noticed in the book is '*z*' for '2' on page 164, line 26.

Of the articles in the Irish Academy Centenary volume, Taylor's is a biographical sketch, a part of which controverts some features of E.T. Bell's account in *Men of Mathematics*. Rhees's is a series of quotations from manuscripts giving contemporary views of Boole's pedagogic abilities and methods. Hackett's is an account, based on manuscript writings by Boole, of his theory of education, and compares it with that of Whitehead. Thomas expounds clearly and concisely Boole's philosophy of science, including mathematics and logic, as set out in *Laws of Thought* and in the volume edited by Rhees. The author largely refrains from criticism or comment. He draws attention to Boole's traditionalist view that the concern of logic is with the laws governing the operations of the *mind*, and quotes Łukasiewicz as expressing the contrary view.

Feys's first article is concerned with 'placing' Boole in the history of logic. Feys regards Boole as *the* originator of modern logic: it seems to the reviewer, on the contrary, that if (say) the development of logic had stopped short with Schröder, there would have been no clear sense in which contemporary logic would be superior to that of, for example, the Scholastics. Feys starts with a contrast between the logical work of Leibniz and of Boole, which seems to the reviewer to overstress the superiority of Boole's achievement: the most important difference was surely that Boole published and Leibniz did not. The modern algebraist, Feys says, would find more similarity between Boole's work and his own than would the modern logician. 'The supposedly fundamental logic of propositions and classes can be considered as a very special kind of algebra, that of the ... Boolean rings'. What, then, 'becomes ... of the all-pervading rationality we suppose to exist in the laws of nature and in the laws of thought? ... Rationality ... must not necessarily be manifested by a univocal uniformity of these laws, but rather by the formal analogy existing between an infinity of systems': it may however be that we cannot in this world reach 'a final interpretation of these . . . logico-mathematical systems'. It seems to the reviewer that in these concluding remarks Feys is regarding logic from a wholly untenable philosophical position.

In his second article Feys expounds Boole's treatment of the fundamental theorem and his method of solving Boolean equations. Feys, like Rhees, is content to accept Boole's merely partial interpretation of his system (pp. 99, 107); although he does remark (p. 112) that it is possible to define '+' in such a way as to mean 'exclusive alternation'. He also wrongly says that '$\frac{x}{y}$' is interpretable whenever x is included in y ($\frac{x}{y}$ is in fact unique only when $y = 1$). He is anxious to preserve and defend Boole's use of the operation of division and of the symbols '$\frac{0}{0}$' and '$\frac{1}{0}$' in solving equations, and his interpretation of solutions of equations as giving 'conditions for interpretability': these procedures seem to the reviewer quite indefensible. Feys offers two incompatible justifications of them. One is that the system has two interpretations, a logical and a mathematical, and it is only in respect of the former that some of the operations are uninterpretable, so that one can switch from the logical to the mathematical interpretation and back again. It is unclear why *this* process would be justified. Moreover, even on the arithmetical interpretation, division by 0 remains uninterpretable, so that it is still not true that we have even a mathematical interpretation for all Boole's procedures. Feys's alternative justification is to say that we do not need to assign a meaning to all the symbols, provided the system could be formalised and proved consistent. This is clearly not enough when we have a partial interpretation: we should have to prove plausibility with respect to that interpretation for those equations which *are* interpretable. (In fact, Boole's procedures do contain lurking inconsistencies, for example in his handling of the 'v' symbol, and in the doctrine occasionally advanced that after 'development' we may equate terms with positive coefficients to 0.)

In his article Brouwer states that an assertion is *judged* if it has been proved or disproved, and *judgeable* if we know an effective procedure for either proving or disproving it. An unjudgeable assertion may become judgeable, since mathematical entities are subject to change. An assertion is *testable* if its negation is judgeable. An example is given of an unjudgeable assertion which, however, satisfies the law of double negation, and also of an assertion which is unjudgeable but testable. In intuitionistic logical algebra we have $\overline{\sigma+\tau} \equiv \overline{\sigma}\times\overline{\tau}$, but not $\overline{\overline{\sigma}+\overline{\tau}} \equiv \sigma\times\tau$, $\sigma+\tau \equiv \overline{\overline{\sigma}\times\overline{\tau}}$, or $\overline{\sigma\times\tau} \equiv \overline{\sigma}+\overline{\tau}$. Counter-examples are given, and it is remarked that the assumption of the general validity of these equivalences can be shown to be contradictory. On page 116, line 22 read '$<$' for '$\geq$' and '$>$' for '$\leq$'.

Rosser's article describes Boole's contributions in 'On a General Method in Analysis', *Philosophical Transactions of the Royal Society of London*, vol. 134 (1844), and *A Treatise on Differential Equations* (Macmillan & Co., Cambridge,

1859) towards the modern conception of functions and functionals, e.g. by considering as mathematical *entities* such operations as differentiation. Craven's article gives the basic ideas of the application of Boolean algebra to the theory of switching circuits. It seems a pity that the Centenary volume contains no article on Boole's work in probability theory.

New College, Oxford UK

JAMES W. VAN EVRA

A REASSESSMENT OF GEORGE BOOLE'S THEORY OF LOGIC

George Boole's theory of logic has not fared well at the hands of the commentators who have written about it, whether they be his successors in logic itself, or historians of the subject. While there is general agreement that his work occupies an important place in the history of logic, the exact nature of that importance remains elusive. On the one hand, he has been called the originator of mathematical logic,[1] but on the other, that claim has been pointedly disputed.[2] On the one hand, his logic does differ significantly from traditional syllogistic logic, and for this he has been applauded.[3] But on the other, Frege's introduction of quantification theory forms such a complete barrier between paleo- and neologic that any lasting influence from Boole's work, if it is there at all, seems permanently obscured.

Fueling these general concerns about the significance of Boole's work are the many claims that errors abound in it. These center mainly on his supposedly uncritical use of mathematics in logical contexts, which, so the critics suggest, resulted in the appearance of logically uninterpretable expressions in his system of logic. Not surprisingly, many of these same critics suggest that Boole's successors in algebraic logic put things right by providing logical interpretations for these expressions, thus extending the symmetry between logic and mathematics.

Here are a few examples of the sort of criticism I have in mind:

> . . . Boole's quasi-mathematical system [can] hardly be regarded as a final and unexceptionable solution of the problem [of supplying a viable alternative to Aristotelian logic]. Not only did it require the manipulation of mathematical symbols in a very intricate and perplexing manner, but the results when obtained were devoid of demonstrative force, because they turned upon the employment of unintelligible symbols, acquiring meaning only by analogy.[4]

> [The] mathematical character [of his work] is responsible at once for the strength and the weakness of Boole's calculus, as on the one hand, it could hardly have assumed so general a form had not Boole been able throughout to utilise the rules and operations which hold good in a 0-1-algebra, while on the other hand, the mathematical method renders the operations effected and the expressions arising in dealing with the problems more or less obscure, since they cannot be subjected to logical interpretation.[5]

> Boole's mathematicism goes so far . . . that he introduces symbols and procedures which admit of no logical interpretation, or only a complicated and scarcely interesting one. Thus we meet with subtraction and division and numbers greater than 1.[6]

> Although Boole's system lends itself in some ways to easy manipulation, we must admit that it contains . . . some defects of rigour such as the use of the letter v for the expression of existential propositions, the admission of numerical coefficients other than 1 and 0, and the use of the operation of division to which no constant meaning has been assigned in logic. In the course of the half century after the publication of his *Laws of Thought* these shortcomings were all removed by his followers.[7]

J. Gasser (ed.), A Boole Anthology, 87–99.

> L'algèbre de Boole souffrait d'un certain nombre de défauts. Avant tout elle comprenait des expressions énigmatiques, par exemple les coefficients 0/0 ou 1/0 . . . ou encore le coefficient 2/3 qui s'y rencontre de temps à autre, etc.[8]

> Owing to his free use of algebraic operations such as division and subtraction, the success of Boole's general method is offset by the appearance, in his calculations, of numerical and fractional coefficients and of other expressions whose logical significance is, to say the least, uncertain. It was the task of his successors to get rid of these anomalies and so to give the system a greater appearance of being what it claimed, a mathematical logic and not the mere logical mathematics that its opponents suspected it to be.[9]

The view of Boole's theory of logic which I present below centers on the following points:

First, specific criticisms such as those above, i.e., those to the effect that Boole includes logically uninterpretable expressions in his system of logic, are largely undeserved. They turn on a misunderstanding of what Boole called his "general method in logic", which provides for the inclusion of purely mathematical interludes in logical transformations, as a way of putting the formal power of mathematics to work in logical contexts. The questionable expressions serve as components in such interludes only, and were not considered by Boole to be logical expressions at all.

Concentrating on the general method, there is also a way of squaring Boole's explicit rejection of division as a logical operation, with his apparent (implicit) acceptance of the operation as a basis for interpreting coefficients such as 0/0 and 1/0. Though the evidence is largely circumstantial, it is possible that Boole did not count such expressions as being logical either, but again as being parts of the general method only.

Second, suggestions to the effect that Boole's successors in algebraic logic corrected his "mistakes" are misleading. Later algebraic logics were indeed more systematic, complete, and generally understandable than *Laws of Thought* (hereinafter '*LT*'), but in an important sense, they remain inferior to it. Specifically, Boole's successors tended uncritically to assume that logic and algebra fit perfectly together in the sense that they share all of their operations. But that assumption creates genuine problems for logic, the main one being, ironically, the unintelligibility of division used as a logical operator. While Boole's logic is far from perfect, he displays a greater awareness of logic as an independent discipline, as well as more sophistication in the use of mathematics as an aid to logic while avoiding the conflation of the two areas, than did his successors in algebraic logic.

Third, much of the relatively recent criticism of Boole's logic treats him as if he were a contemporary logician. But taking him to task for not, e.g., providing the same sophistication in proofs as one finds in contemporary logic not only overlooks the plain fact that Boole stands on the other side of a tremendous revolution in logic, it obscures much of his true importance in the history of the subject. That importance, it seems to me, is best exemplified by his general method. Regardless of its imperfection by today's standards, Boole's use of an independent formal device as an aid to logical processes marks him as an important figure in the transitional epoch between the traditional, purely syllogistic logics, and the contemporary, post-Fregean logics.

The following discussion deals exclusively with Boole's theory as it occurs in *LT*. After its publication, Boole wrote an essay (in [2], pp. 211ff.) containing views which radically diverge from those contained in *LT*. However, the status of that essay remains unclear, and there is evidence that Boole never adopted the views it contains.[10]

1 Even a cursory glance at *LT* reveals that the formal developments it contains occur in two separate sections of the work. On the one hand, Chapter II contains the basic logic, i.e., the fundamental logical system. On the other hand, Chapters V–VIII deal with the process of development, and its subsidiaries, such as elimination and reduction. The remainder of the logical portion of the work (which comprises the first fifteen chapters) is devoted to a variety of essentially auxiliary topics, such as variant interpretations of the symbolism, and specific examples of inferences using the logic already developed.

Boole apparently considered the separation between these two parts of the work to have more than merely spatial significance. As will be seen below, he considered the later chapters to be dependent on the basis laid in Chapter II, but at the same time to be more than just a simple extension of it. The two parts were, for Boole, of two fundamentally different sorts.

Chapter II is entitled "Of Signs in General, and of the Signs Appropriate to the Science of Logic in Particular; also of the Laws to which that Class of Signs are Subject". As the title suggests, the chapter consists of a virtually self-contained presentation of Boole's basic logic. Everything needed for a system of logic is there: a general discussion of the nature of signs, a specification of the particular symbols used in logic, a collection of examples of legitimate expressions of the system using the vocabulary, together with their interpretations, laws of the system, and a specification of its logical coefficients.

The heart of the basic logic consists of the three well-known binary operators '×', '+' and '–', and the two constant class symbols 0 and 1. In modern terms (which are not to be found in Boole's own work), the symbols may be understood as follows: '$x \times y$' stands for the intersection of the classes x and y, '$x + y$' for a truncated version of class union in which the expression in question is significant except in those cases in which x and y intersect, and '$x - y$' for what might be called logical exception (i.e., "x's except for the y's"), which is significant only if y is contained in x. Of the constants, 0 stands for the null class, and in *Laws of Thought,* 1 stands for the class containing everything in the universe of discourse.

Equally well-known is the relationship which Boole set up between his basic logic and mathematical algebra. Each of the functions serves as an analogue of its arithmetical counterpart, and the laws of logic correspond in like fashion with expressions in mathematics. Boole circumscribes the extent of the similarity by laying particular stress on the law of idempotence, $xx = x$, which holds universally in the logic, but in standard algebra, only for the values 0 and 1. This leads Boole to say that "The two systems of elective symbols and of quantity osculate, if I may use the expression, in the points 0 and 1".[11] That is, "The laws, the axioms, and the processes, of [an algebra in which the symbols x, y, z etc., admit indifferently of the values 0 and 1 and of these values alone] will be identical in their whole extent with the laws, the axioms, and the processes of an algebra of logic. Difference of interpretation alone will divide them".[12]

Toward the end of Chapter II, Boole makes it clear that while each logical operator has an algebraic analogue, the reverse is not the case. There the question of whether division is to be counted as a logical operation arises, and Boole meets it by saying,

> . . . the axiom of algebraists, that both sides of an equation may be divided by the same quantity, has no formal equivalent [in basic logic]. I say no *formal equivalent,* because, in accordance with the general spirit of these inquiries, it is not even sought to determine whether the mental operation which is represented by removing a logical symbol, *z*, from a combination *zx*, is in itself analogous with the operation of division in arithmetic. That mental operation is indeed identical with what is commonly termed abstraction, and it will hereafter appear that its laws are dependent upon the laws already deduced in this chapter.[13]

In other words, division is not even considered as a possible logical operation since abstraction can be handled with the logical apparatus already at hand, which does not include division. While the question still remains as to whether Boole slips division in later on, his remarks in Chapter II strongly suggest that he was committed to excluding division from consideration as a logical operator.

2 While Chapter II deals with Boole's basic logic, Chapters V ff. introduce a fundamentally different topic. It is within these chapters that virtually all of the expressions with which the critics are concerned appear, and it is here that Boole lays out what he calls a "general method in logic".[14] The method differs from the basic logic in the sense that Boole considers the earlier logic to be the basis for the generalization which the method will provide:

> The previous chapters of this work have been devoted to the investigation of the fundamental laws of the operations of the mind in reasoning; of their development in the laws of the symbols of Logic; and of the principles of expression, by which that species of propositions called primary may be represented in the language of symbols. These inquiries have been in the strictest sense preliminary. They form an indispensable introduction to one of the chief objects of this treatise—the construction of a system or method of logic upon the basis of an exact summary of the fundamental laws of thought.[15]

As the chapter unfolds, the difference between the basic logic and the general method in logic becomes clear. While the basic logic was restricted to dealing with completely interpreted logical operations, the general method involves the application of more generalized formal devices (development being the primary one) in logical contexts. Boole's apparent aim was to be able to use non-logical transformational devices in logical contexts in order to strengthen logical transformation.

In describing the general method, Boole begins by asking ". . . whether it is necessary to restrict the application of [symbolical laws of logical processes] by the same conditions of interpretability under which knowledge of them was obtained".[16] That is, (using his own example) since the expression '$x + y$' is uninterpretable if x and y do not represent distinct classes, must one only use '+' when it is certain that an interpretable expression will result? Boole goes on to say that if such restriction is necessary, no general method is possible. But, "On the other hand, if such restriction is unnecessary, in what light are we to contemplate processes which appear to be uninterpretable in that sphere of thought which they are designed to aid"?[17] Not surprisingly, Boole holds such restriction to be unnecessary, and then goes on to answer the question just posed.

He begins by pointing out that in ordinary reasoning, there is always a correspondence between what he refers to as process and interpretation, which means that the various steps in an ordinary reasoning process are all interpretable, i.e., none are merely formal moves. Accordingly, there are those who hold that:

> . . . as the laws or axioms which govern the use of symbols are established upon an investigation of those cases only in which interpretation is possible, we have no right to extend their application to other cases in which interpretation is impossible or doubtful, even though (as should be admitted) such application is employed in the intermediate stops of demonstration only.[18]

Such an objection, says Boole, is fallacious. "It is an unquestionable fact", he says, "that the validity of a conclusion arrived at by any symbolical process of reasoning, does not depend on our ability to interpret the formal results which have presented themselves in the different stages of the investigation".[19] This statement forms the foundation of the general method in logic. In concrete terms, Boole reduces the method to the following rule:

> We may in fact lay aside the logical interpretation of the symbols in [a] given equation; convert them into quantitative symbols, susceptible only of the values 0 and 1; perform upon them as such all the requisite processes of solution; and finally restore to them their logical interpretetation.[20]

The process Boole is here advocating is of a distinctly different sort from the basic logic of Chapter II. He is suggesting that any logical symbol may be treated as its mathematical counterpart in the manner laid down in Chapter II. Then *any* available mathematical operation may be used on it, whether that operation is logically interpretable or not. The final (mathematical) expression in the sequence must again be one which corresponds to a logical expression. With the purely mathematical interlude lying between, the sequence may then be treated as the inference of the final (logical) expression from the initial one.

Development plays a central role in this method. Boole considered it to be a process by which one can construct the bridge between logically interpretable expressions in accordance with the above rule. While it performs this logical function, Boole did *not* consider it to be solely a logical device at all; in fact, he initially defined it as encompassing much more than just logical application:

> *Definition*—Any function $f(x)$ in which x is a logical symbol, or a symbol of quantity susceptible only of the values 0 and 1, is said to be developed, when it is reduced to the form $ax + b\,(1 - x)$, a and b being so determined as to make the result equivalent to the function from which it was derived.
>
> This definition assumes, that it is possible to represent any function $f(x)$ in the form supposed.[21]

This definition covers *any* function, logical or not, so long as the variable x may be restricted to the values 1 and 0. As he puts it later on, the results of development are "true and intelligible for quantitative symbols (which meet the restrictions mentioned in the definition) *always;* for logical symbols, *always when interpretable*".[22] In other words, development need not yield logically significant expressions.

It is at this point that some of the critic's complaints can be dispatched, for some of the allegedly uninterpretable logical symbols to which they point can now be seen to be nothing of the sort. The very first example of development which Boole provides contains several of the expressions in question. In the example,[23] Boole sets out to develop the function $\frac{1+x}{1+2x}$, the result being $\frac{1+x}{1+2x} = \frac{2}{3}x+(1-x)$. But Boole does not claim that this expression is logically interpretable, or is a logical expression in *any* sense. Rather, he makes it abundantly clear that x is here being treated as a quantitative symbol restricted to the values 0 and 1, by prefacing the example with the remark that "By the principle which has been asserted in this chapter, it is lawful to treat x as a quantitative symbol, susceptible only of the values 0 and 1".[24] The example is properly one of development, but definitely not of development in a logical context.

Excluding consideration of 1/0 and 0/0, to be considered separately below, the remaining instances of expressions in *LT* which the critics claim to be both logical and uninterpretable[25] fall into one of two classes. First, they may be non-logical expressions which occur in the steps of an inference which intervene between logically interpretable expressions in accordance with the rule of the general method, or second, like the example just considered, they may be constituents of expressions which serve as non-logical examples of processes which also have logical applications. In neither case has Boole done anything obscure, complicated, or unintelligible.

There remains the possibility, however, that Boole illegitimately introduced division as a logical operator. For instance, in providing an example of the development of a function of two variables, Boole uses the function $\frac{1-x}{1-y}$, which, when developed, is $\frac{1-x}{1-y} = \frac{0}{0}xy + 0x(1-y) + \frac{1}{0}(1-x)y + (1-x)(1-y)$. Neither $\frac{0}{0}$ nor $\frac{1}{0}$ have any standing in basic logic, hence we might expect them to be treated as purely mathematical expressions in accordance with the general method as outlined above. Instead, he says that "It will in the next chapter be seen that the forms 0/0 and 1/0, the former of which is known to mathematicians as the symbol of indeterminate quantity, admit, in such expressions as the above, of a very important logical interpretation".[26]

But, in a manner entirely typical of the period, Boole's use of 'logical' is incurably vague and ambiguous. He may have meant that these symbols are, in effect, to be added to basic logic. If this is the proper interpretation, criticisms to the effect that Boole introduced division into logic are justified. But there is an equal likelihood that these expressions were intended to be restricted to the general method, i.e., were meant to apply to, but not be a part of, the expressions of the basic logic. In my view, the latter is the correct interpretation.

The latter interpretation can best be understood by considering another of his examples. Beginning with the statement "Men not mortals do not exist", symbolized $y(1-x) = 0$, Boole wishes to provide a "reverse definition" of mortals in terms of men. The initial equation yields, mathematically, $y - yx = 0$, or $yx = y$. Of this result he says,

> Were this an ordinary algebraic equation, we should, in the next place, divide both sides of it by v. But it has been remarked in Chapter II that the operation of division

> cannot be *performed* with the symbols with which we are now engaged. Our recourse, then is to *express* the operation, and develop the result by the [general method in logic].[27]

Although this passage is somewhat unclear, owing to the fact that Boole does not say in what the difference between performance and expression consists, presumably he is reaffirming the principle of the general method, whereby division may be used as a step in the transformation process, as long as the process ends with a logically significant expression. Thus "expressing" division yields $x = y/y$ which, again, is not interpretable in basic logic. Boole then develops this equation, the result being

$$x = y + 0/0(1 - y).$$

This expression, too, seems to be uninterpretable in basic logic, for it contains 0/0, which is not a part of basic logic. But Boole gives 0/0 the following "logical interpretation":

> Let [a] coefficient be in the form 0/0. Now, as in arithmetic, the symbol 0/0 represents an *indefinite* number, except when otherwise determined by some special circumstance, analogy would suggest that in the system of this work the same symbol should represent an *indefinite class*.[28]

In the particular example under consideration, this interpretation results in the following meaning being assigned to '$x = y + 0/0\ (1 - y)$': Mortals consist of all men together with an indefinite remainder (some, none, or all) of mortals who are not men.

Before proceeding further, note should be taken of Boole's use of 'some'. Symbolized 'v', he first introduces its use in Chapter IV. There he refers to it as an elective of a peculiar sort: ". . . it is indefinite in all respects but this, that it contains some individuals of the class to whose expression it is prefixed . . ."[29] Thus 'vx' means "some x" or, "x is not empty". While Boole's intent is clear enough, i.e., to provide a device for the expression of existential propositions, the device itself suffers from a fatal flaw, namely, the impossibility of a class whose only defining characteristic is that it have members. The flaw is due largely to the restrictive nature of algebraic logics generally.

With 'some' thus understood, the question now arises, does Boole surreptitiously slip division into his logic via the expression 0/0? I am inclined to take his disclaimer concerning division seriously, and suggest that he does not. There are two ways in which the "indefinite remainder" represented by '0/0' might be understood. '0/0' might be conceived of as an elective standing for a truly indefinite class, i.e., a class with a literally indefinitely large membership. Were this the intended interpretation, '0/0' would indeed have to be counted as another (and quite peculiar) variety of elective. Also, on such an interpretation, division would become (implicitly) a part of basic logic.

But it is not at all clear that Boole intended 0/0 to function as a distinct coefficient in basic logic. On the contrary, his use of the symbol in Chapter VI suggests that he thought of 0/0 as being, in effect, a higher level variable ranging over the coefficients already available in basic logic. Specifically, Boole seems to use '0/0' not to stand directly for an indefinite class, but as a way of indicating the irrelevance of class size, in certain cases, for the validity of inferences in which they occur. For instance, concerning the above example, he asks, ". . . what remainder of not-men is implied by the premise ["men not mortal do not exist"]"? His answer is that ". . . whether those beings which are not men are *all, some,* or *none,* of them mortal, the truth of the premise that all men are mortal will be equally unaffected, and therefore the expression here indicates that *some, all,* or *none* of the class to whose expression it is affixed must be taken".[30] Although this explanation of the use of '0/0' is less than totally clear, I take him to be saying, in effect, that the class is indefinite only at a higher level. That is, he is saying that the class is composed of some, or none, or all of the mortals, and that there are no other alternatives. Hence at the basic level, the proposition in question is perfectly determinate: it simply falls under one of the three coefficients which the system already provides. The indefiniteness occurs at a higher level, i.e., at the level at which it is determined which of the three possible propositions the class actually contains. In other words, '0/0' is not simply an addition to 'v', '1', and '0', but is a way of indicating that class size, though definite, need not be specified.

On this interpretation of '0/0', it need not be considered to be a part of basic logic at all. Rather, it remains a part of the general method only. However, unlike the other constituents of the general method, it is neither simply a purely mathematical part of the method, nor does it have a unique logical counterpart. Instead, it operates at the level of the general method as a variable which ranges over the three logical coefficients.

In summary, in this section I have shown that although there are rough spots in it, and although it defies systematically clear understanding, Boole's logic does not merit much of the criticism aimed at it. Charges that Boole includes logically uninterpretable expressions in his logic have been for the most part cleared, in the sense that most of these expressions were understood by Boole to be parts of the general method only, and not logical expressions. While the situation is less clear concerning division, there is at least a way of understanding Boole's use of expressions such as '0/0' which is consistent with his claim that he does not introduce division as an operation in (basic) logic.

3 Did Boole's successors in algebraic logic really remove the "shortcomings" and "anomalies" from his logic? In the sense that the logics of, e.g., Schröder, Peirce and Whitehead were more fully and systematically developed than Boole's, the answer is "yes". Also, in the later algebraic logics, one finds fewer of the relatively minor technical gaffes of the sort which appear in Boole's work. Thus Peirce and Jevons showed how the restriction which Boole placed on '+' could be ignored, and the equivocation on 'v' which occurs in *LT* is not found in later algebraic logics.

In a different sense, however, the matter is not as clear. The algebraic logicians in Boole's posterity tended to overlook Boole's device that permits the use of mathematical operations in logical contexts, while avoiding the general conflation of mathematics and logic. Instead, they tend to treat logic and mathematics as being so related that every operation in the one has its unique counterpart in the other, despite the fact that this poses serious problems for logic. They seem to be more interested, that is, in forcing the "fit" between logic and mathematics than in the problems this creates for logic.

Nowhere is this tendency clearer than in the treatment division received at the hands of Boole's logical descendants. While Boole, as indicated earlier, explicitly rejected division as a logical operation in *LT*, and on the present interpretation avoided even the implicit introduction of the operation into basic logic through such coefficients as 0/0 and 1/0, his successors show no reticence at introducing the operation directly into logic. The remainder of this section consists of a few brief examples of how division was handled by some of these successors. In each case, the result of admitting logic as a basic operation is the same, namely, the generation of fundamentally indeterminate classes into (basic) logic.

In a paper entitled "On an Improvement in Boole's Calculus of Logic", C.S. Peirce sets out to ". . . exhibit Boole's system in a modified form". After introducing some minor notational changes, he turns his attention to the topic of logical division:

> Let $a;b$ be read a logically divided by b, and be defined by the condition that
>
> $$\text{If } b, x \mp a \qquad x \mp a;b$$
>
> x is not fully determined by this condition. It will vary from a to $a + \bar{b}$ and will be uninterpretable if a is not wholly contained under b.[31]

The inclusion of division in logic has the unfortunate result that if treated as logical abstraction, it yields logically indeterminate classes within what amounts to basic logic itself, which is the result Boole seemingly avoids. That is, $x \mp a;b$ is problematic because there is no determination of b with respect to x. From the equation, we know that x contains a, but how many members of b are members of x, if any at all, is left completely undetermined by the operation. Since the analogous indeterminacy in Boole's system was seemingly restricted to a higher level, the same problem did not arise there.

Schröder's *Der Operationskreis des Logikkalküls* consists in large part in an at tempt to clean up Boole's logic (". . . nicht deutungsfähigen Symbolen wie 2, –1, 1/3, 1/0 gerechnet werden muss".[32]) In developing his own calculus of logic, he recognizes four basic operations:

> Auch in dem Kalkül der Logik gibt es, wie in der Arithmetik, 4 *Species* oder Grundrechnungsarten, welche jedoch, wie sich zeigen wird, *endgültig* auf *drei* verschiedenartige Elementaroperationen reducirt werden können. *Nichts hindert, jene 4 Grundoperationen mit denselben Namen zu benennen und mittels derselben Rechenzeichen auszudrücken, wie sie in der Arithmetik gebräuchlich sind.*[33]

The four operations which Schröder recognizes are (as he sets them out):

1°)	Die *Multiplikation*, genannt *Determination*.	1')	Die *Addition* oder Collektive Zusammenfassung (*Collektion*).
2°)	Die *Division* oder *Abstraction*.	2')	Die *Subtraktion* oder *Exception* (Ausschliessung).[34]

Schröder's approach is somewhat different from Peirce's in that he attempts to avoid actively *using* division in his logic (in [12] he speaks of the "inverse" operations (i.e., division and subtraction) as being primarily of "theoretical interest"[35]). Nevertheless, he still considers them to be full-fledged operations of the calculus, even if they are "keine unbedingt ausführbaren Operationen", due to the fact that unless the usual restrictions are followed, meaningless expressions occur.

By incorporating division into logic, Schröder meets the same fate as Peirce. Beginning with $c = a \div b$, Schröder determines the value of c in terms of a and b in the following way:

> Als Werth von c findet sich nun in der genannten Weise:
> $a \div b = ab_1 + ub$
> $\qquad a(b_1 + u)$
> $\qquad ab_1 + uab$
> für ein unbestimmtes u . . . [36]

Immediately thereafter, he refers to the class u as ". . . eine völlig beliebige, *arbitrare*". Here again, the result of using division as a logical operator is the generation of fundamentally indeterminate classes.

Of Boole's successors, John Venn was particularly familiar with his work. Venn's *Symbolic Logic* contains perhaps the most complete discussion of Boole's work to be found in the works of any of his successors. Of Venn's work, for instance, Whitehead remarked, "The task of giving thorough consistency to Boole's ideas and notation, with the slightest possible change, was performed by Venn in his 'Symbolic Logic'".[37] Early in the book Venn comments that Boole ". . . attempts no explanation whatsoever of his use of the sign of division in logic",[38] and then goes on to ask whether there can be a logical interpretation for the operation which is the inverse of multiplication.

His view, like that of Peirce and Schröder, is that division has a place in logic proper. Venn defines 'x/y' in terms of what he calls "restriction" in the sense that 'x/y' is regarded as identifying a class which, when the y restriction is placed on it, yields x. The class is hence composed of all x, and, of all things which are ". . . neither x nor y, it may take as much or as little as it pleases, that is, a perfectly indefinite proportion".[39] Shortly thereafter he says that it is essential that x/y be regarded as identifying a class, even though it may be ". . . indefinite in regard to its limits . . ., but it is never anything else than a true logical class or conceivably assignable group of individuals".[40]

While Boole never gives reasons for his exclusion of division as a logical operation, it is tempting to think that it was occasioned by a desire to avoid such completely indeterminate classes in logic. But regardless of Boole's grounds for rejecting division, suggestions to the effect that Boole's successors did a better job of developing algebraic logic are not entirely convincing. Later logics may have been more tidy, but they lack an important variety of sophistication found in *LT*. That is, as the preceding discussion indicates, while later logicians were intent on mathematicizing logic, Boole's approach is much more sensitive to the requirements of logic as an independent discipline.

4 There remains the question of Boole's importance within the history of logic. The task of fixing his proper historical place has been largely obscured by critics who measure his achievements, and shortcomings, by contemporary standards. Instances of this sort of criticism can be found among the examples at the beginning of the paper; another is Michael Dummett's review of [2]. In his review, he characterizes Boole's achievements by saying,

> As had Leibniz two centuries earlier, he devised a general theory of classes under Boolean operations, a theory which of course contained the traditional theory of the syllogism. This move gained its importance for logic rather from the novelty of *any* extension of logical theory than from the magnitude of the extension itself; and anyone unacquainted with Boole's works will receive an unpleasant surprise when he discovers how ill-constructed the theory actually was and how confused his explanations of it.[41]

Dummett then proceeds to recite the standard criticisms, i.e., those concerning the difficulty of symbolizing statements in algebraic logic, the problems surrounding 'v', '+', and '–', etc. He also criticizes Boole for not *proving* that, in effect, the general method is always truth-preserving, and that the process of development really does, as Boole claims it does, follow from Taylor's theorem.

Such criticism is at best uninformative. Boole is simply not a contemporary logician, and measuring him by our current standards serves only to underscore that fact. Since Boole stands on the other side of a major revolution in the science of logic, holding him up to standards used on our side of that revolution is singularly unproductive. That Boole's logic is not as rigorously developed as ours is undeniably true, but also not very significant. Conversely, what *is* significant are attempts to fix, e.g., the relationship between Boole's work and the logical environment during the period which just preceded the full impact of Frege-*Principia* logic.

Instead of following the more recent variety of criticism, I have concentrated on a feature of Boole's logic which makes him, I think, an important contributor to the transitional period in logic which occupied much of the nineteenth century. His development of the general method, regardless of its imperfections by our standards, goes well beyond the simple formalization of ordinary language, which had characterized logic from Aristotle's day. In the method, we see one of the first important attempts to bring mathematical methods to bear on logic while retaining the basic independence of logic from mathematics. In this sense, the method constitutes a further step away from the dependence on ordinary language which had characterized earlier logics.

Boole thus stands with a foot in each of two logical epochs; the general method links him with contemporary logic, and the still-present pull of the syllogism with the older epoch. In the transition, moreover, Boole's role is pivotal. After him, algebraic logic, by overly stressing the connection between logic and mathematics, headed toward extinction by losing sight of logic as an independent discipline altogether.

NOTES

[1] By R. Feys in "Boole as a Logician", *Proceedings of the Royal Irish Academy*, vol. 57, Sec. A, No. 6, pp. 97–106.

[2] E.g., "Boole cannot correctly be called the 'father of modern logic'". Michael Dummett in [11], p. 205.

[3] A particularly engaging assessment of Boole's general contribution to logic was given by Hugh MacColl:

> [*The Mathematical Analysis of Logic* and *Laws of Thought*] excited much admiration in the mathematical world, and, it may be added, caused no small trepidation among logicians, who saw their hitherto inviolate territory now for the first time invaded by a foreign power, and with weapons which they had too much reason to dread. ("Symbolical Reasoning", *Mind,* vol. V (1880), p. 46.)

[4] [5], p. 113. As early as 1864, Jevons displayed a clear understanding of Boole's general method in logic, explained below. Thus in [6], (§176, pp. 67–68), he accurately describes Boole's process, but rejects it as not corresponding "to the laws and conditions of thought in reality".
[5] [7], p. 115.
[6] [3], p. 298.
[7] [8], p. 422.
[8] [9], p. 171.
[9] [4], p. 543.
[10] Boole never published the essay in question and Rhees ([2], p. 12) quotes De Morgan as saying that Boole would not have wished such an essay published posthumously, since it might give the impression that *LT* had been rejected. Also, in the essay, Boole seems to accept division as a logical operation; however, in a later published work *(Differential Equations,* Supplementary volume, MacMillan & Co., Cambridge (1865)) he again denies (p. 377) that inverse symbols stand for operations. In addition, Venn was of the opinion that Boole never included inverse operations in his logic. (*Cf.* [15], p. 77n.)
[11] [2], p. 119.
[12] [1], pp. 37–38.
[13] [1], pp. 36–37.
[14] Boole refers to the process in a number of ways, and never gives it a definite name. I have chosen "general method in logic" as a matter of convenience. Boole uses the term in [1], p. 70, when he says that if logical symbols were restricted to logically interpretable contexts only, ". . . the inquiry after a general method in Logic would be a hopeless quest".
[15] [1], p. 66.
[16] [1], p. 67.
[17] *Ibid.* Jevons held such a view, though it is unlikely that Boole was referring to him.
[18] [1], p. 67.
[19] [1], pp. 67–68.
[20] [1], p. 70. Boole italicized the entire passage.
[21] [1], p. 72.
[22] [1], p. 77.
[23] [1], p. 73.
[24] [1], p. 72.
[25] A fairly complete list of these as they occur in Chs. V–XV of *Laws of Thought* is as follows: (in each pair the first number is the page, and the expression following is the supposedly uninterpretable symbol occurring there) 94, –1; 95, ½; 97, ½; 115, 2; 123, 2; 125, 2; 131, 2.
[26] [1], p. 74.
[27] [1], p. 89.
[28] *Ibid.*
[29] [1], p. 63.
[30] [1], pp. 89–90.
[31] [12], §3.6 (p. 6). Here 'a, b' stands for the logical multiplication of a and b. '$a \mp b$' stands for the "numerical identity" of the classes a and b.
[32] [13], p. iii.
[33] [13], p. 2.
[34] *Ibid.*
[35] *Cf.* [14], Vol. I, pp. 478 ff.
[36] [13], p. 30.
[37] [16], p. 115.
[38] [15], p. viii.
[39] [15], pp. 77, 78.
[40] [15], p. 79.
[41] [11], p. 205.

REFERENCES

[1] Boole, George, *An Investigation of the Laws of Thought,* Dover Publications, Inc., New York (1951). A reprint of the London edition of 1854.

[2] Boole, George, *Studies in Logic and Probability,* edited by R. Rhees, Watts & Co., London (1952).

[3] Bocheński, I.M., *A History of Formal Logic,* University of Notre Dame Press, Notre Dame, Indiana (1961). A revised translation (by I. Thomas) of *Formale Logik,* Verlag Karl Alber, Freiburg, Munich (1956).

[4] Heath, P.L., "Boole", in *The Encyclopedia of Philosophy,* The MacMillan Co. and The Free Press, New York; Collier-MacMillan, London (1967).

[5] Jevons, W. Stanley, *The Principles of Science, a Treatise on Logic and Scientific Method,* Dover Publications, Inc., New York (1958). A reprint of the London edition of 1874.

[6] Jevons, W. Stanley, "Pure Logic" (1864) in *Pure Logic and Other Minor Works,* Burt Franklin, New York (1971). A reprint of the London edition of 1890.

[7] Jørgensen, Jørgen, *A Treatise of Formal Logic,* Russell and Russell, New York (1962). A reprint of the Copenhagen edition of 1931.

[8] Kneale, William, and Martha Kneale, *The Development of Logic,* Oxford University Press, Oxford, England (1962).

[9] Kotarbiński, Tadeusz, *Leçons sur l'histoire de la logique,* Presses universitaires de France (1964). A translation of *Wykłady z dziejów logiki,* PWN, Warsaw (1964).

[10] Lewis, C.I., *A Survey of Symbolic Logic,* Dover Publications, Inc., New York (1960). A revision of the University of California Press edition of 1918.

[11] Dummett, Michael, Review of [2], *The Journal of Symbolic Logic,* vol. 24 (1959), pp. 203–209. [Reprinted in this volume, pp. 79–85.]

[12] Peirce, Charles S., *Collected Papers of Charles Sanders Peirce,* edited by Charles Hartshorne and Paul Weiss, Harvard University Press, Cambridge (1965).

[13] Schröder, E., *Der Operationskreis des Logikkalküls,* Teubner, Leipzig (1877).

[14] Schröder, E., *Vorlesungen über die Algebra der Logik,* 3 vols., Chelsea Publishing Co., Bronx, New York. A reprint of the Leipzig edition of 1890–1905.

[15] Venn, John, *Symbolic Logic,* Chelsea Publishing Co., Bronx, New York (1971). A reprint of the London edition (2nd) of 1894.

[16] Whitehead, A.N., *A Treatise on Universal Algebra,* Hafner Publishing Co., New York (1960). A reprint of the Cambridge edition of 1897.

University of Waterloo, Waterloo, Ontario, Canada

JOHN CORCORAN AND SUSAN WOOD

BOOLE'S CRITERIA FOR VALIDITY AND INVALIDITY

> The mathematics we have to construct are the mathematics of the human intellect.
> —Boole, 1847

It is one thing for a given proposition to follow or to not follow from a given set of propositions and it is quite another thing for it to be *shown* either that the given proposition follows or that it does not follow.* Using a formal deduction to show that a conclusion *follows* and using a countermodel to show that a conclusion *does not follow* are both traditional practices recognized by Aristotle and used down through the history of logic. These practices presuppose, respectively, a criterion of validity and a criterion of invalidity each of which has been extended and refined by modern logicians: deductions are studied in formal syntax (proof theory) and countermodels are studied in formal semantics (model theory).

The purpose of this paper is to compare these two criteria to the corresponding criteria employed in Boole's first logical work, *The Mathematical Analysis of Logic* (1847). In particular, this paper presents a detailed study of the relevant metalogical passages and an analysis of Boole's symbolic derivations.

It is well-known, of course, that Boole's logical analysis of compound terms (involving 'not', 'and', 'or', 'except', etc.) contributed to the enlargement of the class of propositions and arguments formally treatable in logic. The present study shows, in addition, that Boole made significant contributions to the study of deductive reasoning. He identified the role of logical axioms (as opposed to inference rules) in formal deductions, he conceived of the idea of an axiomatic deductive system (which yields logical truths by itself and which yields consequences when applied to arbitrary premises). Nevertheless, surprisingly, Boole's attempt to implement his idea of an axiomatic deductive system involved striking omissions and Boole does not use his own formal deductions to establish validity. Boole does give symbolic derivations, several of which are vitiated by the fallacy of supposing that a solution to an equation is necessarily a logical consequence of the equation. This fallacy seems to have led Boole to confuse equational calculi (i.e., methods for generating solutions) with proof procedures (i.e., methods for generating consequences). The methodological confusion is closely related to the fact, shown in detail below, that Boole had adopted an unsound criterion of validity.

It is also shown that Boole totally ignored the countermodel criterion of invalidity. Careful examination of the text does not reveal with certainty a test for invalidity which was adopted by Boole. However, we have isolated a test that he seems to use in this way and we show that this test is ineffectual in the sense that it does not serve to identify invalid arguments.

J. Gasser (ed.), A Boole Anthology, 101–128.

This paper goes beyond the simple goal stated above. Besides comparing Boole's earliest criteria of validity and invalidity with those traditionally (and still generally) employed, this paper also investigates the framework and details of *The Mathematical Analysis of Logic* (*MAL*). The investigation is carried out using the context and methods of modern logic. The main object of the investigation is the discovery and exact articulation in modern terms of the objective logical and mathematical content of Boole's first logical work. It is worth emphasizing that no attempt has been made here to characterize a distinctly Boolean doctrine that pervades Boole's logical works as a whole nor has any attempt been made to establish Boole's proper role in the historical development of logic. It is true, however, that some references have been made not only to Boole's later writings on logic and mathematics but also to the views of his predecessors, his contemporaries and his successors. With few exceptions, references to material outside of *MAL* are made solely for purposes of comparison. In a few places, all clearly noted, ambiguity and vagueness of passages in *MAL* are tentatively treated by comparison with his later works.

Boole wrote several things on logic after *MAL* and the later writings contain significant changes (see e.g., [15], p. 205). It is not clear how long Boole spent developing the ideas reported in *MAL* but it is clear that Boole did not spend more than a few months at the most on writing it (see *MAL* [3], Preface). In the preface to *MAL* Boole says that he had thought about the subject previously "at different periods" but in the preface to *Laws of Thought* (1854) he says that *MAL* "was written within a few weeks after its idea had been conceived". In an 1851 article ([6], p. 252) Boole refers to *MAL* as "a hasty and (for this reason) regretted publication". Thus, there is reason to believe that *MAL* does not represent Boole's mature, complete, or final views, and criticisms of *MAL* cannot be taken as criticisms of Boole's work as a whole.

Nevertheless there are good reasons for seeking understanding of the objective logical and mathematical content of *MAL*. In the first place, an understanding of the development of Boole's thought throughout his logical career requires understanding of his thought at each stage. In the second place, without a clear statement of the facts involved one cannot assume that every change in Boole's thinking was objectively an improvement. In the third place, even *if* one's interest is limited to Boole's mature and final logical views there is reason to believe that an understanding of what these *are* and what these *are not* will be enhanced by comparison with his earliest thinking. The latter is particularly likely because, although later works do incorporate changes, Boole never identifies what he takes to be errors in *MAL* and, in most cases, he does not even identify the changes.

One device repeatedly used in this article to illustrate and characterize Boole's logic is comparison of it with Aristotle's logic. Such a comparison may appear to modern logicians as far-fetched, irrelevant, and perhaps even reactionary. This initial negative response is largely justified from a strictly modern and nonhistorical point of view. Moreover it is also partly justified from the point of view of historians whose interest is to see what Boole contributed to modern logic. However if one's interest is to understand Boole's place in the development of logic and to prepare the evidence necessary to assess Boole as a logician then it is Aristotelian logic which is *par excellence* the system to which Boole's logic is to be compared. There are several reasons for this.

In the first place, Aristotelian logic so dominated logical thinking up to Boole's time that the subject of logic was widely considered to be virtually identical to the Aristotelian doctrine. It was precisely within Boole's time, and partly a result of Boole's work, that Aristotle's influence began to be replaced with a more modern outlook. In the second place, Boole (not to mention DeMorgan, Venn, and even Whitehead) devoted much effort to reformulating Aristotelian logic. Roughly one-third of *MAL* concerns categorical propositions and syllogisms. Besides, Boole himself (*MAL,* p. 13) literally invites the comparison. In the third place, it is now clear ([2], [10], [21]) that Aristotle's basic outlook is very close to the modern viewpoint especially as formulated by Church [7] and Tarski [22], and, thus, without serious distortion, it might be said that a contemporary examination of Boole's logic cannot avoid comparison, at least indirectly, with Aristotle's logic.

1. PRELIMINARIES

A set of propositions *P* is said to *logically imply* a single proposition *c* if it is logically impossible that all propositions in *P* be true while *c* be false. Saying that *c* *follows from P,* or is *a consequence of P,* is the same as saying that *P* implies *c*. To connect these ideas to the traditional terminology of arguments, premises, and conclusions, define a *premise-conclusion argument* to be a set, *P*, of propositions called the *premises* together with a single proposition called the *conclusion.* For purposes of abbreviation let (*P*, *c*) be the argument with premises *P* and conclusion *c*. To say that (*P*, *c*) is *valid* is to say that *P* logically implies *c*, and, of course, to say that (*P*, *c*) is *invalid* is to say that *P* does not logically imply *c*.

If one considers only trivial arguments such as the two-premise syllogism then the problem of producing plausible criteria for validity and invalidity does not seem demanding or even interesting. But the situation is otherwise if one considers more complex arguments, such as those found in philosophical or mathematical discourse. For example, if one takes as premise sets various subsets of the axioms of Euclidean geometry and as conclusions various geometrical theorems and the negations of various axioms, then in raising the problem of criteria for validity and invalidity one is implicitly raising a host of traditional questions, including the question of criteria for the independence of the axioms of geometry.

At any rate, Aristotle noticed that the validity of an argument (*P*, *c*) can be established by means of a step-by-step deduction interpolating between *P* and *c* intermediate propositions which clearly link the conclusion to the premises. Thus by presenting a set of rules of inference for inserting intermediate propositions on the basis of those already set down and by indicating how these rules are used to construct deductions, Aristotle presented a criterion for validity.

Let us call Aristotle's rules *A-rules.* Consider a list of propositions beginning with all or some of the members of *P*, whose subsequent lines are obtained by applying A-rules to previous lines, and which ends with *c*. Such a list is called a *direct A-deduction* of *c* from *P*. An *indirect A-deduction* of *c* from *P* is a direct *A*-deduction which begins with members of *P* and the contradictory of *c* and which ends in a contradiction. Aristotle took the existence of an *A*-deduction of *c* from *P*, direct or indirect, as a criterion for validity.[1] Using modern terminology, Aristotle's criterion of validity can be restated as follows: if (*P*, *c*) is *A*-deducible then (*P*, *c*) is valid.

It is, of course, possible to agree with Aristotle that deducibility is a criterion of validity while disagreeing with his specific definition of "deduction". For example, one might wish either to delete rules or to add rules or both. In particular, the logicians who believe that "Some animals are dogs" does not follow from "All dogs are animals" would want to delete one of Aristotle's rules. Likewise, since "All nonanimals are nondogs" is not deducible in Aristotle's system from "All dogs are animals", one might want to add a rule. Indeed, this is one of the changes suggested by Boole (*MAL*, pp. 26, 28, 34).

Aristotle also noticed that the invalidity of an argument (P, c) can be established by exhibiting another argument (P^*, c^*) which (1) has the same *form* as (P, c) but which (2) has premises known to be true and a conclusion known to be false. For example, the second argument below (A2) establishes the invalidity of the argument immediately below (A1).

A1 All maples are plants.
All maples are trees.
? All trees are plants.

A2 All maples are trees.
All maples are plants.
? All plants are trees.

Aristotle took two arguments to have the same form if one can be converted into the other by replacing in a one-one fashion the "content words" of one with those of the other. The replacement (maples → maples, plants → trees, trees → plants) converts A1 into A2.

This method has been used down through the history of logic. For example, it was used in the medieval period by Anselm (with particular dexterity in *De Grammatico* [1]). In the 19th century it was used by Beltrami to establish that the parallel postulate is not implied by the other axioms of geometry (Church [7], p. 328), and in this century it was used by Cohen to establish the fact that the continuum hypothesis is not a consequence of the axioms of set theory [8].

A *counterargument* to an argument (P, c) is an argument (P^*, c^*) where (1) (P^*, c^*) has the same form as (P, c) and (2) the premises P^* are all true and the conclusion c^* is false. Aristotle took a counterargument to establish the invalidity of a given argument. Aristotle's criterion for invalidity can be restated as follows: if (P, c) has a counterargument then (P, c) is invalid.

The striking asymmetry of the two criteria has been remarked on before (e.g., Tarski [22], pp. 117–125). To establish validity one uses a logical deduction; to establish invalidity one appeals to facts (uses a counterargument or, equivalently, a countermodel[2]).

Deduction is usefully construed as a chaining together of evidently valid arguments. The method of refutation by counterargument is virtually implicit in the so-called principle of logical form: an argument is valid or invalid according to its form, i.e., two arguments in the same form are both valid or both invalid.

2. BOOLE ON ESTABLISHING VALIDITY

As indicated above, the traditional approach to logic focuses on the problem of classifying a given argument as valid or invalid. The premises and the conclusion are given, not *as true,* but simply as constituting a premise-conclusion argument, and the role of logic, so to speak, is to determine whether this given argument is valid or invalid, whether the conclusion follows or does not follow from the premises.

In regard to validity Boole presented a deductive system which he took as *a* criterion of validity. But there are several qualifications to be made.

In the first place, he seemed to take the primary role of logic to be, not the determination of validity and invalidity of given arguments, but rather the determination of (unknown) consequences of given premises. Instead of classifying given arguments (whose conclusions are given) he focuses on methods of "generating" the consequences of given premises. It is as if Boole took deducing to be the producing of *consequences* rather than the producing of *deductions* (cf. [6], pp. 218, 232, 238, 239). Boole's paradigm, then, is not to be found in the logical practice of producing deductions, given the premises (or axioms) *and* given a conjectured conclusion. Rather, Boole looked toward the mathematical practice of finding a solution to a given set of equations (*MAL,* pp. 32ff).

Boole did not, however, ignore the traditional concern for producing deductions ([6], p. 142). His deductive system (*MAL,* pp. 15ff) can be used for producing deductions of given conclusions from given premises. In later works he refers to "the synthetical method" ([6], pp. 231, 232) wherein "we begin with the premises . . . and . . . arrive by a direct process at the conclusion . . . ".

But Boole did not emphasize the deductive system as a criterion of validity. In deed, he sometimes employs a different and incorrect criterion (see below) even when he could have used his own deductive method correctly and straightforwardly. Moreover, the presentation of the deductive system is perfunctory, and the system has deficiencies which Boole himself could have recognized easily.

2.1. The Character of Boole's Deductive System

For Boole "logical propositions", including the categorical propositions studied by Aristotle, are identities (equations) between classes *(MAL,* pp. 8, 31, 32). He used 0 to denote the null set and 1 for *the* Universe of all entities. His "elective symbols", x, y, z, u, v, and w, were used to indicate arbitrary but fixed classes, empty and nonempty alike; but, as will be seen below, he often reserves the symbol v for a special use in connection with nonempty classes.

Boole nowhere explicitly asserts that the elective symbols *can* indicate empty sets, nor does he ever explicitly say that they *cannot* indicate empty sets. In passages only indirectly related to present concerns he writes "$x = 0$" (*MAL,* p. 51). However, there are many passages, to be mentioned below, where fallacies exist when elective symbols are able to have empty sets as values. On the other hand, if Boole is to be consistently interpreted as having restricted the values of the elective symbols to nonempty sets then there are many laws involving the empty set which he has not stated or implied (e.g., $0 + 0 = 0$, $0 \cdot 0 = 0$), and the device involving v which is widely supposed to have been invented to explicitly state nonemptiness seems redundant ([15], p. 205).

It is particularly important in this connection to note that Boole later believed that the formal laws of logic are those principles which hold of the null set and the universal set ([6], pp. 20, 211, 218). Thus even if the elective symbols are not allowed to indicate the null set, his propositions and inferences involving the elective symbols may have been supposed by Boole to hold when 0 and 1 are arbitrarily substituted for them.

In *MAL*, + *seems* to resemble union, and – *seems* to resemble relative complement. Juxtaposition (or concatenation) was taken to indicate intersection. Unions and complements are often (but not always, e.g., *MAL*, pp. 43, 53, 58) grouped with parentheses.

It is to be emphasized both that Boole gave nothing which could be interpreted as a formal grammar and that determination of his semantics is highly problematic. Not only is the use of the symbol v unclear but there may be no interpretation for + and – which is consistent with *MAL* as a whole.[3]

The vagueness and incompleteness of his definition of his semantics contrasts sharply with his philosophical comments[4] that ". . . symbols are used with a full understanding of their meaning, with a perfect comprehension of that which renders their use lawful . . . ".

Boole recognized *logical truths*, formulas which, like $x = x$, are true regardless of which classes are denoted by their elective symbols. Although he has no term in *MAL* for logical truth *simpliciter* it is virtually certain that he had the idea: he speaks of relations among the operations ". . . whether as respects the repetition of a given operation or the succession of different ones, or some other particular, which are never violated" (*MAL*, p. 5). In later writings ([6], p. 215), Boole refers to "propositions true in consequence of their form alone" and he gives as examples "A man is a man" and "A man is either a tree or not a tree". Among the logical truths he singles out some which are "obvious" (*MAL*, p. 6). His obvious logical truths are later referred to as *laws* (*MAL*, pp. 16–18) and are used as logical axioms.

Accordingly, Boole's deductions are not merely sequences of propositions derived from premises by means of rules but they also contain logical axioms. This is a striking qualitative improvement on Aristotle's direct deductions which were constructed using eight rules (the three conversions, the four "perfect syllogisms", and the rule of repetition) but no logical axioms (Corcoran [10], [11]). Boole was aware that he was breaking new ground by his discovery of certain logical truths (*MAL*, p. 6) but in *MAL* he did not seem to realize the novelty of using them in deductions.

Nevertheless in later works (cf. [6], pp. 215, 216) Boole is clear about the fact that deductions are constructed by interpolating logical axioms *and* by applying rules of inference. He says "the laws of thought include . . . formal laws of judgment and formal laws of reasoning".[5] (Also see [4], pp. 101–102.)

Speaking loosely for the moment, one can say that Boole presented a system of deductions based on one rule (a version of equals-substituted-for-equals) and four logical axioms (or axiom schemes). Boole was explicitly aware of only *some* contrasts between his own system and that of Aristotle. He compares his own rule with the so-called Aristotelian *dictum de omni* (which had been supposed by some to be *the* basic rule of Aristotle's logic) and he refers to his system as "another mode of considering the subject" (*MAL*, p. 18). In later works (cf. [6], pp. 25, 237, 238) he argues that rules of inference alone would not be sufficient for reasoning and that "we are compelled to have recourse to . . . canons . . . which enable us to add to the premises, whose truth is only assumed, other propositions whose truth is not assumed but necessary".

One thing that Boole did not explicitly notice (and probably did not notice at all) is that Aristotle's system included indirect deductions whereas his own did not. It is important to realize that Boole's deductive system necessarily lacked indirect deductions because his language lacked negation.[6] Likewise, it is important to realize that Aristotle's deductive system necessarily lacked logical axioms because *his* language lacked logical truths.

As far as the analysis of the forms of deduction is concerned, Boole's system was a step forward in one respect and a step backward in another; logical axioms were added but indirect deductions were dropped.

Before going on to consider the details of Boole's deductive system there is one final point of comparison between Boole and Aristotle which should be mentioned. Aristotle emphasizes the epistemic or knowledge-producing property of deductions: by means of a deduction one comes to know that the conclusion follows. Aristotle says that a deduction "makes clear" the logical connection of the conclusion to the premises. In *MAL*, however, there is no *explicit* reference to the epistemic nature of deduction.

2.2. The Details of Boole's Deductive System

Boole presents his deductive system in four pages (*MAL*, pp. 16–19). After stating his logical axioms (or "laws", to use his term) he says that they are "sufficient for the basis of a Calculus" (*MAL*, p. 18).

For convenience we restate his logical axioms and give them names.
Axioms:

I	$x = x$	identity
LDM	$x(u + v) = xu + xv$	left distribution of multiplication
CM	$x \cdot y = y \cdot x$	commutativity of multiplication
IM	$x \cdot x = x$	index law

It is to be emphasized that these are *all* of the axioms which Boole *explicitly* lists. Perhaps it was his casual attitude concerning parentheses which led him to overlook the two associative laws.

AM	$x \cdot (y \cdot z) = (x \cdot y) \cdot z$	associativity of multiplication
AA	$x + (y + z) = (x + y) + z$	associativity of addition

Rule: Boole introduces his rule, which he calls "an axiom", with the following statement: "The one and sufficient axiom . . . is that equivalent operations performed upon equivalent subjects produce equivalent results". It is clear that this is to be understood as a single rule encompassing the following three subrules (a point which Boole himself makes in later works ([6], pp. 235, 236 and cf. [4], p. 36).

FA	$x = y, z = w / x + z = y + w$	functionality of addition
FS	$x = y, z = w / x - z = y - w$	functionality of subtraction
FM	$x = y, z = w / x \cdot z = y \cdot w$	functionality of multiplication

This is *the only* rule of inference that Boole explicitly states *as* a rule of his system in *MAL*. Moreover, as mentioned above, he refers to this rule as "the one and sufficient axiom". It would seem then that Boole is overlooking symmetry and transitivity of identity. However in a footnote (*MAL*, p. 18) he refers to two other rules[7] in a way which suggests that he thinks that they are already somehow included in the above. One of the latter two rules is irrelevant to the system because it presupposes that the language contains negation. This rule is, of course, never used and it is never referred to again.

The other rule mentioned in the footnote is: "If two terms agree with one and the same third, they agree with each other". We take this to be represented by the following:

R1	$x = z, y = z / x = y$	equal to a third
R2	$x = z, z = y / x = y$	transitivity

Boole's statement[8] of this rule presupposes symmetry of identity, i.e., Boole would not have used the above wording unless he thought that identity is symmetrical. Moreover, one can derive symmetry using R1 and I ($x = x$).

S	$x = y / y = x$	symmetry

Although the present authors do not wish to draw the following conclusion, it must be noted that from the language used in *MAL*, together with passages in later works (e.g., [6], pp. 231, 235, 236], some logicians may wish to infer that *the* role which Boole intended for *MAL* was not any of the above but rather was the more general principle of substitution of identicals, viz., from $x = y$ and a sentence S infer any sentence obtained from S by erasing all occurrences of x and y and then filling the blanks *ad lib* with occurrences of x and y. Boole seemed to view such identity inferences in a formally different, but equivalent, way ([6], p. 127).

There is little doubt that Boole believed that his system embodied a complete analysis of reasoning involving his fundamental operations. He uses the term "sufficient" twice and in the footnote mentioned above he states:

> Any account of the process of reasoning is insufficient which does not represent . . . the laws [= logical axioms] of the operation which the mind performs in that process . . .. It is presumed that the laws in question are adequately represented by the fundamental equations of the present Calculus. The proof of this will be found in its capability of expressing propositions, and of exhibiting in the results of its processes, every result that may be arrived at by ordinary reasoning.

2.3. *Omissions from Boole's Deductive System*

The inadequacies of Boole's deductive system are obvious to a casual observer.

Firstly, there are no laws specifically governing the null set and the universal set. Thus one *might* want to add the following:[9]

N1	$0 \cdot x = 0$
N2	$0 + x = x$
U1	$1 \cdot x = x$
U2	$1 + x = 1$

Especially notable is the absence of a law that Boole calls "the principle of contradiction" in *Laws of Thought,* viz.:

NU	$x \cdot (1 - x) = 0$	principle of contradiction

Secondly, the only axiom governing a reiterated argument and explicitly mentioned by Boole is $x \cdot x = x$. Thus the following are missing.

IA	$x + x = x$	index for addition
IS	$x - x = 0$	index for subtraction

Thirdly, Boole gives no axioms involving addition alone. We have already mentioned that the index law and the associative law are missing. (The commutative laws are stated only verbally (*MAL,* pp. 5–6).)

Fourthly, the only distribution law mentioned is that given above. Thus, e.g., the following are missing:

LDA	$x + (y \cdot z) = (x + y) \cdot (x + z)$	left distribution of addition
LDMS	$x \cdot (y - z) = (x \cdot y) - (x \cdot z)$	left distribution of multiplication over subtraction

Finally, we mention that Boole omits the laws that have come to be named after DeMorgan[10] and the so-called absorption laws.

For later reference it should be observed that Boole does not reject any of these omitted laws. Indeed, in *MAL* there are no formulas that he explicitly refers to as *not* being laws. (But, cf. [4], p. 50.)

The absence of the associative laws can be somewhat explained by Boole's inattention to parentheses (e.g., *MAL,* pp. 43, 53, 58), but the other omissions, taken together, seem to indicate that Boole did not systematically investigate deduction. Boole's remarks concerning the "sufficiency" of his system must be understood as revealing casual beliefs not based either on serious reflection or on organized experience. As we will see below, Boole did not *use* his own system and he did not expend any energy in *studying* it prior to publication. Indeed, in the whole of *MAL* there is not one example of a deduction in the above system.

In the course of his symbolic derivations there are places where he makes inferences which could be derived in a few steps by means of his stated rules. For example the following inferences are all made:

$x = y \,/\, x \cdot z = y \cdot z$	multiplying an equation by z
$x = y \,/\, z \cdot x = z \cdot y$	multiplying z by an equation
$x = y \,/\, x - z = y - z$	subtracting z from an equation
$x = y \,/\, z - x = z - y$	subtracting an equation from z

In regard to the production of *formally* correct deductions and in regard to extensive organized experimentation, Boole's work contrasts sharply with Aristotle's. Aristotle gives many examples of deductions meticulously constructed according to his rules. In fact, in the course of hundreds of applications of rules there are at most a handful of missing lines. In addition, Aristotle experimented with his system in a highly organized fashion. In contrast to Boole, Aristotle's belief in the completeness of his system was not only true but it was based on experimentation and serious reflection—even though it was not completely justified according to modern standards (Corcoran [10], [11]). On the other hand, as we have seen, Boole's belief in completeness was false and its falsity would have been obvious to Boole if he had seriously considered the question. Naturally, Boole would have added more axioms and rules, and if he had still failed to have a complete system he would have had a system which at least would not be so obviously incomplete.

As Boole indicates (*MAL*, p. 7), his logical investigations were intended to provide a reformulation of "the received logic". Indeed, over one-third of *MAL* (measured in pages) is devoted to the reformulation of traditional conversions and syllogisms. Moreover, even though Boole's formal language can formulate arguments involving compound terms such as "Europeans or Asiatics" (*MAL*, p. 58), Boole does not give a single example of such an argument. This seems to show that Boole was not fully aware of the fact that he had discovered a framework essentially richer than Aristotle's.[11] In this connection it is worth noting that fifty years later when Whitehead [25] came to apply algebra to logic he derived only the traditional two-premise syllogisms omitting derivations of arguments involving compound terms.

At any rate, given that Boole had a new system of deduction and given that he intended to reformulate traditional logic, one would expect to find detailed deductions of some Aristotelian syllogisms. Below are two examples of deductions of syllogisms using axioms and rules used by Boole (but not necessarily stated by him).

Example 1

All *Y*s are *X*s	1.	$y = yx$	
All *Z*s are *Y*s	2.	$z = zy$	
? All *Z*s are *X*s	?	$z = zx$	
	3.	$z = z$	I
	4.	$zy = z(yx)$	FM, 1, 3
	5.	$z(yx) = (zy)x$	AM
	6.	$zy = (zy)x$	R2, 4, 5
	7.	$x = x$	I
	8.	$zx = (zy)x$	FM, 2, 7
	9.	$zy = zx$	R1, 6, 8
	10.	$z = zx$	R2, 2, 9
		QED	

Example 2

All *X*s are *Y*s	1.	$x = xy$	
No *Z*s are *Y*s	2.	$zy = 0$	
? No *Z*s are *X*s	?	$zx = 0$	
	3.	$z = z$	I
	4.	$zx = z(xy)$	FM, 3, 1

5.	$xy = yx$	CM
6.	$x = yx$	R2, 1, 5
7.	$zx = z(yx)$	FM, 3, 6
8.	$z(yx) = (zy)x$	AM
9.	$zx = (zy)x$	R2, 7, 8
10.	$x = x$	I
11.	$(zy)x = 0x$	FM, 2, 10
12.	$zx = 0x$	R2, 9, 11
13.	$0x = 0$	N1
14.	$zx = 0$	R2, 12, 13
	QED	

These deductions are given for several reasons. In the first place, not only are these what one would expect to find in Boole's writings given what Boole himself says on pages 17 and 18 of *MAL* where he describes his system but also deductions of this sort are strongly suggested to the modern reader of some of Boole's later works ([6], pp. 231–239). In the second place, they give some idea of what is possible using a system like the one that Boole described and they give some idea of the glaring omissions from Boole's system. In the third place, as will be seen below, they are useful in providing a sharp contrast with the way that Boole actually "established" the validity of the two Aristotelian syllogisms in question.

2.4. Solutions and Consequences of Equations

In the next section we show that Boole repeatedly committed a fallacy involving taking a solution to an equation to be a consequence. In order to make the issues clear, this section is devoted to articulation of some distinctions which are widely presupposed in mathematical discourse but which, to the best of our knowledge, have never been spelled out in print. At the end of this section we explain how Boole's fallacy can be seen as a rather natural outcome of overly informal uses of the terms 'deduce' and 'imply' which were common in the mathematical literature of his day and which are relatively common even today.

The phrase "a solution to an equation" has been used in several senses, four of which are explicated in this section. For purposes of illustration we consider only equations in two variables x and y taking values in the domain $\mathbf{Z}$ of all integers. For simplicity we limit our explications to equations in two variables. Generalization to arbitrary equations is obvious.

In *one* sense, the ordered pairs (0, 0) and (2, 2) are the only solutions to the equation $x + y = xy$ and the ordered pairs (1, 3) and (–1, 3) are the only solutions to the equation $x^2y = 3$. What is meant here, of course, is more fully expressed by saying, e.g., that –1 assigned to x and 3 assigned to y (together) form a solution to $x^2y = 3$. The ordered pair locution presupposes a prior ordering of the variables: (3, –1) is not a solution to $x^2y = 3$ (unless y is linked with the first member and x with the second).

This sense of solution may be explicated by defining a *solution* to an equation to be an assignment (of objects in its domain to its variables) which satisfies the equation. In the case of $x + y = xy$, the assignments, α and β, are defined as follows:

	α
'x'	0
'y'	0

	β
'x'	2
'y'	2

Let us call solutions in this sense *root solutions* because they involve what are often called the "roots of an equation". Let $p(x, y)$ and $q(x, y)$ be terms (e.g., polynomials) involving only the variables x and y. When one asks for a root solution to the equation $p(x, y) = q(x, y)$, the variables are taken as names of "unknown objects" and the equation is regarded hypothetically as a true statement about those objects. The problem is to find out exactly which objects the statement is "true of".

In many cases, of course, an equation has a countably infinite number of root solutions. For example, every assignment γ such that

$$[\gamma\text{'}x\text{'}]^2 + [\gamma\text{'}y\text{'}]^2 = [\gamma\text{'}z\text{'}]^2$$

is a root solution to $x^2 + y^2 = z^2$. Also, the root solutions to $x^2 - y^2 = 0$ are the assignments γ such that γ'x' = γ'y' or γ'x' = $-\gamma$'y'.

Now the *set* of solutions γ with γ'x' = γ'y' can be "combined" into a (total) function, viz., $g(n) = n$. Likewise all solutions γ with γ'x' = $-\gamma$'y' can be combined into another function, viz., $h(n) = -n$. Each root solution to $x^2 - y^2 = 0$ is a "member" of exactly one of the two functions g and h but uncountably many other combinations are possible, e.g., the solutions γ with γ'x' = γ'y' where γ'x' is odd combine with the solutions γ where γ'x' = $-\gamma$'y' for γ'x' even to form the function j such that $j(n) = n$ for n odd and $j(n) = -n$ for n even. Notice that no such combining into total functions is possible with the solutions to $x + y = xy$ because there are not enough solutions, e.g., there is no solution assigning 1 to 'x'.

In another sense of the term 'solution', one considers the equation *not* as a condition on objects but rather as a condition on functions. In this sense when one asks for a solution to $x + y = xy$ one is really asking for a (total) function $x(y)$ such that

$$x(y) + y = x(y)y, \text{ for each integer } y$$

or, to put it more precisely, one seeks a function f such that

$$f(y) + y = f(y)y, \text{ for each integer } y.$$

In this sense $x + y = xy$ has no solutions and $x^2 - y^2 = 0$ has an uncountably infinite number of solutions, two expressible by the equations $g(y) = y$ and $h(y) = -y$. In general we call a function f a *functional solution* ("for x as a function of y") *to an equation* $p(x, y) = q(x, y)$ iff for all y, $p(f(y), y) = q(f(y), y)$.

In some cases, as we have seen, functional solutions can be expressed by equations. Sometimes mathematicians are not as explicit as possible in this regard. For example, instead of expressing one of the functional solutions to $x^2 - y^2 = 0$ by "$h(y) = -y$ for all y" this sometimes is expressed simply by the equation "$x = -y$"

(understood as "$x(y) = -y$, for all y"). In either case the term on the right can be called the *defining term* of an expression of a functional solution.

This brings us to a third sense of 'solution' because sometimes the defining terms are called solutions. In this sense 'y' and '$-y$' could be said to be solutions ("for x as a function of y") to the equation $x^2 - y^2 = 0$. These we call *term solutions.* Note that when a polynomial is said to be a solution to an equation the word "solution" is being used in the sense of a term solution.

Finally we note that equations are sometimes said to be solutions of equations. In particular, an equation expressing a functional solution is often itself called a solution to an equation. For example, $x = y$ is sometimes said to be a solution to $x^2 - y^2 = 0$. For this sense we use the phrase *equational solution:* where $t(y)$ is a term involving only the variable y, the equation $x = t(y)$ is said to be an *equational solution* ("for x as a function of y") to an equation when the function f such that for all y, $f(y) = t(y)$, is a functional solution.

It is probably worth noting that there are equations which are called solutions in the literature which are *not* equational solutions in the above sense. For example, in Boole's book on differential equations ([5], p. 7) $xy = \sin x$ is referred to as "a solution" to the differential equation resulting from differentiating it with respect to x. Boole's "solution", of course, is equivalent to $y = x^{-1} \sin x$ which is, in our sense, an equational solution (for y as a function of x) to the same equation.

Let $p(x, y)$ and $q(x, y)$ indicate terms (e.g., polynomials) involving the variables x and y. The following principles summarize the above distinctions.

1. The assignment γ is a root solution to an equation $p(x, y) = q(x, y)$ iff the sentence $p(\gamma\text{`}x\text{'}, \gamma\text{`}y\text{'}) = q(\gamma\text{`}x\text{'}, \gamma\text{`}y\text{'})$ is true.
2. The function f *is* a functional solution to $p(x, y) = q(x, y)$ iff the sentence "For every y, $p(f(y), y) = q(f(y), y)$" is true.
3. The term $t(y)$ is a term solution to $p(x, y) = q(x, y)$ iff the sentence "For every y, $p(t(y), y) = q(t(y), y)$" is true.
4. The equation $x = t(y)$ is an equational solution to $p(x, y) = q(x, y)$ under the same conditions as in 3, i.e., to verify that $x = t(y)$ is an equational solution it is sufficient to substitute $t(y)$ for x in the equation and then to ascertain that the result is true for *all* values of y.

For example, to verify that $x = -y$ is a solution to $x^2 - y^2 = 0$ it is sufficient to ascertain that $(-y)^2 - y^2 = 0$ is true for all values of y in the domain of integers. This trivial point has special significance in interpreting Boole's procedures.

The root solutions are relevant when the problem is simply to find "the roots" of an equation and the other three sorts of solutions are relevant when the problem is "to describe one variable as a function of the others", which, from a foundational point of view, is the same as the problem of combining "roots" into a function in the strict sense. The last three sorts of solutions are very closely related but there is a way that the notion of a functional solution is opposed to the notions of term solution and equational solution. For example, there is an uncountably infinite number of functional solutions to $x^2 - y^2 = 0$ but only a countable number of term solutions and equational solutions. But each term solution and, therefore, each equational solution can be regarded as an expression of a functional solution. We remind the reader that the above remarks apply literally only to equations in two variables x and y, and that modifications are in order when other equations are considered. For example, $x = 2$ can be said to be an equational solution to $x^2 = 4$ and there is no question of the equation having a functional solution.

Of special importance in this article are the equational solutions, and serious confusion is likely to occur unless it is noted that equations are, strictly speaking, ambiguous. They have one sort of meaning when used in problems about "unknown" objects (numbers, classes, etc.) and a completely different sort of meaning when presented as solutions to equations. For example, when $x = y + z$ is used in a problem about numbers it expresses an identity (of the number x to the sum of y and z) but when used to express a solution (e.g., to $x - z = y$) it is read as a definition of a function, i.e., as "for every y and z, $x(y, z) = y + z$". In the problem 'x' is an "arbitrary name", in the solution 'x' is a "dependent variable", i.e., a disguised function symbol.

There are three important things to notice about the relationships between an equational solution and the equations that it solves. Here all equations are read as equations expressing identities. First, every equation is a consequence of each of its equational solutions. For example, $x^2 - y^2 = 0$ is a consequence both of $x = y$ and of $x = -y$. Second, an equational solution is not necessarily a consequence of an equation which it solves. For example, $x = y$ does not follow from $x^2 - y^2 = 0$, nor does $x = -y$ follow. In this case, of course, the disjunction of the two solutions does follow, but the disjunction of two equational solutions is not itself an equational solution. This implies the third point, viz., that a consequence of an equation is not necessarily a solution.

It is clear then that it would be a mistake to infer that an equation is a consequence given only that it is a solution. Below we explain that Boole made this mistake, but here we briefly indicate how this mistake might be traced to overly informal uses of the terms "deduce" and "imply".

In former times (Boole [5], p. 8) and to some extent even recently (Langer [19], p. 359), mathematicians used the word "deduction" vaguely to indicate any process by which a solution was extracted from an equation regardless of whether the process was limited to logical inference. Langer defines a solution to a differential equation in x and y to be any equation in x and y (involving no derivatives) which is deducible from the equation. If one speaks of solutions as being *deduced* from equations then it is more or less natural to refer to solutions as *consequences* of those equations.

Moreover, as Boole himself states ([5], p. 6), there is a sense in which a solution is *implicit* in an equation. An equation describing a physical process does, in some sense, *imply* that the "quantities" involved are related in accord with the solutions. But unless one is clear that this is not necessarily *logical implication* of a solution by the equation, one is liable to mistake solutions for consequences.

2.5. Boole on Conversion of Propositions

Two pages after presenting his deductive system Boole introduces his notorious "v". The symbol v is supposed to indicate a nonempty class, so that $v = xy$ expresses "Some *X*s are *Y*s" (*MAL*, p. 21). There is considerable difficulty in deciphering Boole's intentions in regard to v (Dummett [15], p. 205; van Evra [23], p. 370), and we will not make an attempt here. For our purposes it is enough to know that Boole believed "Some *X*s are *Y*s" to be expressible as $v = xy$, as $vx = vy$, and as $vx(1-y) = 0$ and that he believed "Some *X*s are not *Y*s" to be expressible as $v = x(1-y)$, as $vx = x(1-y)$ and as $vxy = 0$. It is possible to make a certain amount of sense of this if, as Boole suggests, one reads 'vx' (and in some cases 'v' alone), as "some *X*s". For example, $vx = vy$ is read (by Boole) "Some *X*s are some *Y*s", and $v = xy$ is read sometimes as "Some *X*s are *Y*s" and sometimes as "Some *Y*s are *X*s" (*MAL*, p. 21). In addition Boole reads $v(1-y)$ as "Some not-*Y*s" (*MAL*, p. 25).

Boole gives no special rules or axioms for making deductions involving the symbol v. He does however claim to have derived, e.g., "Some *Y*s are *X*s" from "All *X*s are *Y*s", and *Boole's "derivations" of the Aristotelian conversions involved him in the fallacy of taking a solution to an equation as a consequence,* as is shown by the following passage (*MAL*, pp. 26–28).

> The primary canonical forms already determined for the expression of Propositions, are
>
> | | | |
> |---|---|---|
> | All *X*s are *Y*s, | $x(1-y) = 0,$ | A. |
> | No *X*s are *Y*s, | $xy = 0,$ | E. |
> | Some *X*s are *Y*s, | $v = xy,$ | I. |
> | Some *X*s are not-*Y*s, | $u = x(1-y),$ | O. |
>
> [. . .]
>
> The equations A and E, written in the forms
>
> $$(1-y)x = 0,$$
> $$yx = 0,$$
>
> give on solution the respective forms
>
> $$x = vy,$$
> $$x = v(1-y),$$
>
> the correctness of which may be shown by substituting these values of x in the equations to which they belong, and observing that those equations are satisfied quite independently of the nature of the symbol v. The first solution may be interpreted into
>
> Some *Y*s are *X*s,
>
> and the second into
>
> Some not-*Y*s are *X*s.
>
> From which it appears that universal-affirmative, and universal-negative Propositions are convertible by limitation, or, as it has been termed, *per accidens*.

There should be no mistake about our claim that Boole's sole ground for asserting that $x = vy$ follows from $(1-y)x = 0$ is given above. This is so despite the fact that it should have been obvious to Boole that $x = 0$ is also a solution and that if $x = 0$ and y is any nonempty set then the equation is true but the solution $x = vy$ is false (for v any set not disjoint with y).

One should also be clear about the fact that $x = vy$ is *not* deducible from $(1-y)x = 0$ using the rules that Boole explicitly states. The reason is simply that Boole's explicitly stated rules are all satisfied when interpreted as in ordinary class algebra.

For convenience we use the phrases "solutions fallacy" and "the fallacy of solutions" to indicate the inference of a solution of an equation from the equation itself, *when the solution doesn't follow.* Strictly speaking it is always a fallacy to infer a solution from an equation on the ground that it is a solution, even when the solution actually follows. But we use the term in the narrower sense which does not have reference to the ground of the inference.[12]

It is clear enough that Boole is guilty of the solutions fallacy in the above case. It has not been shown, however, that Boole believed that *each* equation used to express a particular proposition ("Some *X*s are *Y*s" or "Some *X*s are not-*Y*s") is deducible in his system from *each* equation used to express any implying general proposition ("All *X*s are *Y*s" or "No *X*s are *Y*s"). This question is a bit more complicated than it appears because Boole permitted a single given proposition to be expressed by as many as three different equations. But after presenting his deductive system and correlating the equations with propositions he states (*MAL*, p. 23):

> . . . all the equations by which particular truths are expressed, are deducible from *any one* general equation, expressing any one general Proposition, from which those particular Propositions are necessary deductions. (emphasis ours)

Of course Boole is using the term "necessary deduction" to indicate either "necessary consequence" or "necessary deduction of the Aristotelian system". Moreover, at the very beginning of the first chapter of *MAL* (p. 6) Boole says that in the system to be presented the laws of conversion and of syllogisms will be deducible.

2.6. *Boole's Derivations of the Syllogisms*

On pages 31 through 47 of *MAL* Boole gives his treatment of the syllogisms. From earlier pages one might expect that Boole would use more or less straightforward deductions except in the cases where Aristotle had deduced a particular conclusion from universal premises. In these cases one would expect to find the solutions fallacy. In turning to the text one does indeed find the solutions fallacy bridging the gap between universal premises and particular conclusions. However, not even in the other cases does one find straightforward deductions.

In order to understand Boole's method of dealing with syllogisms one must recall that, from two equations in three variables, it is sometimes possible to deduce a third equation involving only two of the variables. For example $x + y = z$ and $5x + 2y = 3z$ imply $3x = z$. Boole knew that in certain cases it is possible to use a general rule for "eliminating" y from the equations. On page 32 he notes that the following elimination scheme is valid in arithmetic.

E15
$$ay + b = 0$$
$$a'y + b' = 0$$
$$?\ \ ab' - a'b = 0.$$

To apply this to the above case the equations must be rewritten in the required forms:

$x + y = z$	is rewritten	$1y + (x - z) = 0$
$5x + 2y = 3z$	is rewritten	$2y + (5x - 3z) = 0.$

Then applying the elimination scheme one has:

$$1(5x - 3z) - 2(x - z) = 0.$$

From here one can deduce $3x = z$.

Boole also seems to have noticed that E15 is also valid in class logic.

The above elimination scheme E15, which Boole referred to as "(15)", is the only such scheme symbolically stated in *MAL* but it is not the only one mentioned. On page 34, Boole writes:

> A convenient mode of effecting the elimination, is to write the equation of the premises, so that y shall appear only as a factor of one member in the first equation, and only as a factor of the opposite member in the second equation, and then to multiply the equations, omitting the y.

His example (*MAL*, p. 35) is quoted below.

All *X*s are *Y*s,	$x(1 - y) = 0,$	or $x = xy$
No *Z*s are *Y*s,	$zy = 0,$	$\underline{zy = 0}$
		$zx = 0$
	$\therefore$ No *Z*s are *X*s.	

Boole evidently intends the premises to be written $b = ay$ and $a'y = b'$. There are two forms of the conclusion compatible both with his words and with the example, viz., $a'b = ab'$ and $aa' = bb'$. The first is valid in general but the second is not [e.g., take $a = a' = 1$]. Thus we represent Boole's second elimination scheme as follows.

E $\quad b = ay$
$\quad a'y = b'$
? $a'b = ab'$

Boole gives no derivations for either of these two schemes. However both are easily seen to be valid.

Deduction of E

1.	$b = ay$	
2.	$a'y = b'$	
?	$a'b = ab'$	
3.	$a'b = a'(ay)$	multiplying a' times (1)
4.	$a(a'y) = ab'$	multiplying a times (2)
5.	$a'(ay) = (a'a)y$	associativity
6.	$a(a'y) = (aa')y$	associativity
7.	$a'b = (a'a)y$	things equal to the same (5, 3)
8.	$(aa')y = ab'$	things equal to the same (6, 4)
9.	$aa' = a'a$	commutativity
10.	$(aa')y = (a'a)y$	multiplying (9) times y
11.	$a'b = (aa')y$	things equal to the same (7, 10)
12.	$a'b = ab'$	things equal to the same (8, 11)
	QED	

A derivation of E15 is an elaborate triviality not worth citing in detail. One way to proceed is to first derive $b' = 0$ from $a'y + b' = 0$ and then from $b' = 0$ derive $ab' - a'b = 0$. This reasoning shows that Boole's E15 is very weak and that it is a special case of the following.

SE $ay + b = 0$
$a'y + b' = 0$
? $zb' - w = 0$

The point to notice is that the conclusions, both of E15 and of SE, follow from the second premise ($a'y + b' = 0$) alone. Moreover, the only thing that is relevant to the conclusion is the fact that the second premise implies $b' = 0$. Thus whenever Boole uses E15 he is using only the second premise.[13]

Boole believed that elimination schemes provide an essential part of derivations of syllogistic conclusions. Just prior to stating E15 he wrote:

> The equation by which we express any Proposition concerning the classes X and Y, is an equation between the symbols x and y, and the equation by which we express any Proposition concerning the classes Y and Z is an equation between the symbols y and z. If from two such equations we eliminate y, the result, *if it do not vanish,* will be an equation between x and z, and will be interpretable into a Proposition concerning the classes X and Z. *And it will then constitute the third member, or Conclusion, of a Syllogism, of which the two given Propositions are the premises.* (*MAL*, pp. 31, 32, italics added)

Shortly after this statement he explains the qualification "if it do not vanish". Notice that if b and b' are both 0 then the elimination scheme yields, in effect, $0 = 0$. Boole explains that in such cases one does not apply a scheme to the two equations directly but one must first *solve* one of them. And he repeats his observation that $y = v(1 - x)$ is a solution to $xy = 0$ and that $y = vx$ is a solution to $(1 - x)y = 0$. (He seems to think that these are the only solutions, but for the present this can be ignored.)

We quote in full Boole's treatment of the first Aristotelian syllogism (*MAL*, p. 34).

> All Ys are Xs, $y(1 - x) = 0$, or $(1 - x)y = 0$.
> All Zs are Ys, $z(1 - y) = 0$, or $zy - z = 0$.
>
> Eliminating y by (15) we have
>
> $z(1 - x) = 0$,
> $\therefore$ All Zs are Xs.

That Boole did not correctly apply E15 to get his conclusion will become clear presently. What is immediately clear, however, is that Boole thinks that he used E15 to deduce $z(1 - x) = 0$ from $(1 - x)y = 0$ and $zy - z = 0$. But this is a fallacy because $zy - z = 0$ is a logical truth: zy must be a subset of z, and the complement of z with respect to one of its own subsets is necessarily 0. In other words, no matter what classes y and z may be, $zy - z = 0$ is true. Thus $z(1 - x) = 0$ is being deduced from $(1 - x)y = 0$, which does not imply it, of course (take $x = y = 0$ and $z = 1$).

This and other passages *suggest* that Boole may *not* have been using "–" to indicate relative complement. But in *MAL* he never explicitly states what he takes "–" to indicate. Compare ([15], p. 205). In later works (e.g., [4], p. 33) he takes "–" to indicate a certain partial function. Regardless of whether Boole is using "–" to indicate a partial function as in *Laws of Thought,* in the course of "applying" E15, there are other fallacies involved in the above quoted passage.[14]

The inescapable conclusion is that Boole is not using E15 as he says he is. Moreover, the important point is not that there is a gap between what he claims to be doing and what he is doing, but rather that no care has been taken to refer his derivations to the deductive system that he has set forth several pages earlier.

Boole's second example was quoted in full above in connection with the discussion of elimination scheme E. Likewise in that example the connection, if any, with the deductive system is not referred to.

In the next four examples (*MAL*, pp. 35, 36) Boole derives particular conclusions from universal premises. In each of these cases he uses the fallacy of solutions to get intermediate particular propositions and then he applies elimination scheme E. For example, in the first of these he infers the solution $y = vx$ from $y(1 - x) = 0$, failing to notice that the solution is not a consequence ($y(1 - x) = 0$ does not imply that y is nonempty).

In the final three (*MAL*, p. 37) of his nine examples he deduces particular conclusions from a pair of premises each including a particular premise. In these cases, of course, no fallacy is needed to get a particular conclusion. Nevertheless in the second of these his second step seems, in effect, to deduce "Some *Y*s are not *Z*s" from "Some *Z*s are not *Y*s" (e.g., "Some dogs are not animals" from "Some animals are not dogs"). See the example on page 124.

In order to summarize Boole's treatment of the syllogisms we make the following abbreviations.

Axy	All *X*s are *Y*s
Nxy	No *X*s are *Y*s
Sxy	Some *X*s are *Y*s
$Sx(1 - y)$	Some *X*s are not-*Y*s
$S(1 - x)y$	Some not-*X*s are *Y*s
$S(1 - x)(1 - y)$	Some not-*X*s are not-*Y*s.

1. (1) Ayx (2) Azy ? Azx	On page 34, E15 is misapplied to inferences from the premises.
2. (1) Axy (2) Nzy ? Nzx	On page 35, E is applied to inferences from the premises.
3. (1) Ayx (2) Nzy ? $Sx(1 - z)$	On page 35, E is applied to a solution of (1) taken together with an inference from (2).
4. (1) Ayx (2) Nzy ? $S(1 - z)x$	On page 36, E is applied to (1) taken together with a solution of (2).
5. (1) Ayx (2) Ayz ? Sxz	On page 36, E is applied to a solution of (1) taken together with an inference from (2).
6. (1) Nyx (2) Nzy ? $S(1 - z)(1 - x)$	On page 36, E is applied to an inference from (1) taken together with a solution of (2).

7.	(1) Ayx	On page 37, E is applied to (1) and (2).
	(2) Szy	
	? Szx	
8.	(1) Axy	On page 37, E is applied to inferences from (1) and (2).
	(2) $Sz(1-y)$	
	? $Sz(1-x)$	
9.	(1) $Sy(1-x)$	On page 37, E is applied to (1) and an inference from (2).
	(2) Nzy	
	? $S(1-x)(1-z)$	

In his treatment of the syllogism Boole does not explicitly employ the deductive system which he has described on pages 17 and 18. Instead he treats syllogistic inference as a case of eliminating a variable from two equations. To effect this he gives two elimination schemes; one, E15, is stated symbolically and the other, E, is stated ambiguously. The one which is stated symbolically is applied once, and then incorrectly. The one which is not stated symbolically is used eight times. In one case it seems to be used in conjunction with an elementary conversion fallacy, and in four other cases it is used in conjunction with the fallacy of solutions. Since E was *not* deduced, none of the derivations are proper deductions.

If it is not discouraging enough that the fallacy of solutions occurs in four of his nine examples, in the discussion (*MAL*, p. 42) following the examples he says that it would have been allowable to make the same step in the other cases as well. We quote.

> We have found it, in a certain class of cases, to be necessary to replace the two equations expressive of universal Propositions by their solutions; and it may be proper to remark, that it would have been allowable in all instances to have done this . . .

The inattention to logical detail exhibited in Boole's derivation of the syllogisms is clear enough so that further comment is not needed. It is remarkable, however, that Boole had the idea of a sound, equational deductive system but that he made no attempt to develop it or to apply it. It is clear that Boole's fallacious derivations of the syllogisms drawing particular conclusions from universal premises are all based on the fallacy of solutions. But one should not overlook the fact that other fallacies are also present.

3. BOOLE ON ESTABLISHING INVALIDITY

If a deductive system is "strong enough", i.e., if it contains a deduction of the conclusion from the premises whenever the conclusion is implied by the premises then it is said to be *complete*. This could have been what Boole had in mind by his term "sufficient". But in view of Boole's failure to clearly distinguish consequences from solutions he could have been thinking of his system as a calculus, i.e., as a method of deriving solutions to equations (and sets of equations). It would be natural to call a calculus "complete" if it contains a derivation of a given equation from a given set of equations whenever the given equation is a solution to the set of equations.

Of course, one must also take account of the possibility of having "unsound deductions", i.e., deductions which permit inference from given premises of conclusions which do not follow. In such a case one says informally that the deductive system is "too strong". If a deductive system is *not* "too strong", i.e., if it permits deduction of a conclusion from premises *only* in those cases where the argument is valid, then it is said to be *sound.* For a deductive system to be complete and sound is for it to be "just strong enough". In this case one can use it to deduce each valid argument but one cannot use it to deduce any invalid arguments. It apparently never occurred to Boole to wonder whether his system (or his method) was unsound. Indeed, he paid only the most casual notice to the concept of invalidity.

As indicated above, the usual criterion of invalidity is refutability by a counterargument. One knows that "All men are happy and Joe is happy" does *not* imply "Joe is a man" because one knows that "All dogs are animals and Felix is an animal" is true whereas "Felix is a dog" is known to be false. Boole notes the invalidity of many arguments but he *never* uses the method of refutation by counterargument (even implicitly) to establish his observations. In his discussion of conversions (*MAL*, pp. 26–30) he does not discuss, or even mention, invalid *conversions* such as "All *X*s are *Y*s" with "All *Y*s are *X*s". Moreover in his discussion of syllogisms, although invalid arguments are mentioned, there is not one counterargument to be found in the entire fifteen pages (*MAL*, pp. 31–47).

As mentioned above, the method of refutation by counterargument is virtually implied by the so-called principle of form, viz., that arguments are valid or not in virtue of form, i.e., that two arguments in the same form are both valid or both invalid. Boole indicates some awareness of form as opposed to content in *MAL.* And in later works ([6], pp. 216, 217) Boole emphasizes the fact that validity does not depend on the particular meanings of the terms employed but he never seems to see this fact as leading to a method of refutation. It is as if establishing invalidity by counterargument has never occurred to him.

It might be suspected that refutation by counterargument was somehow out of favor in the time during which Boole wrote. But this suspicion is neither confirmed nor put to rest by consulting the several works on logic written in the British Isles just prior to Boole's work. For example, Whately's *Elements of Logic* [24], which was supposedly the most popular logic book of the period and which was quoted or referred to four times by Boole (*MAL*, pp. 7, 18, 20, 28), uses the method of counterarguments repeatedly. But there is no discussion of it and there is no systematic use of it that would enable one to infer that a principle was being tacitly appealed to. Some of Whately's counterarguments are quoted below.

1. All birds are animals. [24], p. 82
   ? All animals are birds.

2. Some animals are beasts. [24], p. 89
   Some animals are birds.
   ? Some birds are beasts.

3. Some animals are sagacious. [24], p. 90
   Some beasts are not sagacious.
   ? Some beasts are not animals.

The method of refutation by counterargument is conspicuously absent from Mill's famous *System of Logic* [20] (originally published in 1843 and referred to in

MAL, p. 2). Despite the fact that Mill seems to be unaware, not only of the traditional criterion of invalidity but also of the traditional deductive criterion of validity, it must be admitted that he seems to refer the reader elsewhere for these ideas. After displaying the traditional valid two premise syllogisms, he writes:

> The reason why syllogisms . . . are legitimate, that is, why if the premises are true the conclusion must inevitably be so, and why this is not the case in any other possible combination . . . of propositions . . . may be presumed to have [been] . . . learned from the common school books of syllogistic logic . . .. The reader may . . . be referred . . . to . . . Whately's *Elements of Logic*, where he will find stated with philosophical precision, and explained with remarkable perspicuity, the whole of the common doctrine of the syllogism. ([20], p. 191)

In many respects DeMorgan's *Formal Logic* [14], published in 1847, the same year as *MAL*, is a breath of fresh air compared to Whately, Mill, and Boole. DeMorgan is much more sensitive both to language and to the conventions which had become traditional in logic but which deviated from those common in ordinary discourse. In addition he is much more systematic and his attention to detail is what one would expect from a mathematician. *However* there is nothing about organizing deduction into a system and, although the method of refutation by counterargument seems just below the surface in many places, one could not argue that DeMorgan was consciously aware of it as a method.

3.1. Boole's Criterion of Invalidity

In order to grasp Boole's treatment of invalidity it is useful to recall a few elementary ideas from analytic geometry. An equation in three variables $f(x, y, z) = g(x, y, z)$ describes a set of points in three-dimensional space. For example, $z = x + y$ describes the plane containing the three points (0, 0, 0), (1, 0, 1), and (0, 1, 1). This plane cuts each plane (z = constant) parallel to the xy-plane in a line $y = z - x$ (which for $z = 0$ is the line $y = -x$). For another example, $z^2 = x^2 + y^2$ describes the cone obtained by rotating the line $z = y$ about the z-axis. This cone cuts each plane parallel to the xy-plane in a circle whose radius $(x^2 + y^2)^{1/2}$ is the distance $|z|$ of the given plane from the xy-plane.

The set of points described by a law of arithmetic, e.g., $x \cdot (y \cdot z) = (x \cdot y) \cdot z$, is the set of all points in the space. An equation which is "universally false", e.g., $x \cdot (y \cdot z) + 1 = (x \cdot y) \cdot z$, describes the null set of points. And the set of points described by several equations taken together is the set of points common to the sets described by the individual equations. For example the set of points described by $z = x + y$ and $z^2 = x^2 + y^2$ taken together contains two points $(0, z, z)$ and $(z, 0, z)$ in each plane (z = constant) parallel to the xy-plane and thus consists of the two lines $z = x$ (in the plane $y = 0$) and $z = y$ (in the plane $x = 0$).

If an equation c is implied by a set P of equations then the set of points described by the members of P (taken together) is a subset of the set of points described by c.

The equation $0 = 0$ is implied by every set of arithmetic equations: $p = q$ implies $p - p = q - q$ or $0 = 0$. Moreover, $0 = 0$ is equivalent to $(x - x) + (y - y) = (z - z)$, so $0 = 0$ can be regarded as describing the whole space.

The situation in Boole's class algebra is exactly analogous. An equation in three variables describes a set of triples of classes. Several equations taken together describe the triples of classes common to the sets of triples described by the individual equations. The set of triples described by a consequence of an equation contains as a subset the class of triples described by the equation. Each Boolean law, including $0 = 0$, describes the set of all triples of classes. And $0 = 0$ is implied by each and every set of equations.

The important point to bear in mind here, both with respect to arithmetic as well as in Boole's algebra, is that nothing whatever can be concluded about a set of equations from the fact that $0 = 0$ has been deduced from it. The above discussion was designed to put this point in a context that would make it completely clear and unproblematic.

Boole first makes note of invalidity on page 35 where he makes the assertion that there is no inference possible from "All *X*s are *Y*s" and "All *Z*s are *Y*s". Of course, no valid Aristotelian syllogism has these as premises but clearly these two premises imply "All (*X* or *Z*)s are *Y*s", "All (*X* and *Z*)s are *Y*s", "All *X*s except[15] *Z*s are *Y*s", and so on, all of which are easily representable in Boole's notation. One must assume that Boole intended to restrict himself to the Aristotelian cases here even though he considers valid but non-Aristotelian inferences on the very next page.

What is somewhat baffling is that the only ground that Boole gives for asserting that no inference is possible in the above case is that $0 = 0$ is deducible from the result of applying elimination scheme E to the premises $x = xy$ and $zy = z$. But he doesn't state that that is *the* ground.

On page 38 he does something similar with each of two pairs of Aristotelian premises which do not yield syllogistic conclusions. But in these cases, instead of deducing $0 = 0$ from the result of applying the elimination scheme E, he deduces $0 = 0$ from the result of applying E taken together with what he calls "the auxiliary equation" for the second premise. To understand what he has in mind by "auxiliary equation" one should recall tht v, the "auxiliary symbol", is supposed to mean "Some *X*s" in certain contexts. In these contexts the auxiliary equation is $v(1 - x) = 0$. In contexts where v is supposed to mean "Some not-*X*s" the auxiliary equation is $vx = 0$ (see *MAL*, pp. 24, 25). Although Boole does not explicitly say so, it is clear that the auxiliary equation is "attached" to a given equation when used to express a given proposition and it is clear that the auxiliary equation states that v is a subset of the class that Boole's usage *takes* it to be a subset of. According to Boole's usage, e.g., "All *Y*s are *X*s" is expressed by $y = vx$ with $v(1 - x) = 0$ as auxiliary. Of course, it would have been much clearer to say that "All *X*s are *Y*s" is expressed by the *pair* of equations $y = vx$ and $v(1 - x) = 0$.

On the basis of these three examples one might suppose that Boole thinks that the deduction of $0 = 0$ from the equations of premises (or from their equations taken together with their auxiliary equations) establishes that no Aristotelian conclusion follows. And this supposition is somewhat supported by Boole's own words after the third example.

> Indeed, in every case in this class, in which no inference is possible, the result of elimination is reducible to the form $0 = 0$. Examples therefore need not be multiplied. (*MAL*, p. 38)

One should be clear about the fact that by "reducible to the form $0 = 0$" Boole could not have meant "equivalent to $0 = 0$" because in the example immediately preceding his quoted remark the result of elimination is $vx = vxz$ (and the auxiliary equations are $v(1 - y) = 0$ and $v(1 - z) = 0$).

On page 38 he considers his final class of "unlawful cases" which he divides in two. The first division is "when the result of elimination is reducible by auxiliary equations to the form $0 = 0$". The second is "when the result of elimination is not reducible by auxiliary equations to the form $0 = 0$". The first division is compatible with our supposition. His example of the latter division is the following, quoted in full (*MAL*, p. 39). See middle of page 119 above.

Some *Y*s are *X*s,	$vy = vx$,	$vy = vx$
Some *Z*s are not *Y*s,	$v'z = v'(1 - y)$,	$\underline{v'(1 - z) = v'y}$
		$vv'(1 - z) = vv'x$

Now the auxiliary equations being

$$v(1 - x) = 0, \quad v'(1 - z) = 0,$$

the above reduces to $vv' = 0$. It is to this form that all similar cases are reducible. Its interpretation is that the classes v and v' have no common member, as is indeed evident.

It is not at all clear why Boole thinks that this example is different from the others. Given the auxiliary equation $v'(1 - z) = 0$, it follows that the left side, $vv'(1 - z)$, of the result of elimination is 0. Thus the right side is 0. Thus $0 = 0$ follows.

However the rest of the page in question, the next page, and the following page (*MAL*, pp. 39–41) make it plain that Boole believes that the last example is different.

In the whole of *MAL* there is nothing on invalidity except the material reviewed above taken from pages 35 through 41 and some scattered remarks on pages 44 through 47. There is not one counterargument in the entire book. No criterion of invalidity, correct or incorrect, is explicitly offered as such. And the only test which comes close to being used as such a criterion is deducibility of $0 = 0$ from the equations and auxiliary equations of the premises. But this has been seen to apply to every set of premises and thus does not serve to distinguish any argument from any other.

4. CONCLUSIONS CONCERNING *MAL*

Boole had an idea of a deductive system which, if developed, would have been superior to the system handed down by Aristotle. In the first place it permitted the use of logical axioms (instead of being restricted to rules of inference). In the second place, it could be used in connection with negative terms and with conjunctions and disjunctions of terms.[16]

But Boole did not seem to realize the importance of the ideas. He omitted many obvious axioms. He noticed two "derived rules" (E15 and E) but he gave no derivations of them. He did not give a single application of the system, not even easy and obvious ones such as derivations of the easiest of the Aristotelian syllogisms. In deriving the syllogisms using other methods (involving the "derived rules") fallacies have been noted in six of the nine examples. In four cases, the fallacy of solutions is involved. In a comment following the examples, Boole thinks that the fallacious move involving solutions is allowable in all cases. (The syllogisms drawing particular conclusions from universal premises are all obtained *via* the fallacy of solutions.)

The fact that Boole did not notice his fallacies may be partly explained by the fact that he was oblivious to the need for a criterion of invalidity. He mentions invalidity only within the space of twelve pages (*MAL*, pp. 35–47). When he mentions the goals of logic he usually mentions the generation of valid inferences—never the identification and classification of invalid ones. He gives no criterion of invalidity and the test that he does seem to employ is absurdly inadequate.

Boole's standing in the history of logic does not, of course, rest on his achievements in *The Mathematical Analysis of Logic,* the work here being discussed.[17] And it would not be fair to Boole to compare his understanding of logical methodology or his attention to logical detail,[18] as represented by *MAL,* to that of Aristotle as represented by Book I of *Prior Analytics.* It is worth pointing out, however, that Aristotle clearly stated his deductive criterion of validity and applied it with meticulous accuracy in twenty-odd deductions. Perhaps because Aristotle did so many examples, he managed to discover and state enough rules to have a complete deductive system (Corcoran [10]). It is also worth noting that although Aristotle did not *state* his counterargument criterion of invalidity, he applied it with precision and ingenious conciseness to over one hundred cases, making it possible to infer with considerable certainty what the criterion was.

Aristotle's work included the following experiment. He considered a rather simple class of almost two hundred arguments supplying a deduction for each valid case and supplying a counterargument for each invalid case. Of course, the arguments considered by Aristotle were syllogisms containing exactly one or exactly two premises. Boole would have discovered a complete set of axioms had he run an analogous experiment on the equations between the following terms (where $k = +, \cdot$ or $-$): 0, 1, x, xky, $xk(ykz)$, $(xky)kz$, $(xky)k(xkz)$ (a total of 51 terms or 2601 equations). What we have in mind, of course, is supplying a deduction for each logical truth and a counterexample (in 0 and 1) for each of the others. We can excuse Boole from conducting the whole of such a huge and repetitious exercise. But whether Boole should be excused for running no such experiment is a matter to be judged by those with a greater appreciation for the state of logic and mathematics of the time.

Although we hereby register our doubts, we have no adequate ground for disagreement with Dummett [15] who wrote:

> There can be no doubt that Boole deserves great credit for what he achieved, in the sense that in those historical circumstances what he did must have been very difficult to do.

NOTES

*The authors thank James van Evra (University of Waterloo) for help over a period of years. Some of the ideas in this paper are from the second author's doctoral dissertation [26] which was supervised by the first author. Van Evra was the outside reader. A National Science Foundation grant enabled the second author to read the unpublished Boole manuscripts on deposit in the Library of the Royal Society of London. Wendy Ebersberger, John Kearns, William Parry, Charles Lambros, and, especially, Michael Scanlan (all of SUNY/Buffalo) gave suggestions, criticisms, and friendly support and cooperation. Earlier versions were presented to the Buffalo Logic Colloquium (November 1977), to the Association for Symbolic Logic (April 1978, see [12]), and to the Philadelphia Logic Colloquium (April 1979). Finally, the authors are grateful for conscientious and penetrating criticism to the editor and to anonymous referees.

[1] This interpretation of Aristotle's logical system was discovered independently by Smiley [21] and Corcoran [10]. Some related ideas were discovered earlier by Bacon [2].

[2] The intended interpretation of the language in which the counterargument to (P, c) is stated is, of course, a countermodel for (P, c). Moreover, reinterpreting (P, c) in one of its known countermodels

produces a counterargument. Thus having a known countermodel and having a counterargument are the same thing. Boole, of course, uses the traditional terms 'argument', 'valid', 'train of reasoning', etc., and he has the idea of different interpretations (or models) of his formal language, but he does not introduce a special term for them.

[3] It is very likely that Boole sometimes thought of + in *MAL* as union. In *Laws of Thought* he explicitly takes + to be union on disjoint sets and undefined on overlapping sets, but such seems not to be the case in *MAL*. For example, in *MAL*, p. 58, he speaks of adding the condition of disjointness in a situation where disjointness could be inferred were + partial. In this connection, and for other reasons as well, it is important to notice (1) that Boole *never* did interpret his operation symbols as truth-functions and (2) that in *MAL* Boole never conceived of logic except as class logic. Propositions that are now treated by propositional logic were treated by Boole as equations between classes in a section called "Of Hypotheticals". There the elective symbol x corresponding to the proposition X indicates the class of "conceivable cases" in which X is true. Thus $x + y$ is a term (not a sentence), it indicates a class of "cases" (not a truth value), and, so, + must be an operation on classes (not a truth-function) (*MAL*, pp. 48ff). References to Boole's "Theory of truth-functions" (e.g., Kneale and Kneale, [18], p. 420) are at best misleading. And reference to $x + y$ as expressing the truth-value of the exclusive disjunction of X and Y (e.g., [18], p. 413) is wrong for two reasons. First, it does not express a truth-value. Second, if + is union then $x + x = x$ so $x + y$ corresponds to the nonexclusive disjunction and if + is partial union then (for $x \neq 0$) $x + x$ is *undefined* so $x + y$ does not correspond to exclusive disjunction.

Boole introduces the minus sign (*MAL*, p. 20) in the context $1 - x$ and he uniformly uses it this way (as a *binary* function symbol in the expression of the nonrelative complement, which is a *unary* function, of course) for twelve pages (*MAL*, p. 32) where he first uses it without 1 in the context $ab' - a'b$. Then (*MAL*, p. 43), without comment, he uses it as a unary function symbol in the term $-v$. Thereafter he uses it as a binary function symbol and as a unary function symbol (e.g., *MAL*, pp. 46, 55).

[4] This quotation is taken out of context (*MAL*, p. 10), but it is clear from the broader context that it expresses Boole's view. Corroborating passages are also present (e.g., [6], pp. 17, 217, 218).

[5] The distinction between logical axioms ("laws of judgment") and rules of inference ("laws of reasoning") is particularly crisp in *MAL* because there is no "if . . . then" in the object language. (This point is missed by Kneale and Kneale ([18], p. 412).) The logical axioms are stated using symbolic formulas and the rules of inference are given in English. Since "conducting an argument" ([6], p. 237) is clearly "reasoning" as opposed to "judgment", Boole feels it necessary to emphasize the necessity of using "laws of judgment" in deductions ([6], pp. 144, 237, 238). There are some passages which might appear to a modern reader to be attempts to articulate the axiom/rule distinction (e.g., [6], p. 25). But because the distinction is unproblematic for Boole, other interpretations are more likely.

[6] Complementation should not be confused with negation. Not only did Boole's language lack negation but it also lacked the ability to express negations. For example, there is no Boolean formula involving x and y which is true if and only if x and y are distinct. Thus $x \neq y$ cannot be expressed. In a later work ([6], p. 128, fn) Boole explicitly considers "All *X*s are not-*Y*s" and "the proposition 'Some *X*s are *Y*s' is false". He says that his preference of the former (complement) reading is "not a useless refinement but a necessary step, in order to make the proposition truly a *relation between classes*" (Boole's emphasis). Moreover, when he deals with what is now taken as negation of a proposition he again avoids negation by taking an elective symbol for a proposition to indicate *not* a truth-value but rather a *class*, viz. the class of "cases" in which the corresponding proposition is true (*MAL*, p. 49). Thus even in this context negation is supplanted by complementation. In *Laws of Thought* (pp. 164–165), Boole understands the elective symbol x corresponding to the proposition X to indicate "that portion of time for which the proposition X is true", still avoiding negation.

[7] Although Boole does not say so, the same two rules in exactly the same wording are found in Whately ([24], II, iii, 2).

[8] In the above we have followed Boole's practice of using elective symbols in metalinguistic statements of axiom schemes and rules as if they were syntactic variables. The alternative, in regard to axioms, is to suppose that Boole implicitly assumed a general substitution rule (from S infer any sentence obtained from S by replacing all occurrences of a single elective symbol by one and the same term). To preserve consequence general substitution must be restricted so that it applies only to sentences derived from logical axioms (e.g. Corcoran [9], p. 155). This virtually entails that the elective symbols are free variables, but there is no evidence in *MAL* to support that conclusion. On the contrary, the elective symbols are repeatedly used to refer to "arbitrary but fixed" classes. In effect we are claiming that Boole's object language had no variables and thus is not an algebraic language in the usual sense (e.g., Craig, [13], Hailperin, [16]).

[9] The additions suggested below are based on the supposition, made for illustrative purposes only, that + and – are interpreted as union and relative complement. Of course, if these two symbols are interpreted differently then other additions would be appropriate. For example, if + is taken as partial union then Boole's distributive law is not among the "propositions true in consequence of form alone" because if $x = \{1, 2, 3\}$, $u = \{1, 4\}$ and $v = \{3, 4\}$ the left side is *undefined* and the right side is $\{1, 3\}$. The main point is that no matter how the two symbols are interpreted major additions are necessary. Incidentally, contrary to Kneale and Kneale ([18], p. 406), important additions *are* made in *Laws of Thought*. Also, in connection with U1 see *MAL*, p. 17n.

[10] Wood ([26], pp. 195–196) has shown that if + and – are interpreted as the partial functions used in *Laws of Thought* then DeMorgan's Laws do not hold. If DeMorgan's laws are taken as conditions on two classes x and y then they imply that x is the complement of y.

[11] Boole was also not aware of the fact that Aristotle's rules do not admit of easy generalization into the richer framework. For example, Aristotle inferred "Some *A*s are *B*s" from "All *A*s are *B*s" because his terms were "logically simple" universals which are grasped through perception of individuals coming under them and which are, therefore. inherently nonempty. But, of course, terms which are compounded out of logically simple universals are not logically simple and thus may be empty. Thus for Aristotle, "Some *CA*s are *B*s" does not follow from "All *CA*s are *B*s". For example, "Some circular squares are rectangles" does not follow from "All circular squares are rectangles" (Corcoran [11], pp. 103–104).

[12] If one is solving equations $p(x, y) = q(x, y)$ one is asking *either* (1) whether there are objects x and y which satisfy the equation and if so what they are or (2) whether there are functions $x = f(y)$ which satisfy the equation (for all values of y) and if so what they are. In neither case can one *assume* that such objects or functions exist. If one is considering equations as expressive of propositions for purposes of drawing inferences then the letters x and y are taken as "names" of objects (or classes) and in this case it is *absurd* to speak of x as a function of y. Could 7 be a function of 8? Could the class of men be a function of the class of animals? Nevertheless Boole confuses these distinct roles of letters (*MAL*, pp. 70, 71). In this location he thinks of an equation involving x and y as if it gives "y as a function of x" (*MAL*, pp. 24, 70 and cf. Langer [19], p. 359). This confusion has unfortunate consequences for Boole's "analytic method" which, however, is not to be discussed in this article.

In the passage quoted from *MAL*, pp. 26–28, Boole derives "Some *Y*s are *X*s" $[x = vy]$ from "All *X*s are *Y*s" $[x(1 - y) = 0]$ taking the latter as a condition on x as a function of y. But if he had taken it as a condition on y as a function of x he would have seen that $y = x$ is a solution and thus that the equivalence of the *X*s to the *Y*s could be derived by his reasoning from "All *X*s are *Y*s".

[13] A similar difficulty obtains if Boole is taking "–" as partial complement. In this case one must use both premises but the conclusion still depends only on the fact that $c + d = 0$ implies $d = 0$. Thus the strengthened scheme would be SPE: from $ay + b = 0$ and $a'y + b' = 0$ infer $zb' - wb = 0$. Here the conclusion is again independent of the coefficients of y.

[14] If one is to apply E15, as stated by Boole, to $(1 - x)y = 0$ and $zy - z = 0$ then one must construe the two as $(1 - x)y - 0 = 0$ and $zy + -z = 0$ (and one must construe the result as $z(1 - x) - z0 = 0$). Now, $zy + -z = 0$ implies that $zy = 0$ and that $-z = 0$, neither of which follows from the original premises.

[15] Boole read "–" as "except". He did not take up the question of whether "All *A*s except *B*s are *C*s" implies "Some *A*s except *B*s are *C*s".

[16] As indicated above, Boole had one logical system which he tried to use to treat both syllogistic and truth-functional reasoning. But, it does not seem to have occurred to him that there are arguments which involve a combination of the two, e.g., "All *A*s are *B*s" implies "Either All *B*s are *C*s or Some *A*s are not –*C*s". It is well-known that such arguments cannot be treated in Boole's system.

[17] Boole's claim ([6], pp. 20, 211, 218) that the "laws" of his algebra are exactly the sentences which are true on arbitrary substitutions of 0 and 1 seems to qualify him for mention in any history of decision procedures and in any other history of logic which emphasizes attempts to reduce reasoning to computing. Boole's discussions of the role of logical axioms in deduction (*MAL*, pp. 5, 6, 18; [6], pp. 144, 215) qualify him for mention in any history of formal deduction.

[18] The reader should notice that most of our criticisms of Boole's inattention to logical detail are independent of which interpretation Boole had in mind for + and – in *MAL*. Moreover, each particular charge which was supported by argument based on one interpretation applied to passages which contain mistakes relative to the other interpretation as well.

REFERENCES

[1] Anselm, *De Grammatico* [XIth c.], Latin text and English translation in D. Henry, *Commentary on De Grammatico*, Reidel Publishing Co., Dordrecht, Holland, 1974.

[2] Bacon, J., "Natural-deduction Rules for Syllogistic" (abstract), *The Journal of Symbolic Logic*, vol. 31 (1966), pp. 686–687.

[3] Boole, G., *The Mathematical Analysis of Logic, Being an Essay Towards a Calculus of Deductive Reasoning*, Macmillan, Barclay & Macmillan, Cambridge and George Bell, London, 1847, also reprinted 1948 by Basil Blackwell, London.

[4] Boole, G., *An Investigation of the Laws of Thought on Which Are Founded the Mathematical Theories of Logic and Probabilities*, Macmillan, London, 1854. Pagination referred to here is from the 1954 Dover reprint.

[5] Boole, G., *A Treatise on Differential Equations*, Cambridge 1859. Pagination referred to from reprint of the fifth edition by Chelsea Publishing Co., New York.

[6] Boole, G., *Studies in Logic and Probability*, edited by R. Rhees, Open Court Publishing Company, LaSalle, Illinois and Watts & Co., London, 1952.

[7] Church, A. *Introduction to Mathematical Logic*, Princeton University Press, Princeton, N.J., 1956.

[8] Cohen, P. and R. Hersh, "Non-Cantorian Set Theory", *Scientific American*, vol. 217 (1967), pp. 104–116.

[9] Corcoran, J., "Three Logical Theories", *Philosophy of Science*, vol. 36 (1969), pp. 153–177.

[10] Corcoran, J,. "Completeness of an Ancient Logic", *The Journal of Symbolic Logic*, vol. 37 (1972), pp. 696–702.

[11] Corcoran, J., "Aristotle's Natural Deduction System" in *Ancient Logic and Its Modern Interpretations*, ed., J. Corcoran, Reidel Publishing Co., Dordrecht, Holland, 1974.

[12] Corcoran, J. and S. Wood, "Validity and Invalidity: Boole's Criteria" (abstract), *The Journal of Symbolic Logic*, vol. 44 (1979), p. 662.

[13] Craig, W., *Logic in Algebraic Form*, North-Holland, Amsterdam, 1974.

[14] DeMorgan, A., *Formal Logic*, ed., A.E. Taylor, Open Court Co., London, 1926, orig. publ. Taylor and Walton, London, 1847.

[15] Dummett, M., Review of [6], *The Journal of Symbolic Logic*, vol. 24 (1959), pp. 203–209. [Reprinted in this volume, pp. 79–85.]

[16] Hailperin, T., *Boole's Logic and Probability*, North-Holland, Amsterdam, 1976.

[17] Henry, D., *Commentary on De Grammatico*, Reidel Publishing Co., Dordrecht, Holland, 1974.

[18] Kneale, W. and M. Kneale, *The Development of Logic*, Oxford University Press, Oxford 1962.

[19] Langer, R.E., "Differential Equations", *Encyclopedia Brittanica*, 1953 edition, vol. 7, pp. 358–363.

[20] Mill, John Stuart, *A System of Logic*, Longmans, Green and Company, London, 1879, originally published in 1843.

[21] Smiley, T., "What Is a Syllogism?", *Journal of Philosophical Logic*, vol. 2 (1973), pp. 136–154.

[22] Tarski, A., *Introduction to Logic and to the Methodology of Deductive Sciences*, tr., O. Helmer, Oxford University Press, New York, 1941.

[23] Van Evra, J., "A Reassessment of George Boole's Theory of Logic", *Notre Dame Journal of Formal Logic*, vol. 18 (1977), pp. 363–377. [Reprinted in this volume, pp.87–99.]

[24] Whately, R., *Elements of Logic*, James Munroe & Co., Boston, 1845, originally published in London, 1825.

[25] Whitehead, A.N., *Universal Algebra*, vol. 1, Cambridge University Press, Cambridge, 1898.

[26] Wood, S., *George Boole's Theory of Propositional Forms*, Ph.D. Dissertation in Philosophy, SUNY/Buffalo, 1976.

State University of New York at Buffalo, Buffalo, New York, 14260 USA

THEODORE HAILPERIN

ALGEBRAICAL LOGIC: LEIBNIZ AND BOOLE

An association of algebraic symbols and their properties with those of logic would hardly have been conceivable before the 16th century. Systematic use of letters for numbers in general, and symbols for operations on them, were a development of the 16th and 17th centuries. Although Aristotle's theory of the syllogism used letters of the alphabet for arbitrary general terms, i.e., used variables, there was no formal notion of the composition of terms nor of a negative term, hence no notion of an operation on terms. An early, perhaps the earliest, mention of such an algebraic-like operation occurs in Jacques Bernoulli's *Parallelismus ratiocinii logici et algebraici* . . .[1]

In this 'open disputation' Bernoulli alludes to a parallel between compounding general terms (e.g., *Virtus et eruditio*) and an indicated algebraic literal sum $a + b$; likewise that of the logical operation of removing a term from a compound and the algebraic difference. He also remarks on the parallel between algebraic and logical inferences, citing the examples:

	$a = b$		$b = a$		$a = b$
	$c = a$		$c = a$		$a = c$
therefore	$c = b$	therefore	$c = b$	therefore	$c = b$

as well as (using Bernoulli's symbols '$\subset$' and '$\supset$' for 'less than' and 'greater than')

	$a \subset b$		$b \supset a$		$a \subset b$
	$c = a$		$c = a$		$a = c$
therefore	$c \subset b$	therefore	$c \subset b$	therefore	$c \subset b$

which, he points out, correspond respectively to the first, second and third syllogistic figures.

These few brief remarks of Bernoulli's are in marked contrast to what Leibniz had accomplished by that time. Unfortunately Leibniz's achievements were effectively unknown, residing in manuscripts that were not published until the 19th and early 20th centuries. By then symbolic logic had developed independently of him. In these Leibniz manuscripts, antedating Boole's work by more than 150 years, we find expressed a clear conception of a formal deductive system for (term) logic with algebraic features. These ideas of Leibniz were developed over the course of decades and took various forms. I shall here focus on his more mature writings.[2]

We begin with his manuscript of 1686 *Generales inquisitiones de analysi notionum et veritatum* (Couturat 1903, 356–99. English translation in Parkinson 1966,

J. Gasser (ed.), A Boole Anthology, 129–138.

47–87). Much of it is concerned with analysis of concepts and of truths, and to matters of a grammatical, philosophical or metaphysical nature. We shall be ignoring these as not being germane to our topic. As his lengthy essay progresses there is a gradual transition from an essentially verbal form of logic to one quite algebraic in character. In its later portion, and in two short notes written some four years later, his efforts were directed towards the construction of a logical calculus—for the time a most remarkable idea.

Leibniz's concept of a formal calculus was quite up to present day standards, but his conception of what constitutes logic was, from a modern perspective, quite meager—hardly more than the Aristotelian-medieval-scholastic syllogistic theory of general terms. He knew, and gave examples to the effect, that terms could be interpreted intensionally, as a composition of attributes, as well as extensionally, as a collection of individuals. Habitually, though, he chose to express himself intensionally; for example his 'A is B' was often written 'A contains B', meaning that when A and B are analyzed into a composition of terms, those of B would be among those of A. For example, 'rational animal' contains 'animal'. Now for his quite remarkable innovations. To maintain historical authenticity we shall be using, with minor exceptions, his language and his notation.

First, there is the introduction of symbols for logical notions. The relation of 'same as' or 'coincides with' is replaced by '=', and its negation by '*non* =' (we shall use '≠'). The composition of two terms A and B is written 'AB'. Though not named, the operation is occasionally referred to as 'multiplication'. The privative (absence of, negation of) A is written '*non A*' (we shall write 'not-A'). Leibniz says a term is *positive* (i.e. not privative) "unless advised that it is privative". He says *non-entity* (no symbol for it) is the privative of everything. Although he uses 'A not-A' as coinciding with non-entity, there is no expression for *everything*. Finally, there is the introduction of indefinite or undetermined terms, for which he uses letters at the end of the alphabet:

> For by the sign Y I mean something undetermined, so that BY is the same as 'some B', ... Thus 'A is B' is the same as 'A is coincident with some B', or $A = BY$.

Although Leibniz has no notion of a quantifier, his use of the 'indefinite' letter is equivalent to that of a *free* (*existential*) *variable*. This notion, not normally in use, is the analogue of the free (universal) variable, i.e., one understood to be universally quantified with a quantifier of *widest* possible scope. In the case of the free (existential) variable this quantifier would have the *narrowest* possible scope. As an example of the use of his indefinite letter Leibniz cites the inference from '$A = AA$' to '$A = AY$'—a special case of what we would now write as an inference from $\Phi(A, A)$ to $\exists Y\Phi(A, Y)$.

Before listing Leibniz's "primary bases" for his logical calculus we call attention to his restriction of the substitution of a compound term for a single letter to possible terms, i.e., terms not resolvable into a compound containing non-entity (A not-A, B not-B, etc.) as a component (assuming commutativity and associativity).

The following is a compilation of logical principles abstracted from Leibniz's *Generales inquisitiones* and two later notes relating to it. Although the order of arrangement is ours, the language (Englished) is Leibniz's. The title is taken from the first of his two notes.

PRIMARY BASES OF A LOGICAL CALCULUS

(1) $A = B$ means that one can be substituted for the other, B for A or A for B, i.e., that they are equivalent.
(2) $A = A$. [Add: If $A = B$, then $B = A$. If $A = B$ and $B = C$, then $A = C$.]
(3) $A = AA$.
(4) A = not-(not-A).
(5) $AB = BA$. [Add: $(AB)C = A(BC)$.]
(6) $A = AB$ and not-B = not-B not-A coincide [are equivalent].
(7) That in which there is [has as a component] B not-B is a non-entity or false term; for example if $C = AB$ not-B, C would be a non-entity. [Since entities A, AB, etc., are not non-entities, Leibniz uses this principle to justify having $A \neq AB$ not-B and $AB \neq ACB$ not-B.]

From these Leibniz *formally* derives the following:

(8) If $A = B$, $AC = BC$.
(9) $B \neq$ not-B; more generally, $AB \neq C$ not-(EB). [Erroneous. Delete the 'E' and use $AB \neq C$ not-B.]
(10) $A \neq B$ and $B \neq A$ are equivalent.
(11) $A = B$ and not-A = not-B are equivalent.
(12) not-B = not-B not-(AB).
(13) $A \neq B$ not-A.
(14) If $A = A$ not-B, then $A \neq AB$.
(15) If $A = B$, it follows that $A \neq$ not-B.
(16) If $A = AB$, one can assume a Y such that $A = YB$.
(17) If $A = YB$, it follows that $A = AB$.

Some brief comments are in order. In (1) it looks as though Leibniz is giving a definition of $A = B$. But as he uses it it is a rule of inference to the effect that if $A = B$ then $S[A]$ is equivalent to $S[B]$, where $S[A]$ is a formula and $S[B]$ one obtained from it by replacing one or more occurrences of A by B. The bracketed additions in (2) and (5) are principles which Leibniz does make use of although he has not included them. As for (3) we have, more than a century and a half before Boole a statement of the 'index law' for logical product. In (4) we have the involutary property of negation, while (6) states the equivalence of 'A is B' and 'not-B is not-A'. After presenting (8) Leibniz emphasizes that $AC = BC$ does not imply $A = B$.

Unquestionably, this logical calculus of Leibniz's is a remarkable accomplishment. As primitive notions there is the relation of equality '$\alpha = \beta$', the unary operation 'not-α' and the binary '$\alpha\beta$'. Additionally there is a tacit use of existential quantification, as in items (16) and (17). Although developed with a specific interpretation in mind, it is formulated as an abstract deductive system in which, from (1)–(7), consequences are derived without appeal to meanings. Leibniz had no notation for variables ranging over terms in general—as we have noted, his variables $A, B, C, \ldots$ are restricted to range over *possible* terms. If A is a possible term so is not-A. But the composition of possible terms need not be—for example 'A not-A'. The principal interpretation which Leibniz had in mind for his calculus was one in which the letters stood for intensions of concepts. Since modern logicians are unpracticed in intensional ways of thinking I shall present an interpretation of Leibniz's 1686–90 logical calculus, that is, a model for his axioms, in more familiar extensional terms. Let

(i) 'term' denote a subset of some non-empty set U
(ii) 'not-α' denote the complement of α with respect to U
(iii) '$\alpha\beta$' denote the set intersection of subsets α and β of U, and
(iv) $A, B, C, \ldots$ be variables restricted to range over subsets of U *excluding both U and the empty set.*

Then with this interpretation Leibniz's axioms (1)–(7) are satisfied.[3]

One readily sees that syllogistic doctrine *in the traditional version*, in which terms carry existential import, holds. For example, as Leibniz shows, '(All) A is B' implies 'Some A is B'.

Sometime after 1690 Leibniz wrote a pair of studies on logic, printed as items XIX and XX in Gerhardt (1890), which are also remarkable. The following two brief remarks give some indication of Leibniz's algebraic way of thinking in these mss.

(i) In manuscript XIX, instead of juxtaposition, the symbol '+' is used for compounding terms and an operation, '–', inverse to it is admitted. Its title "Non inelegans specimen demonstrandi in abstractis" was struck out. Could it have been that something like $A + A - A$ being susceptible to two conflicting values, e.g.

$$A + A - A = (A + A) - A = A - A = \text{Nothing}$$

and

$$A + A - A = A + (A - A) = A,$$

made Leibniz realize that $A + A = A$ together with an inverse to '+' is unworkable?

(ii) In manuscript XX the symbol '$\oplus$' is used to form compound terms but now there is no mention of an inverse operation for it. The formal properties of the relation expressed verbally by *est in*, are developed in detail; axioms and postulates ascribe to it (in modern terminology) properties of a partial order having algebraic (semi-lattice) connections with $\oplus$ (e.g., if A is in L and B is in L, then $A \oplus B$ is in L). The notion of 'privative' does not occur, only that of terms being disparate—A and B are *disparate* if 'A is not in B and B is not in A'. (In present day extensional language, A and B are disparate if 'A not-B' and 'B not-A' are both non-empty.)

I turn now to our other logician.

At the time Leibniz was writing the two manuscripts just referred to, Aldrich's *Artis logicae compendium* was published in Oxford. It contained a concise account of standard topics of the day and, except for a somewhat original organization of syllogistic doctrine, is otherwise unremarkable. What *is* remarkable is that it (including also its various editions) was the last book emphasizing the formal aspects of logic that appeared in England for over a hundred years. An abrupt change, and a renewed interest in formal logic, occurred with the publication in 1826 of Richard Whately's *Elements of Logic*. It became extremely popular, going into many editions. Whately's book included a vigorous defense of logic as an abstract science connected with linguistic structure, as opposed to its being an art of reasoning, as with Aldrich, or as an instrument in search for truth as with the British empiricist school, exemplified by J.S. Mill. Throughout Whately's book syllogistic principles are explained solely in extensional terms, that is, in terms of classes, no mention being made of 'ideas', or 'concepts'. This viewpoint became widespread among English logicians. Mentioned by Boole in his *Laws of Thought* as a reference source, Whately's book exemplifies the *logical* milieu in which Boole's ideas for an algebra of logic were developed.

Another influencing factor came from mathematics. As with logic, a change took place early in the 19th century, especially in England, in the perception of the nature of algebra. Although these ideas, referred to as symbolical algebra, are no longer current, being supplanted by modern abstract algebra, they nevertheless were instrumental in suggesting to Boole a first step in the construction of an algebra of logic. I expand a bit on this last statement.

To provide a set of principles for algebra, and in particular to provide a rational basis for operating with negative and imaginary numbers, George Peacock (1830, 1834) introduced the idea of a symbolical algebra extending arithmetical algebra.

Peacock's symbolical algebra was a science of symbols and their combinations "suggested" by arithmetical algebra [of whole numbers] in which "forms" in arithmetical algebra such as $a(b - c)$ and $ab - ac$, though meaningless when b is less than c, are nevertheless taken to be equivalent in symbolical algebra. For Peacock the interpretation of such extended equivalences will *follow* not precede the adoption of such laws. He also included this restriction: the adopted laws should reduce to arithmetical ones when the symbols are given arithmetical meaning. A considerable extension of the scope of symbolical algebra was introduced by D.F. Gregory (1840) who dropped Peacock's restriction when applying its principles to mathematical operators considered separately from their operands. For example, if D stands for the operation of differentiation (of a function of one variable) then there is the "algebraic" law $D^nD^m = D^mD^n = D^{m+n}$, the exponent indicating the number of successive applications of the operation.

Boole, an active follower of Gregory's ideas, made important use of operator calculus in some of his earliest research papers. Being a tool of broad and flexible applicability foreshadowing modern abstract algebra, it is not surprising that he should make use of it in his initial venture into logic. Boolean algebra was not yet in existence, and indeed did not make its appearance until Schröder (1877). So how did Boole do logic by algebra thirty years earlier?

In his *Mathematical Analysis of Logic* (Boole 1847) capital letters X, Y, Z, . . . were used to represent (logical) classes as well as general names for their elements, and lower case x, y, z, . . . as operators ("elective symbols"), where the "symbol" x

operating on a "subject" selects all the Xs it contains. Elective symbols were assumed to obey the laws

1. $xy = yx$
2. $x(u + v) = xu + xv$
3. $x^n = x \quad [x^2 = xx = x]$

No doubt the ordinary algebraic properties $x1 = 1x = x$ and $x0 = 0x = 0$ suggested the use of 1 to stand for the 'Universe', 0 for 'Nothing' and $1 - x$ for the operator selecting not-X. Although De Morgan's contemporaneously appearing *Formal Logic* makes mention of a *universe*, there is no symbol for it; and although for him each term had a complement with respect to the universe there was no symbol for the *operation* of forming the complement. Distinctively different from De Morgan's *Formal Logic*, Boole's *Mathematical Analysis of Logic* carries out logical inferences by algebraic operations which are assumed to preserve validity: for example

$$x = xy \quad \text{(All } X\text{s are } Y\text{s)}$$

leads, by common algebraic operations, to

$$x(1 - y) = 0 \quad \text{(No } X \text{ is not-} Y\text{).}$$

With these few remarks I take leave of Boole (1847). Boole himself referred to it as "a hasty and (for this reason) regretted publication ...".

Seven years later in *The Laws of Thought* there is a marked change in conception: the lower case $x, y, z, \ldots$ no longer are selection operators but stand directly for the classes themselves. On the basis of generalizing from specific instances Boole concludes that an algebra for classes would be the same as "common" algebra if the variables take on only the values 0 and 1. But other quantitative values do come in—e.g. $x + x$ is written as $2x$. Not until Hailperin (1976) [2nd edition 1986, §1.4] was a complete list made of the algebraic properties actually used in *Laws of Thought*. Viewed as a set of axioms I called it an algebra of signed multisets. A multiset is a set in which duplication of elements can occur, and a signed multiset is one in which elements can occur positively or negatively—like assets and liabilities in a firm's accounts. The following is a list of the algebraic properties which Boole used, some explicitly and some implicitly.

"BOOLE'S" ALGEBRA OF $\times, +, -, 0, 1$

1. $A + B = B + A$	1′. $AB = BA$
2. $A + (B + C) = (A + B) + C$	2′. $A(BC) = (AB)C$
3. $A + 0 = A$	3′. $A1 = A$
4. $A(B + C) = AB + AC$	
5. $A + X = 0$ has a solution for $X(= -A)$	

6. $1 \neq 0$
7. $A^2 = 0$ only if $A = 0$
8. $nA = (1 + 1 + ... + 1)A = 0$ only if $A = 0$

It can be shown (Hailperin 1986, §2.3) that Boole's algebra is equivalent to Boolean algebra when A, B, C, ... are restricted to range over idempotents (A is idempotent if $A^2 = A$). When x and y are idempotent then

$$(x + y)^2 = x + 2xy + y,$$

so $x + y$ is not idempotent except if $xy = 0$. However, as is readily shown, $(1 - x)$, xy and $x + (1 - x)y$ are idempotent. For Boole only when an expression satisfied the idempotency condition—such an expression he referred to as being "independently interpretable" [in logic]—was it taken to be a class. But to maintain algebraic freedom of operations in his algebra he made use of non-idempotents (e.g. $x + x$). He then was faced with the task of producing meaningful (in logic) results. This he does by clever and ingenious devices, devices decried by some, like Jevons, as mysterious and incomprehensible. One of these devices is replacing an algebraic expression by its "expansion" and, if equal to an idempotent, setting those terms in the expansion whose coefficient is not idempotent equal to zero. We illustrate this with some simple examples.

If $f(x,y)$ is an algebraic expression in (idempotent) variables x and y, then as its expansion Boole gives

$$f(x,y) \stackrel{?}{=} f(1,1)xy + f(1,0)xy' + f(0,1)x'y + f(0,0)x'y',$$

where the coefficients $f(1,1)$, $f(1,0)$, $f(0,1)$ and $f(0,0)$ are the results of substituting 1 and 0 for x and y, and where x', y' are short for $(1 - x)$, $(1 - y)$. The "?" over the equality is our insertion to indicate that Boole, though using '=' did not actually justify the equality of the two sides and, indeed, doesn't really use the right-hand side as being *equal* to the left side.

Example 1. From $w = x + y$ Boole would obtain

$$w = 1 \cdot (xy' + x'y) + 2xy$$

which he would interpret as equivalent to the *pair* of interpretable equations

$$\begin{cases} w = xy' + x'y \\ xy = 0. \end{cases}$$

Example 2. From $xw = y$ he would solve for w and have

$$w = \frac{y}{x} \stackrel{?}{=} 1 \cdot xy + 0 \cdot xy' + \frac{0}{0} x'y' + \frac{1}{0} x'y$$

interpreted by him as

$$\begin{cases} w = xy + vx'y', & v \text{ indefinite} \\ x'y = 0. \end{cases}$$

We have shown (Hailperin 1986, Chapter 2) that this queer method of Boole's for doing term logic is sound and, moreover, can be carried out without appeal to division, that is, remaining within the algebra of signed multisets. Of course, as we now know, even the simple (Boolean) algebra of idempotents suffices to do term logic.

In the following parallel columns we compare, mainly highlighting differences, the respective approaches of our two logicians to mathematizing logic.

Chez Leibniz (1685, 1690)	*Chez Boole* (1847, 1854)
1. Defines =, and derives its properties.	1. = is assumed as part of algebra.
2. Makes use of 'non =' (≠).	2. No use of ≠, <, or >.
3. Has only two algebraic operations: logical compound (*AB*) and privative (non *A*).	3. Has the full panoply of operations: +, −, ×, ÷; also numerals.
4. All simple terms taken *with* existential import; likewise compound terms except if '*B* not-*B*' is part of the expression. Non-emptiness expressed by '≠ *B* not-*B*' (or ≠ *AB* not-*B*, etc.).	4. All terms are *without* existential import. Has no way of formally expressing non-emptiness.
5. Refers to 'Nothing' but has no formal symbol for it. No mention of a universal term.	5. Uses '0' for 'Nothing' and '1' for 'Universe', with appropriate (logical) algebraic properties assigned to them.
6. Expresses '(All) *A* is *B*' by $A = AB$, or [for some *Y*] $A = YB$, where *Y* is positive (i.e. can't be Nothing).	6. Expresses 'All *X* is *Y*' by $x = xy$, or $x = vy$, *v* indefinite (but can be 0).

7. Disjunction not used.	7. Disjunction expressed either by $x + (1 - x)y$ (*non-exclusive*) or by $x(1 - y) + y(1 - x)$ (*exclusive*).
8. In ms XX the relation *est in* is endowed with algebraic (semi-lattice) properties.	8. Has no formal *inclusion* though in use his $w = A + vC$, v indefinite is equivalent in meaning to $A \subseteq w \subseteq A + C$.
9. Has a carefully constructed abstract formal system.	9. No actual formal system. Logic is viewed as applied mathematics using a variant of 'common' algebra.

One might be curious to know the effect that Leibniz's ideas would have had on Boole. Indeed, about a year after publication of his *Laws of Thought* Boole was informed of Leibniz's anticipation of an algebraic approach to logic—in particular of his having the law $aa = a$. The information doesn't seem to have been very much.[4] In any case, Boole had unshakably firm views of his work. For example, although a good friend and correspondent of De Morgan's one detects next to no influence on him of De Morgan's ideas. As another example, Jevons tried in vain to convince Boole that $x + x$ should be x, rather than $2x$.

But what if the manuscripts were published in Leibniz's lifetime and there then ensued 150 years of development before Boole? That's *quite* a different question, to which there can hardly be any reasonable answer.

NOTES

[1] Bernoulli (1685). According to the editor of Bernoulli (1969): Obwohl die beiden Brüder Jakob und Johann gemeinsam für diese Arbeit als Autoren zeichnen, ist kein Zweifel, daß der 19jährige Johann hier nur eine sekundäre Rolle spielte, zumal er nur als Respondent fungierte, weshalb Johann später auch auf die Aufnahme dieser Disputation in seinen eigenen *Opera omnia* verzichtet hat.

[2] The book Peckhaus (1997), which I came to know only after the Boole 97 conference, is worthy of special mention. It has, in its Chapter 2 (Die Idee der *mathesis universalis* bei Leibniz) and Chapter 5 (Leibniz und die englische Algebra der Logik), a wide-ranging and in-depth description of the historical and philosophical background relevant to the topic of our paper, as well as extensive coverage of the secondary literature.

[3] The existence of such a model for Leibniz's axioms is not surprising. See Lenzen (1984), where it is shown that "Leibniz's conjunction-negation logic, with 'est Ens', i.e. 'is possible' as an additional (although definable) logical operator, is provably equivalent (or isomorphic) to Boolean algebra".

[4] For details see Peckhaus (1997), §5.2.1.

BIBLIOGRAPHY

Bernoulli, Jacques. 1969. *Die Werke von Jakob Bernoulli*, Band 1. Basel: Birkhäuser.

Bernoulli, Jacques and Jean Bernoulli. 1685. *Parallelismus ratiocinii logici et algebraici*, ... Reprinted in *Jacobi Bernoulli, Basileensis, opera* vol. 1 (Geneva 1744) 211–24 and in Bernoulli 1969, 263–74.

Boole, George. 1847. *The mathematical analysis of logic, being an essay towards a calculus of deductive reasoning*. Cambridge: Barclay & Macmillan; London: George Bell.

———. 1854. *The laws of thought*. London: Walton and Maberley. Reprinted 1916, 1951.

Couturat, Louis, ed. 1903. *Opuscules et fragments inédits de Leibniz. Extraits des manuscrits de la Bibliothèque royale de Hanovre*. Paris: Presses Universitaires de France. Reprinted 1961, Hildesheim: Olms.

Gerhardt, C.I., ed. 1890. *Die philosophischen Schriften von Gottfried Wilhelm Leibniz*, Band 7. Berlin (English translation of XIX and XX in Parkinson 1966, 122–44).

Gregory, D.F. 1840. On the real nature of symbolical algebra. *Transactions of the Royal Society of Edinburgh* 14: 208–16.

Hailperin, Theodore. 1976. *Boole's logic and probability*. Amsterdam: North-Holland.

———. 1986. Revised and enlarged second edition of Hailperin 1976.

Leibniz, Gottfried Wilhelm. 1768. *G.G. Leibnitii opera omnia*. Edited by L. Dutens. Geneva.

———. 1890. *Die philosophischen Schriften von Gottfried Wilhelm Leibniz*, Band 7. Berlin. Reprinted 1961, Hildesheim: Olms.

———. 1971. *Sämtliche Schriften und Briefe*. Sechste Reihe, Erster Band. Berlin: Akademie-Verlag.

Lenzen, Wolfgang. 1984. Leibniz und die Boolesche Algebra, *Studia Leibnitiana* 16 (2). Wiesbaden: Franz Steiner.

Parkinson, G.H.R., ed. 1966. *Leibniz logical papers*. A selection translated and edited with an introduction. Oxford: Clarendon Press.

Peacock, George. 1830. *A treatise on algebra*. Cambridge: J. & J.J. Deighton.

———. 1834. Report on recent progress and present state of certain branches of analysis. *Report on the third meeting of the British Association for the Advancement of Science held at Cambridge in 1833*. London: John Murray.

Peckhaus, Volker. 1997. *Logik, Mathesis universalis und allgemeine Wissenschaft*. Berlin: Akademie-Verlag.

Schröder, Ernst. 1877. *Der Operationskreis des Logikkalküls*. Leipzig: Teubner.

Lehigh University, Bethlehem, PA 18015 USA

MARIE-JOSÉ DURAND-RICHARD

LOGIC VERSUS ALGEBRA: ENGLISH DEBATES AND BOOLE'S MEDIATION

The work of George Boole (1805–64) is often considered a first attempt at mathematization of logic. It is often opposed to that of Gottlob Frege (1848–1925), who is regarded as the true founder of mathematical logic. This way of thinking may be considered as the appraisal of the twentieth-century philosopher, logician or mathematician who is well informed about historical developments in this field since the last century. It lays great stress on the formalization of logic but tends to neglect what could now be called the semantic question, despite its importance for Boole with the problem of the interpretation of symbols, and for Frege with the problem of representation by which he tries to understand the relation between propositions and reality.

My own viewpoint is that of an historian of science. With regard to Boole, I am not primarily interested in knowing whether he was a precursor of modern logic, or a mathematician suddenly inspired by a flash of genious.[1] My purpose is not to reconstruct Boole's ideas from the current state of logic. I do not think of historical evolution as being a necessary process, for which I should have only to indicate the dates of the main events. What is important to me in the history of ideas is that thought is part of action. Ideas do not simply happen, once the conditions of historical development are right, nor are they concerned merely with the description of the world. My own view is that they are actually involved in its transformation, and that they contribute to the solution of problems—of problems pertaining not only to mathematics or logic, but to science and knowledge generally, at specific cultural moments, where they introduce new meanings. I am concerned with how Boole's work is included in a social and cultural context where logic and mathematics are unceasingly cross-examined about their foundations. My main interest here will therefore be to understand how it happened that Boole came to think of his system of logic, in order to show that his work is not an historical accident but that it answered to some fundamental questions about the nature of knowledge that were central to debates taking place in Great Britain in the first half of the nineteenth century.

During that period, publications such as *The Edinburgh Review* and *The Quarterly Review* were filled with the question of whether it was necessary for the only two English universities, those of Oxford and Cambridge, to reform their curriculum in order to include new knowledge in natural philosophy, obtained for the most part by use of analytical calculations in mathematics, and new approaches to knowledge, resulting from the Industrial Revolution, historically located between 1760 and 1830. From the time when Bacon proposed a *Novum organum,* and Newton asserted that the 'Principles of Natural Philosophy' were mathematical, scholastic logic, despite

J. Gasser (ed.), A Boole Anthology, 139–166.

being at the core of university *disputationes*, had been threatened with being unable to maintain intellectual control of what was seen as progress. I intend to clarify epistemological problems which were raised both in logic and in mathematics, and which concerned the foundations of these disciplines as well as their conditions of truth, in order to decide whether they ought to be considered sciences or arts (Richards 1980) and whether they could serve as a basis for further knowledge. I shall have occasion to show how George Peacock, as he developed his symbolical algebra in 1833, and how George Boole, as he developed his symbolical logic from 1847 to 1854, were trying to answer such questions. I show in particular how the symbolical mediates between logic and mathematics, and provides a new way of founding knowledge. For mathematical aspects of these questions, Peacock and Boole turned to the works of their contemporaries; as for philosophical aspects, they made use of the work of John Locke (1632–1704) on the relationship between perceptions, ideas and language. From Boole's work it is apparent that, at the time when non-Euclidean geometries were first being conceived in Central Europe, the symbolical approach to mathematics and logic was making an important contribution to awareness of the distinction between coherence and truth.

1. THE SOCIAL AND CULTURAL CONTEXT

It is necessary to keep in mind some specific features of the main institutions of knowledge at the beginning of the nineteenth century. Before the University of London was created in 1828, and that of Durham in 1837, the only two English universities in existence, those of Oxford and Cambridge, were Anglican institutions. This is a very specific characteristic because Anglicanism was a national religion and there were therefore very close ties between church and state. The university statutes, recognized by Queen Elizabeth I in 1570 and strengthened by the Royal Letters of James I in 1616, survived the two English revolutions of the seventeenth century. They made the universities a branch of the Anglican church, which was directly responsible for clerical learning. In 1841, Peacock wrote that at least half of the students at Cambridge intended to take orders; according to Jenkins and Jones, a third of them followed the same path between 1850 and 1900[2].

Mathematical and logical authors such as Peacock and Boole, because of their direct or indirect involvement in the reformation of scientific institutions, were strongly committed to the long process that led the English universities to become 'national seminaries' rather than 'seminaries of sound learning and religious education'. Numerous and persistent initiatives all through the nineteenth century culminated in the reform of the statutes of Cambridge and Oxford during the 1850s (Durand-Richard 1996), but the complete separation between Anglicanism and the university was to take place only with the new reform of the university in 1871.

1.1. In Search of New Foundations of Knowledge

Neither the structures nor the values of the kind of university that existed in England at that time were conducive to expansion into new branches of learning or new ways of thinking. Without official bodies to promote changes and to adjust programmes and examinations,[3] it was difficult for anything new to be included in courses and therefore in the curriculum (Winstanley 1940, 259; Tillyard 1913, 72–4). It was inevitable, then, that the first discussions about the necessity of these transformations would take place in the press.

At Oxford, teaching was still based on the scholastic logic. At Cambridge, the 'Mathematical Tripos' was *de facto* but not *de jure* the examination for the 'Bachelor of Arts degree'.[4] Since the second half of the eighteenth century, it had become the cornerstone of the examinations, and mathematics the keystone of teaching. Attempts to introduce experimental philosophy, for example, had no real chance of succeeding as long as the examination was unchanged. But if mathematics were regularly celebrated as the 'glory and honour of the University' in Cambridge, there was something paradoxical in the fact that their dominating position was coupled with faithfulness to Euclidean geometrical foundations and to the fluxionary notation used by Newton in his *Principia*. This situation is often imputed to the quarrel between Newton and Leibniz concerning who had been the first to invent the infinitesimal calculus. But it is important to insist on the consistency of this system of teaching, where Euclid's geometry was part of a classical cursus because its logical rigour was considered closer to that of the syllogistic and in every way more solidly grounded than the ill justified operativity of the analytical calculations (Garland 1980, 28–51). Moreover, the teaching of logic was maintained in the colleges until 1839, together with a preliminary examination in Latin known as the *disputationes* or *viva voce examination.* Students had to present or oppose arguments in syllogistic form about questions such as: 'The doctrine of eternity of punishment is inconsistent with the doctrine of the omnipotence of the Deity', or 'Newton's method of prime and ultimate ratios is correct' (Rouse Ball 1889, 78). Even De Morgan (1806–71), fourth wrangler at Cambridge in 1827, had bad memories of them:

> The real disputations were very severe exercises. I was badgered for two hours with arguments given and answered in Latin,—or what we called Latin—against Newton's first section, Lagrange's derived functions, and Locke on innate principles. And though I *took off* everything, and was pronounced by the moderator to have disputed *magno honore*, I never had such a strain of thought in my life. For the inferior opponents were made as sharp as their betters by their tutors, who kept lists of queer objections, drawn from all quarters. (De Morgan 1915, 305)

1.2. The New Issues

At the turn of the century, a wide-ranging discussion began on the role of the universities and on the basic knowledge that would have to be mastered by the governing classes in order to deal with the contradictions of a country that was quickly transforming itself. In a few short decades, it changed from a rural economy to a new society in which the landscape of bridges, railways, ships, steam engines and machine tools bore witness everywhere to the importance of technical production and new values organized by principles such as transformation and operativity.

The Edinburgh Review, founded in 1802, was the principal organ of the Whig society. It was the scene of forceful attacks against every form of conservative government, taste and learning, in diatribes so scathing that the government actually felt threatened for a time (Ponteil 1968, 66). This journal launched a radical criticism of institutions according to utilitarian principles, inspired on a moral level by the natural theology of William Paley (1743–1805), but founded philosophically on ideas of John Locke (1632–1704), Dugald Stewart (1753–1828), Jeremy Bentham (1748–1832) and James Mill (1773–1836). The teaching of the English universities was repeatedly the focus of criticism. In 1808, the mathematician and geologist Rev. John Playfair (1748–1819) began criticism of the English universities by asserting that faithfulness to scholastic logic was inconsistent with experience and observation, as well as with curiosity and innovation, i.e. with any experimental approach. Oxford, where the role of mathematics was not as important as at Cambridge, and where scholastic logic was maintained as the ultimate foundation of knowledge, incurred the severest criticism. About knowledge, Playfair (1810, 161, 185) asserted:

> What is the main object of most branches of human knowledge, if it be not to minister the bodily wants of man? What is the utility of mathematics, but as they are brought to bear upon navigation, astronomy, mechanics, and so upon bodily wants? What is the object of medecine? What of anatomy? What greater purposes have law and politics in view, but to consult our bodily wants—to protect those who minister them—and to arrange the conflicting interests and pretentions which these wants occasion? One of the greatest objects of human wisdom ... is to turn the properties of matter to the use of man.
>
> What was said about the dictates of Aristotle, was not meant of his Physics, but of his Logic and his Metaphysics ... The logic of Aristotle is particularly hostile to inductive sciences. By turning the mind to the syllogistic method, it becomes a very powerful obstruction to that knowledge which is derived, by induction, from experience and observation.

Another motive of Playfair's as he opened the debate was to stigmatize his compatriots' ignorance of Laplace's *Mécanique céleste*. Playfair laid great stress on the fact that the Laplacian theory, which used the Leibnizian form of the differential calculus, provided complete confirmation of the Newtonian mechanics; his analysis held out the possibility of a reconciliation between the two forms of calculus, considered as the calculus of universal motions. Moreover, Playfair argued that the Laplacian theory was not threatening religion, but, on the contrary, that it reinforced both the thesis of the existence of an initial design, which can be nothing else but the design of God, and the Aristotelian doctrine of final causes (Playfair 1808, 278–9).

> When we consider the provision made by nature for the stability and permanence of the planetary system, a question arises, which was before hinted at,—whether this stability is necessary or contingent, the effect of an unavoidable or an arbitrary arrangement? If it is the necessary consequence of conditions which are themselves necessary, we cannot infer from them the existence of a design, but must content ourselves with admiring them as simple and beautiful truths, having a necessary and independent existence. If, on the other hand, the conditions from which this stability arises necessarily, are not necessary themselves, but the consequences of an arrangement that might have been different, we are then entitled to conclude, that it is the effect of wise design exercised in the construction of the universe.
>
> Now, the investigations of La Place enable us to give a very satisfactory reply to these questions; viz. that the conditions essential to the stability of a system of bodies gravitating mutually to another, are by no means necessary, insomuch that systems can eas-

ily be supposed in which no such stability exists. The conditions essential to it, are the movement of the bodies all in one direction, their having orbits of same eccentricity, or not far different from circles, and having periods of revolution not commensurable with one another. Now, these conditions are not necessary; they may easily be supposed different; any of them might be changed, while the others remained the same. The appointment of such conditions therefore as would necessarily give a stable and permanent character to the system, is not the work of necessity; and no one will be so absurd to argue, that it is the work of chance. It is therefore the work of design, or of intention, conducted by wisdom and foresight of the most perfect kind. Thus the discoveries of La Grange and La Place lead to a very beautiful extension of the doctrine of final causes the more interesting the greater the objects are to which they relate. This is not taken notice of by La Place; and that it is not, is the only blemish we have to remark in his admirable work.

In spite of the vehement tone of this criticism, which was intended for both church and state because of the primary function of Cambridge and Oxford Universities, it concerned the legitimacy of neither spiritual nor temporal power:

We have no doubt of the truth of the one (our religion) or of the excellence of the other (our government) and are convinced that both will be placed on a firmer basis, in proportion as the minds of men are more trained to the investigation of truth. (Playfair 1810, 179–80)

This strong criticism was therefore intended not to divide, but to find a way towards reconciliation by means of the reform of institutions. Whatever their motivation, these criticisms always asserted that the most important consideration was to assure stability and permanence—which, however, were taken to mean different things depending on which of the opposing positions was adopted.

1.3. How These Issues Were Faced from Cambridge

In Cambridge, a group of young students of mathematics, led by Charles Babbage (1791–1871), John F.W. Herschel (1792–1871) and George Peacock (1791–1858), met as early as 1812 and soon constituted itself as the 'Analytical Society'. Its leading members were convinced of the need for the university to escape from the lethargy of a provincial institution in order to become a national centre for research in mathematics.[5] In this context, their clearly expressed determination to force the adoption of Leibnizian notation for the infinitesimal calculus in Cambridge examinations was really a political act, in the Greek meaning of the term, i.e. an act constitutive of a school of thought. Although there had been, since the beginning of the century, previous attempts at reforming some aspects of teaching or examinations (Panteki 1987, 124; Wilkes 1990, 207; Becher 1980), none had had such firm and continuous determination. The remarkably learned preface of the sole volume of *Memoirs of the Analytical Society,* published anonymously in 1813, stressed the fruitfulness of the most recent developments in algebraic analysis on the Continent. Written for the most part by Babbage, this preface laid down, by its insistence on the English origin of numerous ideas and the necessity of 'reimporting' their achievements, a research programme of national interest. Babbage, Herschel and Peacock translated the *Traité élémentaire de calcul différentiel et intégral* (1802) of Lacroix in 1816, and each of them published a volume of examples in 1820, thereby providing the tutors of the colleges with suitable materials for their students' use. Peacock, moderator in 1817, 1819 and 1821, formulated all his questions in the differential notation, which was definitively adopted in the Senate-House Examination and consequently in the teaching given in preparation. The same threesome prepared the publication of numerous textbooks in order to update their contents completely.[6] Peacock, who would stay in Cambridge, first as a tutor in Trinity College and then as Lowndean Professor of Geometry and Astronomy, until Lord Melbourne appointed him deacon of Ely cathedral in 1839, showed an enthusiasm for all these projects that never diminished:

> We are proposing College Reforms, which will introduce the true faith[7] in Trinity .... In the meantime, the hall of Saint John's resounds with daily altercations at the deitrs and the dotts, it is wrong symptoms, these of approaching change: for who would have discussed the point five years since .... Trust me: the golden age of the university is approaching: our business is to accelerate the desirable period. You mentioned the want there is of a good Algebra: will you undertake it immediately? ... But after all, there is as much a want of good treatises upon mechanics, hydrostatics, and optiks, as of other books and all reform will be imperfect which does not attend to these points as well as the others.[8]

> I assure you, my dear Herschel, that I shall never cease to exert myself to the utmost in the cause of reform, and that I will never decline any office which may increase my power to effect it. I am nearly certain of being nominated to the office of Moderator in the year 1818–19, and as I am an examiner in virtue of my office, for the next year, I shall pursue even more decided than hitherto, since I shall feel men have been prepared for the change, and will then be enabled to have acquired a better system by the publication of improved elementary books. I have considerable influence as a lecturer, and I will not neglect it. It is by silent perseverance only that we can hope to reduce the many-headed monster of prejudice, and make the University answer her character as the loving mother of good learning and science.[9]

These reformers were to create, from Cambridge, a network of new institutions, so as to reinforce the representativeness of the university in scientific fields, and to develop research and scientific communication in Natural Philosophy. This is the project they carried out with other scientists politically close to Whigs and Radicals. The historian of sciences Walter Faye Cannon named this group of scientific men the 'network of Cambridge' (Cannon 1964). It busily contributed to the establishment of new scientific societies of national importance between Cambridge and London, so as to counterbalance the increased weight of provincial scientific societies that had been created since the beginning of the century. Peacock and Herschel would be two of the five members of the Royal Commission instituted in Cambridge to prepare and then to implement, from 1852 to 1858, the reform of the University.

1.4. How These Issues Were Faced from Oxford

The reply given by Oxford logicians to the criticisms of *The Edinburgh Review* first came, as early as 1808–09, from another group from Oriel College, which Pietro Corsi named 'the Noetics of Oxford'. This group sought to refute *The Edinburgh Review*'s judgment on logic from a philosophical point of view; Sir James Mackintosh (1765–1832) characterized it as the school of speculative philosophy (Corsi 1988). Edward Copleston (1776–1849), Principal of Oriel College from 1814 to

1826, was the author of a series of replies, as well as of an anonymous tract that appeared in 1809, entitled *Examiner Examined*. Dugald Stewart was also to reply in 1814, in the second volume of his *Elements of the Philosophy of the Human Mind.*

Although the Noetics of Oxford agreed to the introduction of some new courses, they did so provided that they be optional and that they be accessible on the basis of epistemology induced by scholastic logic. These conditions were met by a course on political economy introduced in Oxford by Copleston and Richard Whately (1787–1863), Fellow of Oriel College, and then archbishop of Dublin from 1831. Both contributed greatly to the ultimate acceptance of this new field in the university as well as in Anglican circles[10] by presenting political economy in the framework of the natural theology of Paley; this helped to counter the positions of David Ricardo (1772–1823) and Thomas Malthus (1766–1834). Whately's *Elements of Logic*, which Boole would later use as a reference textbook in writing his *Mathematical Analysis of Logic*, was written from that standpoint. Indeed it resulted from a collective enterprise of the Noetics: Copleston entrusted Whately with its writing and gave him some of his notes, while Whately turned to John Henry Newman (1801–90) and William Nassau Senior (1790–1864), first professor of political economy in Oxford (1825–30), for the writing of some parts. Moreover, in its edition of 1826, this work contained an appendix with an analysis of terms in political economy by Nassau Senior, which was influenced by Dugald Stewart's analysis of the role of language in science. But above all, this logic was prepared for the *Encyclopædia metropolitana* of Samuel Coleridge (1772–1834), for whom the specialization of sciences represented the menace of the secularization of knowledge. Coleridge sought therefore to constitute a body of theologians, scholars and scientific men, with the declared aim of preparing, not an alphabetical listing, but a rational exposition of a philosophical system, founded on the nature of the subjects and on the genesis and the harmony of human knowledge (Yeo 1993, 44, 51). The Noetics of Oxford were opposed to Dugald Stewart with respect to his rejection of scholastic logic, but they agreed with his defence of the philosophy of human mind against utilitarians and practical men who considered it unnecessary for the advancement of science and for human happiness.

The debates in *The Edinburgh Review* became vehement again in the 1830s, especially when the Scottish philosopher and logician William Hamilton (1788–1856), educated at Oxford, published a series of articles on philosophy, logic and university teaching.[11] He was strongly opposed to the Cambridge philosopher and mathematician William Whewell (1794–1866), who had just published, in 1836, his *Thoughts on the Study of Mathematics as Part of a Liberal Education.* They disagreed on two fundamental points:

(1) in the discussion of the necessity of a reform of the universities, Hamilton supported the university against the colleges, and Whewell the colleges against the university;

(2) in the discussion of the foundations of human knowledge, understood as the principles of a liberal education, Hamilton supported the preeminence of logic, and Whewell that of geometry, which he conceived

> as the sole permanent form of mathematics, as opposed to algebraic analysis, which he conceived as its progressive form. (Whewell 1836, 1845)

This quest for what must be the permanent forms of institutions and knowledge sought to respond to numerous witnesses of the new dynamics connected with the growth of new industrial towns, which found expression in new scientific societies concerned mainly with science and marked with utilitarian values. All through the eighteenth century outside the universities, the mathematical culture of the *Philomaths* (Pedersen 1963; Wallis 1972) was developed in England, particularly in journals intended for non-academic readers such as *The Ladies Diary* and *The Mathematical Repository,*[12] but also in the Dissident Academies (Rogers 1981), or in the commercial and navigation schools, in towns such as Newcastle upon Tyne, Whitehaven, Durham or Sunderland (Fauvel et al. 1991, 16–9, 27–32). An important part of teaching was devoted to algebraic calculus. Numerous provincial societies were created, and their vitality showed a strong gap between a landed aristocracy with its eighteenth-century culture, where science was part of the leisure of the gentleman amateur, and new provincial elites—merchants, factory owners, capitalists, engineers, new lawyers and doctors—all in search of cultural recognition of their new economic status. Those philosophers and men of science who were conscious of the fact that the inadequacies of the university with respect to new knowledge increased the threat of a social explosion worked to find a new equilibrium, which could lead beyond the chasm between 'learned men', i.e. those with traditional classical learning, and the new 'practical men' (Morrell and Thackray 1981; D. Lardner in Babbage 1989, 2: 176).

2. INDUCTIVE SCIENCES VERSUS FORMAL SCIENCES

From these problems concerning the real nature of the principles of knowledge new works emerged that sought to reorganize both mathematics and logic. In the same way that was happening at Cambridge, new works on logic were initiated around Oxford, and they were numerous enough to constitute what Hamilton would call 'The New Analytics' and refer to as a kind of miracle:

> In these circumstances, when .... logic seemed in Oxford on the eve of following metaphysic and psychology to an academic grave, a new life was suddenly communicated to the expiring study, and hope at least allowed for its ultimate convalescence under a reformed system. ....
>
> This was mainly effected by the publication of the *Elements* of Dr Whately, then Principal of St Alban's Hall, and recently (we rejoice) elevated to the Archepiscopal See of Dublin. ....
>
> The last decade, indeed, has done more in Oxford for the cause of this science than the whole hundred and thirty years preceding. (Hamilton 1833, 199–200)

The crucial epistemological point was to establish the source of legitimacy of the new sciences. That is to say that the nature of this knowledge was very much an issue long before Whewell wrote his *History* (1837) and his *Philosophy of Inductive Sci-*

ences (1840). A systematic investigation was to result from these debates that would ultimately distinguish the formal part of science from the part that is in relation to the real world, and specify the conditions of certitude in both parts.

2.1. Dugald Stewart and Demonstrative Truth

At the very beginning of the nineteenth century, Dugald Stewart, in his *Elements of Philosophy of Human Mind,* sought to identify the difference between mathematics and other sciences in terms which seem very much in accordance with twentieth-century epistemology. His analysis of mathematical certitude follows Locke's conception of demonstrative truth, founded on the fact that, for mathematical objects, nominal and real existence are confounded (Locke 1694, IV-6-13, IV-6-16, IV-12-4). In Stewart's terms, the certitude of mathematical results must reside in the fact that the definition of its objects coincide with their existence. Consequently, mathematical certitude is not truth, but consistency.

> In these extracts from Leibnitz, as well as in that quoted from Condillac, in the beginning of this article, the essential distinction between mathematics and the other sciences, in point of phraseology, is entirely overlooked. In the former science, where the use of an ambiguous word is impossible, it may be easily conceived how the solution of a problem may be reduced to something resembling the operation of a mill, —the conditions of the problem, when once translated from the common language into that of algebra, disappearing entirely from the view; and the subsequent process being almost mechanically regulated by general rules, till the final result is obtained. In the latter, the whole of the words about which our reasonings are conversant, admit, more or less, of different shades of meaning; and it is only by considering attentively the relation in which they stand to the immediate context, that the precise idea of the author in any particular instance is to be ascertained. In those sciences, accordingly, the constant and unremitting exercise of the attention is indispensably necessary, to prevent us, at every step of our progress, from going astray. (Stewart 1814, 141)

In the other sciences, experience and induction are not such naive acts as asserted by the followers of Bacon. Without theory, or without general principles inferred from a discriminating comparison of various phenomena, experience is nothing more than a blind useless guide (Stewart 1814, 329). Every act of observation involves an interpretation of nature. Every act of induction relies on some fundamental belief in the uniformity and constancy of the laws of nature. No enquiry into natural phenomena can be undertaken without conscious or unconscious use of the principles of reasoning, which regulate the activities of the human mind. And to study these principles is as essential to the development of science as to study the object of each scientific discipline.

2.2. Whately and the Process of Operation

Whately's *Elements of Logic* is an important text in its quest to understand the precise role of logic in the constitution of science, because it brings together some topics from philosophy of logic and philosophy of mathematics that go beyond the opposition between Oxford and Cambridge. Whately's view of logic is marked by arithmetical practice and by Locke's specific distinction between reality, ideas and language.

First of all, Whately adopts a conciliatory position on logic, characterizing it both as the science of reasoning, for it analyses the processes of the mind, and as the art of reasoning, for it states practical rules in order to safeguard the mind against errors in its deductions. The difficulties which arise in the right estimation of the role of logic in scientific discovery come, for Whately, from a wrong estimation of the respective roles of observation and experience on the one hand, and of the operation of reasoning on the other hand. These difficulties do not come from logic, but from the data to which it is applied.[13] In the same way, arithmetic will give a right result to a problem, provided the data are correct. And Whately speaks clearly of an analogy between the two sciences. In particular, he insists that the signs used in arithmetic are arbitrary, and takes advantage of the reader's familiarity with arithmetic to make clear to him his own analogous conception of logic.

Whately distinguishes three operations of mind: *apprehension*, which brings the notion or the idea or the conception of an object into the mind, *judgment*, which allows comparison of two ideas in the mind, and *reasoning*, which allows transition from one judgment to another that is 'founded upon it'. These three operations are expressed in language respectively by terms, propositions and syllogisms. The syllogism, which Whately understood as the expression of an act of reasoning, is considered the universal principle of reasoning and the reduced form of every right reasoning (Whately 1849, 23–4). Syllogisms express reasoning about classes of objects, which are designated by proper names and result from generalizations. The generalizations are founded on abstraction and retain only what is essential to reasoning. Whately's interest in abstraction, unlike Locke's, does not extend beyond its function of separating or its function of characterizing the human being. He is in complete agreement with Locke, however, when he observes the same distinction between reality, ideas and words, when he begins its analysis from ideas in the mind rather than from perceptions, and when he manages abstraction through its expression by words in syllogisms.

Having so defined the object of logic, Whately puts much emphasis on its formal aspect, based on the use of 'arbitrary unmeaning symbols', which offer the same advantages as the symbols of arithmetic: reasoning is independent of the truth or falsity of the conclusion, which is accidental and variable (Whately 1849, 6, 14–5).

Finally, Whately is willing to analyse induction in the framework of scholastic logic. He again takes up Locke's distinction between real and nominal definitions, and asserts, as Locke did, that in mathematics, both sorts of definition coincide exactly; he concludes that logic deals only with nominal definitions (Whately 1849, 29–31). In this way, Whately is able to work on induction in the framework of logic, reducing the inductive argument to a syllogism in Barbara. He is also able to distinguish between logical and physical discoveries. The discovery of new truths could not arise from reasoning only, since the conclusion is included in the premisses, unless what is meant is that it is new only in the sense that it was never expressed before. And so Whately is led to conclude that mathematical discoveries are logical ones.

2.3. Hamilton and the Laws of Thought

The review of the 'Recent Publications on Logical Sciences' given by Hamilton in 1833 was mostly devoted to Whately's *Elements of Logic*, considered the most important of the 'recent publications'. It is worth noting that Hamilton says very little about the work of the botanist George Bentham (1800–84), *Outline of a New System of Logic, with a Critical Examination of Dr Whately's 'Elements of Logic'*, published in 1827, in which the quantification of the predicate is used for the first time. As is well known, however, Hamilton began a quarrel of priority with De Morgan on that subject in 1847, and their public discussion induced Boole to write his *Mathematical Analysis of Logic* (Boole 1847, preface).

But Hamilton's review of Whately's work is not without ambiguity. Using the support of scholastic logicians and of Kant's philosophy, he followed Whately in asserting that logic is neither a material nor a real science, but a formal one. Like Whately, then, he rejects truth and falsity outside logical concerns, and extends this rejection to a long series of modalities.

> Logic is a formal science; it takes no consideration of real existce, or of its relations, but is occupied solely about that existence and those relations which arise through, and are regulated by, the conditions of thought itself. Of the truth or falsehood of propositions, in themselves, it knows nothing, and takes no account: all in logic may be held true that is not conceived as contradictory. In reasoning, logic guarantees neither the premisses nor the conclusion, but merely the consequence of the latter from the former. ...
>
> But if truth or falsehood, as a material quality of propstns and syllogisms is extralogical, so also is their modality. Necessity, Possibility, &c., are circumstances which do not affect the logical copula or the logical inference. They do not relate to the connexion of the subject and predicate of the antecedent and consequent, as terms in thought, but as realities in existence; they are metaphysical, not logical conditions. The syllogistic inference is always necessary; it is modified by no extraformal condition; is equally apodictic in contingent as in necessary matter. (Hamilton 1833, 215–6)

Hamilton goes on insisting on the confusions introduced in logic by what he calls these metaphysical notions such as the necessary, the impossible, the contingent, the possible, the true and the false, all of them already in Aristotle's *De Interpretatione*, but also the certain, the probable, the useful, the good, the just—and so on, since the list may be extended at will. The essential problem for Hamilton is to limit the field of logic, considered as a formal science, that is to say as a primary and independent science, to its specific concerns, and so to exclude from it radically the truth problem, since 'the reasoner must transcend the sphere of logic if he would attempt to vindicate the truth of his conclusion'. In that way, Hamilton can use this formal conception of logic to reject any participation of it in scientific invention, or in the choice of the axioms of a science.

This insistence on characterizing logic as a formal science allows Hamilton to correct Whately's analysis of the inductive syllogism. He asserts that logic does not consider really existing things, but 'the general forms of thought under which the mind conceives them'. So he can distinguish between the relation of causality, which concerns existing things, and the relation from reason to consequence, which concerns the whole and its parts, and with which he analyses logical induction. Whole and parts have no real or essential existence; they are but 'creations of the mind itself'. Hamilton, then, understands deductive reasoning as a process of analysis, from the whole to the parts, and inductive reasoning as a process of synthesis, from the parts to the whole (Hamilton 1833, 226–7).

But the most important part of Hamilton's review is certainly his refusal to follow Whately in considering the process of reasoning, rather than the analysis of the laws of thought, as the principal object of logic. What is the difference? Hamilton is resolutely opposed to what may be called an instrumentalistic conception of logic. He repeatedly stands apart from any reference to Locke, to whom he imputes the decline of logic, errors in his philosophy, and the lack of a distinction between logic, psychology and metaphysics. He blames Whately for this limited conception of logic which restricts its study to that of syllogistic and rejects that of apprehension and judgment, which Hamilton considers as constitutive of thought. Moreover, Hamilton considers Whately's view confusing because it involves the analysis of language rather than that of reasoning; this approach goes back not only to Locke but to Hobbes.

> ...We conceive as a very partial conception of the science,—that Dr Whately makes the process of reasoning not merely its principal, but even its adequate object; those of simple apprehension and judgment being considered not in themselves as constituent elements of thought, but simply as subordinate to argumentation. In this view logic is made convertible with syllogistic. (Hamilton 1833, 206)

Hamilton also rejects Whately's definition of logic for not indicating 'the differential quality by which logic is discriminated from the other sciences', namely psychology and metaphysics, which have the same object, but in 'its real nature'.

> Other sciences, as psychology and metaphysic, propose for their object (among the other faculties) the operations of reasoning, but this considered in its real nature: logic, on the contrary, has the same for its object, but only in its formal capacity; in fact, it has, in propriety of speech, nothing to do with the process of operation, but is conversant only with its laws. (Hamilton 1833, 207)

In these criticisms we find the difficulties involved in the whole issue, which had already been identified in Locke's *Essay on Human Understanding* (Durand 1990, 158–9), where a distinction is made between three different meanings of the term 'operation' conceived as an operation of thought:

(1) the faculty of mind, i.e. the potentiality of the operation

(2) the effective process of the action of operation

(3) its effective realization, i.e. its result

This is a crucial distinction, which would lead mathematicians of the English Algebraic School to formulate operations explicitly as laws of combination, and to define them by their properties rather than by their results. But Hamilton refuses it, because he refuses to consider logic as the study of operative processes. He refuses the instrumental character that logic would have if its object were to be limited to such a study.

> Logic considers Thought, not as the operation of thinking, but as its product; it does not treat of Conception, Judgment, and Reasoning, but of Concepts, Judgments, and Reasonings ... In so far as the form of thought is considered in reference to the thinking mind,—to the mind by which it is extended, it is considered as an act, or operation, or energy; and in this relation it belongs to Phænomenal Psychology. Whereas, in so far as this form is considered as the product of such an act, and, in this relation, it belongs to Logic. Thus Phænomenal Psychology treats of thought proper as conceptn, judgment, reasoning; Logic, or the Nomology of the Understanding, treats of thought proper as a concept, as a judgment, as a reasoning. Whately, I have already shown you, among other errors in his determination of the object-matter of Logic, confounds or reverses this; for he proposes to Logic, not thought considered as a product, but reasoning alone; and that, too, considered as a producing operation. He thus confounds Logic with Phænomenal Psychology.[14]

Some years later, Hamilton's rejection of instrumentality would surface again in his criticism of Whewell's *Thoughts on the Study of Mathematics as Part of a Liberal Education*. His defence of university teaching against the colleges takes the form of supporting a general cultivation of the mind against narrow professional training, a personal development of the individual against obedience to national or professional interests:

> (Liberal education is) an education in which the individual is cultivated, not as an instrument towards some ulterior end, but as an end unto himself alone; in other words, an education, in which his absolute perfection as a man, and not merely his relative dexterity as a professional man, is the scope immediately in view. (Hamilton 1836, 409)

Hamilton shared Whewell's rejection of the mechanization of algebraic processes (Hamilton 1836, 412), but did not accept that geometry is superior to logic as a practical means for learning about reasoning. On the contrary, he held that mathematics are concerned only with necessary matter, whereas reasoning has to work on contingent matter as well.

> Logic, by a famous distinction, is divided: —into Theoretical or General Logic, in so far as it analyses the mere laws of thought; and into Practical or Special Logic, in so far as it applies these laws to a certain matter or class of objects. ... Now, as all matter is either necessary or contingent (a distinction which may be here roughly assumed to coincide with mathematical and non-mathematical) we have thus, besides one theoretical or general logic, also two practical or special logics in their highest universality and contrast.

Theoretical Logic

1) Practical Logic As specially applied to Necessary Matter = Mathematical Reasoning	2) Practical Logic As specially applied to Contingent Matter = Philosophy and General Reasoning

> Now, the question which Mr Whewell proposes to handle, is—What is the best instrument for educating men to a full development of the reasoning faculty? and his answer to that question is—Mathematics. But the reasoning faculty of men, being in all principally, in most altogether, occupied upon contingent matter, comprising what Mr Whewell himself calls 'the most important employments of the human mind', he was bound articulately to prove, what certainly cannot be presumed, that Mathematics (the Practical Logic of necessary matter) cultivate the reasoning faculty for its employment on contingent matter, better than Philosophy, the Practical Logic itself of contingent matter (Hamilton 1836, 413).

For Hamilton, the fundamental laws of thought remain those of the Scholastics: the laws of identity, of contradiction, of exclusion or excluded middle, and of reason and consequent or sufficient reason.

To sum up, three characteristics stand out from these debates on logic:

- the distinction between three specific parts of the meaning of the term 'operation': its potential, its process and its result;
- the assertion of the strictly formal character of logic;
- and the distinction between the concepts of logical coherence and physical causality, which would become the distinction between the concepts of consistency and truth.

3. OPERATIONAL SYMBOLISM VERSUS MATHEMATICAL REALISM

With the increasing development of algebraic means of expression, mathematics were also subject to instrumentalism, which involved writing symbols without always keeping in mind the objects signified. Therefore, mathematics, even more than logic, were threatened by the loss of meaning, by the loss of the referent values of which their objects are bearers.

Before the end of the eighteenth century, the same Playfair who would stigmatize his compatriots for their ignorance of the *Mécanique céleste* of Laplace carried out an analysis of the conceptual difficulties raised by the introduction of impossible quantities in operations. But he used a geometrical argument to explain the coherence of the results.

> Those expressions, as is well known, owe their origin to a contradiction taking place in that combination of ideas which they were intended to denote. Thus, if it be required to divide the given line $AB = a$ in C, so that $AC \times CB$ may be equal to a given space b^2, and if $AC = x$, then $x = \frac{1}{2}a \pm \sqrt{\frac{1}{4}a^2 - b^2}$; which value of x is imaginary when b^2 is greater than $\frac{1}{4}a^2$; now to suppose that b^2 is greater than $\frac{1}{4}a^2$, is to suppose that the rectangle $AC \times CB$ is greater than the square of half the line AB, which is impossible. ...
>
> The natural office of imaginary expressions is therefore, to point out when the conditions, from which a general formula is derived, become inconsistent with each other. (Playfair 1778, 319–20)

This is to say that quantities devoid of intelligibility would be equally devoid of utility. But this is far from being the case. Playfair was thus obliged to seek to legitimize them by analogy with geometry. He developed the example of the integration of two differential equations $\dot{a} = \frac{\dot{z}}{\sqrt{1+z^2}}$ and $\dot{a} = \frac{\dot{z}}{\sqrt{1-z^2}}$, which can be considered to have a single common form if, in the second one, $z\sqrt{-1}$ is substituted for z. The solution of the first one is $z = \text{sh}\ a$, whose expression in terms of the exponential does not use imaginary expressions, since $\text{sh}\ a = \frac{e^a - e^{-a}}{2}$. The solution of the second one is $z = \sin a$, whose the expression in terms of the exponential is an imaginary expression, since $\sin a = \dfrac{e^{a\sqrt{-1}} - e^{-a\sqrt{-1}}}{2\sqrt{-1}}$. Playfair refers this analogy of the written forms to a geometrical analogy, between an arc of an equilateral hyperbola, whose length is a, and an arc of a circle whose length is a too, the properties of the two geometrical figures corresponding to each other by a geometrical affinity.

In consequence, he attributes to the results obtained by impossible quantities only heuristic value. Reasoning, for Playfair, remains the monopoly of geometry, in which the meaning of the propositions is not strictly included in the formal logic of propositions. The figures have a meaningful reference and authorize each step of the reasoning to be checked. Algebraic symbols, on the other hand, are artificial characters, which have no link with the idea they represent, and lead to mechanical and blind operations.

At the turn of the nineteenth century, developments in algebraic analysis provided another important example of a calculus that could be carried out on meaningless symbols, which raised the question of the place of analogy in mathematical reasoning. Lagrange's theorem, which was to be referred to as the 'symbolical form of Taylor's theorem' by the followers of Peacock, comes from the analogical manipulation of these two series:

$$u(z+x) = u(z) + \frac{du}{dz}x + \frac{d^2u}{dz^2}\frac{x^2}{2!} + \frac{d^3u}{dz^3}\frac{x^3}{3!} + \text{etc.},$$

which is the development of a function in a Taylor series

$$e^z = 1 + z + \frac{z^2}{2!} + \frac{z^3}{3!} + \text{etc.}$$

which is the series which defines the exponential.

The observed analogy between the forms of these two series leads to the following expression of the finite increment of a function:

$$u(z+x) - u(z) = \Delta u = e^{\frac{du}{dz}x} - 1,$$

$$\text{and then } \Delta u = (e^{x\delta} - 1)u$$

$$\text{and finally } \Delta^n u = (e^{x\delta} - 1)^n u$$

where u, the symbol for the fonction, and $\delta = \dfrac{d}{dz}$, the differential operator, have been separated from the symbol for the variable (Arbogast 1800, 350–1).

As they were excessively conscious of the difficulties involved, the mathematicians of the English Algebraic School did not embrace instrumentalism. On the contrary, they were convinced that practices such as those of analytical algebra are fruitful only because they are founded on reason: efficiency results from the laws of symbolical calculus, which they considered it their job to discover. Because they shared the finalism of Playfair, they wanted to avoid the materialistic aspect of utilitarianism, but without giving up the knowledge acquired. They thus sought to formulate explicitly the principles of a logical and symbolical calculus adequate for founding algebra.

3.1. Woodhouse and the Logic of Operations

In Cambridge, the Rev. Robert Woodhouse (1773–1827), fellow of Caius College (1798–1820) and later Lucasian Professor (1820–22) and Plumian Professor of Astronomy and Experimental Philosophy (1822–27), was far from refusing the instrumentalism of algebra and did not share the intellectual puritanism of the Rev. William Frend (1757–1841), who sought to reconcile the rigour of algebra and its empirical foundation by excluding all negative and impossible quantities. Woodhouse did not attempt, like Playfair, to found the success of algebraic methods solely on their possibility to express truths of a geometrical nature; on the contrary, he supposed that paradoxes in algebra could not originate in the symbolism itself, because it is a rational creation of human mind:

> I (am) convinced in my own mind, that there can be neither paradoxes nor mysteries inherent and inexplicable in a system of characters of our own invention, and combined according to rules, the origin and extent of which we can precisely ascertain ... Demonstration would be defined to be a method of showing the agreement of remote ideas by a train of intermediate ideas, each agreeing with that next it; or, in other words, a method of tracing the connection between certain principles and a conclusion, by a series of intermediate and identical propositions, each proposition being converted into its next, by changing the combination of signs that represent it, into another shewn to be equivalent to it.[15]

This philosophical point of view led Woodhouse to accept the observed analogies, and to found the exactness and the efficiency of the methods on the existence of a logic specific to the operations themselves, which is valid for all quantities, whether they be real or not. In this manner, he defined for instance the exponential of an imaginary quantity using the analytical form of the exponential of a real one, by its development in a series.

His style and his views were very much influenced by Locke's philosophy, particularly with regard to what it is to be a demonstration. Woodhouse had the same confidence as Locke in the practice of using arbitrary symbols to express some truth of the world. This method of separating the system of signs and signified things offers a solution to the problem of abstraction. With it, Woodhouse was able to found an algebraic calculus with the same fundamental properties for polynomials and series, in which utility is compatible with rigour. He considered positive the very things that were condemned by his detractors.

> The arguments that seem to render all operations performed with impossible quantities unintelligible may be included under the following statement. Algebra is a species of short-hand-writing, a language, or system of characters or signs, invented for the purpose of facilitating the comparison and combination of ideas. Now all demonstration by

> signs, must ultimately rest on observations made on individual objects; and all the varieties of the transformation and combination of signs, except what are arbitrary and conventional, must be regulated by properties observed to belong to the things of which the signs are the representatives. Demonstration by signs is shewn to be true, by referring to the individual things the signs represent; and is shewn to be general, by remarking that the operation is the same, whatever is the thing signified, or, in other words, that the operation is independent of the things signified. (Woodhouse 1801, 90)

The specific contribution of the English Algebraic School is to be found in the conviction that operating mechanisms were necessarily justified. Following Woodhouse, but more insistently than he, the members of this school endeavoured to clarify the formal properties of operations, and thereby to change the contents of algebra, which would no longer be concerned primarily with such mathematical objects as quantities, but with laws for combining operations.

3.2. Peacock and His Legitimatization of Symbolical Properties of Operations

Rev. George Peacock was not the most celebrated of the mathematicians of this School. Apart from his prominent role in the introduction of the differential notation of the infinitesimal calculus in Cambridge examinations, his *Treatise of Algebra*, written as a textbook in 1830, and his famous *Report on the Recent Progress and Actual State of Certain Branches of Analysis,* given at Cambridge at the third meeting of the British Association for the Advancement of Science, he did not have prestigious publications. He did not, for example, publish in the *Transactions of the Cambridge Philosophical Society,* despite being one of its founders. His historical essay on *Arithmetic,* written in 1826 for the *Encyclopædia metropolitana,* and published in 1845, was praised in the encyclopedia preface as being of equal interest for the scholar, the mathematician, and the speculator in metaphysics. In the same way, Herschel, in his obituary notice (Herschel 1858), honoured him as the most philosophical of them all.

Nevertheless, he was the most progressive of the first-generation mathematicians of the English Algebraic School in his attempt to legitimatize the instrumental aspect of algebra. Peacock had numerous followers among his students, notably De Morgan and Duncan F. Gregory (1813–44), who would be the first Cambridge mathematician to correspond with Boole and his advisor as he prepared his first writings. The radical change in the conception of algebra brought about by Peacock enabled his followers to think of it as a calculus of operations, and to seek meaningful reference in the operations' own properties. For this reason, it is possible to speak of his work as introducing an epistemological breakthrough in algebra.

In the very first pages of his *Report,* Peacock announced his ambition of developing an algebra that could take the place of geometry; in this he followed Woodhouse, who exclaimed in 1802: *'That the science of geometry was first invented is properly an accidental circumstance'* (Woodhouse 1802, 89)! Like Woodhouse, he had no doubt as to the independence of algebra and the autonomy of the logic of its operations.

The first task of Peacock in his *Report* was to state his position on algebra in the British debates about the foundation of truth in mathematics. He does so by giving the same analysis of what is formal in mathematics as Locke, and then Woodhouse, had before him. He characterized algebra as a speculative, abstract, and demonstrative science, for which completeness and coherence are secured only by the rigour of deductive reasoning from first principles considered as ultimate facts; of these he did not specify the origin, but said simply that unlike those in natural philosophy they are not 'veiled in the mystery of the first cause'.

> We consider those principles and definitions as *ultimate* facts, from which our investigations proceed in one direction only, giving rise to a series of conclusions which have reference to those facts alone, and whose correctness or truth involves no other conditions than the existence of a necessary connection between them, in whatever manner the evidence of that existence may be made manifest.
>
> We merely regard the results of the science and the logical accuracy of the reasoning by which they are deduced from assumed principles; and all our conclusions possess a necessary existence, without seeking either for their strict or for their approximate interpretation in the nature of things. (Peacock 1833, 187)

His work consisted in giving explicit operating laws without excluding the knowledge acquired by experience in mathematics. This acquired knowledge was included, not only with its results, but as an empirical method used to discover the general forms of algebra, in accordance with its historical genesis and development as viewed by Coleridge in the preface to his *Encyclopædia metropolitana*. This empirical part of algebra had however to remain completely subordinate to its operating laws, which constitute its symbolical part. This is why Peacock regarded as two distinct sciences (1) an arithmetical algebra, considered as a science of suggestion, where symbols are general in their form, but specific in their value, and where the conditions of existence of the results are maintained, and (2) a symbolical algebra, which he defined as the 'language of symbolical reasoning'. This symbolical algebra possesses its own logic which is described as a 'system of combination of arbitrary symbols, general in their form as in their value' (Peacock 1833, 194, 208), to which all the practices of what is known in algebra are necessarily subjected. Peacock thus made explicit Woodhouse's own conviction in asserting the existence of a universal structure which guarantees the symbolical validity of all acquired results. Moreover, it is the universality of this validity which constitutes its own meaning. The attention with which Peacock very carefully excluded analogy as a possible legitimatizing factor of the acquired results is evidence of his conviction of the primacy of deductive over inductive reasoning:

> The operations and resulting forms ... expressed in symbols, bear a strict analogy to operations bearing the same names, and to the forms resulting similarly from them in Algebra, where the symbols are perfectly general: but it is by the law of the permanence of equivalent forms, and not by analogy, that we are enabled to pass from one to the other: it is only so far therefore, as Analogy may be considered as a modified expression of this law, that we are legitimately enabled to generalize conclusions deduced by means of it. (Peacock 1830, 108)

The following double statement of the 'law of the permanence of equivalent forms' is the key to the relationship between arithmetical and symbolical algebra. It asserts the double possibility of a transfer from the equalities between the general forms of arithmetical algebra to the equivalences between the general forms of symbolical algebra, and conversely.

> (A): Whatever form is algebraically equivalent to another when expressed in general symbols, must continue to be equivalent, whatever those symbols denote.
>
> (B): Converse Proposition: Whatever equivalent form is discoverable in arithmetical algebra considered as the science of suggestion, when the symbols are general in their form, though specific in their value, will continue to be an equivalent form when the symbols are general in their nature as well as their form. (Peacock 1833, 194)

All the difficulty of Peacock's thought is contained in this double statement: it expresses his determination not only to account for all mathematical knowledge produced by experience but also to maintain its subordination to universal symbolical laws. Because he never gave up a finalistic conception of knowledge, he did not accept that the mathematician might create axiomatically some algebraic structure of his own choosing. Peacock insisted on the fact that these symbolical laws are not generalizations of arithmetical ones, but their necessary prerequisites (Peacock 1833, 194). Statement (A) in the principle of permanence presupposes therefore as natural these universal operating laws. This being supposed, statement (B) asserts the subordination of inductive to deductive reasoning, and indicates how the form of the results obtained by these preexisting laws can be discovered. If these statements appear strange or contradictory now, they are philosophically clarified by the attempt of Peacock to reconcile his finalistic conception of knowledge with his conception of its historical genesis, which he developed essentially in the pedagogical presentation of his *Treatise of Algebra,* as well as in the historical presentation of his *Arithmetic,* where the operations of arithmetic are given, first by words, then by symbols, as a universal experience (Peacock 1845, 383, 395, 397, 442).

Symbols may be arbitrary, and the names of the operations may be arbitrary too, but their structures are not. They express the necessary laws of combination by which faculties of mind operate in algebraic thought. This is a universal experience that has to be finally expressed; it has to be stated as general, as Locke said about principles. Peacock thus stated as general the reciprocal character of addition and subtraction, and also of multiplication and division, by means of a principle of combination, and rules for the concurrence of signs.

> The operations called addition and subtraction are denoted by the signs + and –. They are the inverse of each other. In the concurrence of the signs + and –, in whatever manner used, if two like signs come together, whether + and +, or – and –, they are replaced by the sign +; and when two unlike signs come together, whether + and –, or – and +, they are replaced by the single sign –. When different operations are performed or indicated, it is indifferent in what order they succeed each other.
>
> The operations called multiplication and division are denoted by the signs × and ÷, or more frequently by a conventional position of the quantities or symbols with respect to each other ...
>
> The operations of multiplication and division are the inverse of each other.
>
> In the concurrence of the signs + and – in multiplication or division, if two like signs come together, whether + and +, or – and –, they are replaced by the single sign +; and

> if two unlike signs come together, whether + and –, or – and +, they are replaced by the single sign –.
>
> When different operations succeed each other, it is not indifferent in what order they are taken. (Peacock 1833, 196–7)

This philosophical background for Peacock's thought facilitates understanding of the constancy with which he considered his symbolical algebra as equally valid for finite and infinite computations, such as terminable or interminable quotients, rational or irrational roots, or such as polynomials and series. His constant refusal to distinguish between diverging or converging series, or to consider problems such as the continuity of functions as essential, may be understood in the same way. For Peacock, these problems concern the numerical values of expressions, and consequently, the arithmetical algebra, the interpretation of symbols, and the field of contingent truths.

> It is this confusion of necessary and contingent truth which has occasioned much of the difficulty which has attended the theories of the interpretation of algebraical signs. It has been supposed that a meaning could be transmitted through a succession of merely symbolical operations, and that there would exist at the conclusion an equally necessary connexion between the primitive definition and the ultimate interpretation, as between the final symbolical result and the laws which govern it. So long as the definitions of the meaning of the symbols are sufficient to deduce the results, those results will have a necessary interpretation which will be dependent upon a joint consideration of all those conditions; but whenever an operation allow it to be strictly defined or interpreted, the chain of connexion is broken, and the interpretation of the result will be no longer traceable through its successive steps. (Peacock 1833, 230)

Peacock made effective the distinction that was manipulated by Locke, perceived by Whately, but refused by Hamilton, namely the distinction between the three meanings of 'operation', which correspond to three different levels of its occurrence: potential, process and result. Symbolical algebra is the science of necessary truths, which applies only to processes of operations. However, because Peacock wanted to ascertain these operations not on the basis of an axiomatization of principles but with respect to the faculties of the mind—because he wanted to maintain the primacy of the operating laws over the experience of algebra—he observed a strict hierarchy between arithmetical and symbolical algebra, or between form and content, as a model theorist might say today. Analogy is too much associated with probability to be used as a legitimatization of symbolical laws, and the mathematician's creativity is not recognized in itself.

The ambitions of the English Algebraic School are to be viewed as part of an essential symbolical revolution, where the operating processes become more important in mathematics than the structuring of movement representing action, as was achieved by the calculus. But this new trend of thought does not mean the end of an ontological foundation for operations. It leads to conceptual work on notions such as truth, consistency and interpretation, where necessary truth becomes synonymous with demonstrative certitude, and contingent truth with interpretation. It results in abstract mathematics shifting to a science of mind, which it already was, socially, in Cambridge, and which it was to become, conceptually, with Peacock.

4. BOOLE'S MEDIATION

As George Boole is usually presented as self-taught, and as his logic is usually analysed in itself, or from a twentieth-century point of view, his involvement in these debates is not always clearly perceived. Yet, as I indicated above, mathematical culture was not rare in England outside the universities at this period. Boole's father, who was one of the founders, then later curator and librarian of the Mechanics Institute of Lincoln, where George first learnt higher mathematics, had a telescope at his window and invited passers-by to look into it (Diagne, 1989, 17); he was an example of such a scientifically cultivated and motivated man. Boole was initiated into most continental developments in mathematics by Sir Edward French Bromhead, a close friend of Babbage's and one of the founders of the Analytical Society. His very first interest in mathematics concerned the method of separation of symbols,[16] which was the main topic of his first papers in the *Cambridge Mathematical Journal,* which appeared beginning in 1841, and the subject of his 'On a General Method in Analysis', for which he was awarded the Royal Medal of the Royal Society in 1844. He had a close professional relationship with numerous mathematicians of the second generation concerned with symbolical algebra, such as Gregory, Robert L. Ellis (1817–59), lecturer in Trinity College of Cambridge and co-founder with Gregory of the Cambridge Mathematical Society, William Thomson (1824–1907), the future Lord Kelvin, who would manage the same journal when it became the *Cambridge and Dublin Mathematical Journal,* De Morgan, the Rev. Philip Kelland (1808–79), professor of mathematics in Edinburgh University, who wrote on Peacock's principle of permanence (Kelland 1840), and the Rev. Charles Graves (1812–99), professor of mathematics at Trinity College in Dublin. All of them were to write letters of recommendation in support of Boole's candidacy for a professorship at Queen's College of Cork in 1849. Another interest of Boole's concerned the same problems which led Babbage to develop his analytical engine, namely the resolution of differential equations and of equations on finite differences (Boole 1859, 1860), and their symbolical methods of resolution.

Between Peacock's work of 1833 and Boole's of 1847, the mathematicians just mentioned explored more concretely the conditions of a calculus of operations, and established that these conditions are as crucial as its algebraic forms and transformations. But the general problematics remain: it is possible to separate, in mathematics, the question of necessary truth from that of the interpretation of a calculus. This was clearly articulated by Boole in the course of his involvement in debates about the status of mathematics in the foundation of knowledge.

Although Boole does not refer to Peacock by name in his work, he does rely on Peacock's symbolical algebra from the very beginning of his *Mathematical Analysis of Logic* in stressing the distinction between the interpretation and the laws of combination which concern the processes of analysis. As Woodhouse and Peacock did before him, he repeatedly considers as 'accidental circumstances' the previous lack of full recognition of this separation and the previous attribution of these relations to magnitudes only.

> They who are acquainted with the present state of the theory of Symbolical Algebra, are aware, that the validity of the processes of analysis does not depend upon the interpre-

> tation of the symbols which are employed, but solely upon the laws of their combination. Every system of interpretation which does not affect the truth of the relations supposed, is equally admissible, and it is thus that the same process may, under one scheme of interpretation, represent the solution of a question on the properties of number, under another, that of a geometrical problem, and under a third, that of a problem of dynamics or optics. This principle is indeed of fundamental importance; and it may with safety be affirmed, that the recent advances of pure analysis have been much assisted by the influence which it has exerted in directing the current of investigation. (Boole 1847, 3)

This separation between the symbolical properties and the interpretation of symbols would be firmly reasserted in his *Investigation of the Laws of Thought* in 1854, essentially about parts of the calculus which do not find a direct correspondence with the laws of thought themselves (Boole 1854, 68), and for which Boole takes impossible quantities as an example.

Boole was not reluctant to accept the mechanical aspect of this symbolical calculus. He did not hesitate to quote John Stuart Mill and to agree with him about the mechanization of reasoning—albeit with some reservations[17]—and replies to the objection that this use of symbols might obviate the need for thinking, by emphasizing the new possibilities offered to the exercise of the inventive faculties, particularly to that of generalization (Boole 1847, 10). But in no case does he accept an instrumental conception of logic. He strongly asserts that logic is founded upon 'facts of another order which have their abode in the constitution of the Mind', and insists on 'the speculative truth of its principles'. Although logic may be 'intimately connected with Language', the fundamental reason has nothing to do with technical matters but only with processes of the mind. This is affirmed very clearly in the *Mathematical Analysis of Logic*:

> That which renders Logic possible, is the existence in our minds of general notions,—our ability to conceive of a class, and to designate its individual members by a common name. The theory of Logic is thus intimately connected with that of Language. A successful attempt to express logical propositions by symbols, the laws of whose combinations should be founded upon the laws of the mental processes which they represent, would, so far, be a step toward a philosophical language. (Boole 1847, 4–5)

In both *Mathematical Analysis of Logic* and the *Laws of Thought*, Boole's logic is founded on the mental act of selection in the universe of objects, a mental act which is not considered in its nature, but in the efficiency of its action given its laws of combination and succession (Boole 1847, 16). The elective symbol x first represents the mental operation by means of which the elements are chosen, and then is identified with the class itself. Like the operations of ordinary algebra, these mental operations are commutative and distributive. On this point, Boole is clearly nearer to Gregory than to Peacock, as he uses the very same theorem enunciated by the former (Gregory 1839, 34–5), which authorizes the transfer of mathematical properties from one field to another—in this case, from ordinary algebra to logic—provided that the same expressions be used and that they be submitted to the same conditions. Boole himself already uses this theorem for symbols such as differential operators in his 1844 memoir. He establishes there that the act of selection leads to logical operations with properties such as $xy = yx$, $x(y + z) = xy + xz$, $x^2 = x$, where the product represents the successive selection of two classes from the same universe, and the sum the description of two disjoint classes, and he goes on to assert:

> The laws, the axioms, and the processes, of such an Algebra will be identical in their whole extent with the laws, the axioms, and processes of an Algebra of Logic. Difference of interpretation will alone divide them. Upon this principle the method of the following work is established.[18]

Provided with these symbolical expressions of the laws of thought, the problem of eliminating the middle term from the two premises of a syllogism becomes that of eliminating an unknown from two equations, and can be generalized to any number of equations, which is all the easier given that $x^2 = x$ in this special algebra. In spite of Boole's declared purpose of constructing 'the mathematics of the human intellect' (Boole 1847, 7), it would be anachronistic to speak of his logic as applied mathematics. Perhaps it could be said that logic and mathematics are both applications of a symbolical calculus, but only on condition of never losing sight of the fact that when Boole speaks of the first principles of such a symbolical calculus, he still views them as based on faculties of mind, which are the only means of explaining them. Boole refuses to consider them as a creation of the human mind just as he refuses induction as a method for his logic. It is essential to remember that although Boole considers only the efficiency of the combinations of these laws of thought, he always refers them to some principle of order, of harmony, or of coherence, which the human mind can perceive by analogies, but which always reveals the reign of order and law.

> It is the ability inherent in our nature to appreciate Order, and the concurrent presumption, however founded, that the phænomena of Nature are connected by a principle of Order. Without these, the general truths of physical science could never have been ascertained. [... In logic,] that principle of order or analogy upon which the reasoning is conducted must either be stated or apprehended as a general truth, to give validity to the final conclusion. In this form, at least, the necessity of general propositions as the basis of inference is confirmed,—a necessity which, however, I conceive to be involved in the very existence, and still more in the peculiar *nature*, of those faculties whose laws have been investigated in this work. [...] But besides the general propositions which are derived by induction from the collated facts of experience, there exist others belonging to the domain of what is termed *necessary* truth. Such are the general propositions of Arithmetic, as well as the propositions expressing the laws of thought upon which the general methods of this treatise are founded; and these propositions are not only capable of being rigorously verified in particular instances, but are made manifest in all their generality from the study of particular instances. (Boole 1854, 403–4)

Mathematical expression of laws is viewed as a manifestation of the reign of necessity, for the intellectual operations as well as for movements of the external world.

> [...] the absolute connexion between the initial and final elements of a problem, exhibited in the mathematical form, fitly symbolizes that physical necessity which binds together effect and cause. The necessary sequence of states and conditions in the inorganic world, and the necessary connexion of premises and conclusion in the processes of exact demonstration thereto applied, seem to be co-ordinate. [...] Subjection to laws mathematical in their form and expression, even the subjection of an absolute obedience, would stamp upon the two series one common character. The reign of necessity over the intellectual and the physical world would be alike complete and universal. (Boole 1854, 407–8)

The whole last chapter of the *Laws of Thought* is thus an expression of Boole's direct involvement in the debates about the concurrence between logic and mathematics in the system of human knowledge. Boole refers to Whewell's *Philosophy of the Inductive Sciences,* and to Cournot's *Essai sur les fondements de notre connaissance.* He feels the same difficulty as Peacock with respect to legitimatizing the conceptual jump which separates inductive from deductive processes. He does not try to elaborate explicit intellectual processes, but refers to limiting states of experience, and to the fact that reason is more profound than imagination. Finally, his last argument is that 'there exist in our nature faculties which enable us to ascend from the particular facts of experience to the general propositions [...]; as well as faculties whose office it is to deduce from general propositions accepted as true the particular conclusions which they involve' (Boole 1854, 420).

Boole concludes with a comment on 'the place of Mathematics in the system of human knowledge' (Boole 1854, 422), in which he tries to reconcile the use of logic and the use of mathematics by appealing to the fact that both are parts of the foundation of knowledge, the latter being the expression of the necessity of the former. Like his predecessors, he separates the necessary symbolical part of mathematics from the contingent applied part. In doing so, he reinforces, in logic as in mathematics, the distinction between validity and truth.

> Perhaps the most obviously legitimate bearing of such speculations would be upon the question of the place of Mathematics in the system of human knowledge, and the nature and office of mathematical studies, as a means of intellectual discipline [...] But upon the very ground that human thought, traced to its ultimate elements, reveals itself in mathematical forms, we have a presumption that the mathematical sciences occupy, by the constitution of our nature, a fundamental place in human knowledge, and that no system of mental culture can be complete or fundamental, which altogether neglects them.
>
> But the very same class of considerations shows with equal force the error of those who regard the study of Mathematics, and of their applications, as a sufficient basis either of knowledge or of discipline. If the constitution of the material frame is mathematical, it is not merely so. (Boole 1854, 422–3)

But these mathematics remain a profane knowledge and, as Locke did before him, Boole does not fail to conclude that all knowledge tends towards the knowledge of God—which, in 1847, he qualified as 'Divine Providence'. This is of course the fundamental reason for which Boole could not conceive of truth as the 'creation' of the human intellect (MacHale 1985, 43)—which he also asserted in 1847, in celebration of the reduction of the business day to ten hours in all trades. The sole pretention allowed to human knowledge is to discover the truth and to comprehend it; but for neither Locke, Peacock nor Boole did it belong to man to create truth.

CONCLUSION

The idea that Boole's work, carried out some thirty years before Frege's, is somewhat isolated historically reflects the point of view of a twentieth-century reader. In this paper, it is shown that Boole's work represented an important contribution to debates that took place in Great Britain about the legitimacy of the inductive sciences, about the renewal of mathematics by symbolical methods, and about the status of mathematics and logic as foundations for human knowledge. Boole was among those—so well described by Morrell and Thackray (1981) on the subject of the British Association for the Advancement of Science—who argued that inductive sciences are subordinate to deductive sciences. Like the others in this movement, Boole was no conservative. He sought to grasp and to implement the major changes of his time. His conception of logic represents a mediation between those who maintained their attachment to logic and those who argued for mathematics as the foundation of knowledge that was appropriate to the time. This mediation exploited the new importance given to operating processes resulting from recognition of the distinction between the potential, the process and the result of operations, and by specifically studying the operating processes. Boole's mediation follows up this way of thinking and gives a mathematical form to the processes of logic working on the laws of combination of classes.

Following in the footsteps of the English Algebraic School, Boole sought after a reign of necessity and permanence. Their efforts, founded on belief in the symbolical function of language and its properties, made it possible to do without a theory of abstraction, which had led to considerable difficulties already with Locke. They succeeded in subordinating contingency to necessity, and in deepening the distinction between logical validity and truth—truth retaining a strong religious dimension.

The non-Euclidean geometries were thus far from being the only part of mathematics that led to questioning of the nature of mathematical truth. The introduction of new objects, such as impossible quantities or differential operators, but also new thinking on the nature of inductive and deductive sciences, as well as the comparison of mathematical and logical truth, led to similar questioning. The work of Peacock and Boole transformed mathematics into the study of operations of the mind. The study of the confrontation between validity and truth was to be the next step in the transformation of the links between mathematics and logic.

NOTES

I wish to thank very sincerely Professor James Gasser, who has so kindly and so efficiently accepted to adjust my awkward English.

[1] He recalled that when he was eighteen years old, walking in the open fields, he suddenly felt mystical enlightenment and saw the possibility of writing logical relations in an algebraical form (MacHale 1985, 19).

[2] Peacock (1841, 166). Between 1750 and 1799, 60% of the registered students took orders; the proportion was 62% between 1800 and 1849, and 38% between 1850 and 1899 (Jenkins and Jones 1950, 110).

[3] The Board of Mathematics in Cambridge was created in 1848.

[4] The 'Tripos' would be named 'Mathematical Tripos' only with the creation of the 'Classical Tripos' in 1824.

[5] For example, Herschel refused to accept that they be known as the 'Cambridge' Analytical Society. Cf. Herschel Manuscripts, Royal Society Library Hs.2.6 (Herschel to Babbage: 8 February 1813) and Durand (1985, 201).

[6] Herschel Manuscripts, Royal Society Library Hs.2.69 (Herschel to Babbage: 24 December 1816), Hs.13.246 (Peacock to Herschel: 13 November 1816), Hs.13.249 (Peacock to Herschel: 17 March 1817), Hs.13.250 (Peacock to Herschel: 30 May 1817). Cf. also Durand (1985, 199).

[7] The differential notation of the infinitesimal calculus.

[8] Herschel Manuscripts, Royal Society Library Hs.13.247 (Peacock to Herschel: 3 December 1816).

[9] Herschel Manuscripts, Royal Society Library Hs.13.249 (Peacock to Herschel: 17 March 1817).
[10] Whately gave *Introductory Lectures on Political Economy* in Oxford in 1831.
[11] They were to be translated into French and German, and also came to be known in America (Stephen 1890): Anonymous (W. Ham.), Ed. Review, June 1831, vol. 53, n° 106, pp. 384–427 & December 1831, vol. 54, n° 108, pp. 478–504, 'On the State of the English Universities'. Anonymous (W. Ham.), Ed. Review, April 1833, vol. 58, n° 115, pp. 194–238, art. IX, 'Recent Publications on Logical Science'. Anonymous (W. Ham.), Ed. Review, January 1836, vol. 62, n° 126, pp. 409–455, art. VII, 'Thoughts on the Study of Mathematics as a part of a Liberal Education, from W. Whewell'.
[12] Mathematical exercises were regularly proposed in *The Ladies Diary* (1704–1840), of which Thomas Simpson and Charles Hutton were successive editors, as well as being professors at the Royal Military College in Woolwich, and in *The Mathematical Repository* (1770–1840) edited by Thomas Leybourne, which included problems of sufficient difficulty to be considered useful for Cambridge students preparing their examinations. Cf. Fauvel et al. (1991, 8–9) and Panteki (1987, 125).
[13] Whately agrees with Bacon when he attributes this confusion to the scholastics, who were the first to try to use logic in physical discoveries.
[14] Hamilton (1860, vol. III, lecture V, part I, section I: 'Noetic: On the fundamental laws of thought. Their contents and history', 73–4).
[15] These paradoxes appear, for the most part, in developments in series of functions where impossible quantities occur (Woodhouse 1801, 93, 107).
[16] MacHale (1985, 48–9). Letter of Boole to Bromhead: 24 July 1839.
[17] Mill (1843, 2:292) quoted by Boole (1847, preface).
[18] Boole (1854, 37–8). Similar statements are to be found in Boole (1847, 18; 1854, 31).

REFERENCES

Arbogast, L.F.A. 1800. *Du calcul des dérivations*. Strasbourg.

Babbage, C. 1989. *The works of Charles Babbage*. Edited by M. Campbell-Kelly. London.

Becher, H.W. 1980. Woodhouse, Babbage, Peacock, and modern algebra. *Historia Mathematica* 7: 389–400.

Boole, G. 1844. A general method in analysis. *Philosophical Transactions* 134: 225–82.

———. 1847. *The mathematical analysis of logic, being an essay towards a calculus of deductive reasoning*. Cambridge.

———. 1854. *An investigation of the laws of thought, on which are founded the mathematical theories of logic and probabilities*. London.

———. 1859. *A treatise on differential equations*. Cambridge.

———. 1860. *Treatise on the calculus of finite differences*. Cambridge.

Cannon, W. 1964. Scientists and broadchurchmen: An early intellectual network. *Journal of British Studies* 4: 65–88.

Corsi, P. 1988. *The heritage of Dugald Stewart: Oxford philosophy and the method of political economy*. Florence.

De Morgan, A. 1915. *A budget of paradoxes*. 2d ed. Chicago.

Diagne, S.B. 1989. *Boole, l'oiseau de nuit en plein jour*. Paris.

Durand(-Richard), M.J. 1985. George Peacock (1791–1858): La synthèse algébrique comme loi symbolique dans l'Angleterre des réformes (1830). Thèse de doctorat, E.H.E.S.S.

———. 1990. Genèse de l'algèbre symbolique en Angleterre: une influence possible de John Locke. *Revue d'Histoire des Sciences* 43: 129–80.

———. 1996. L'Ecole algébrique anglaise: les conditions conceptuelles et institutionnelles d'un calcul symbolique comme fondement de la connaissance. In *L'Europe mathématique: Mythes, histoires, identités*, edited by C. Goldstein, J. Gray and J. Ritter. Paris: Editions de la Maison des sciences de l'homme.

Fauvel, J., P. Ransom, P. Wallis and R. Wallis. 1991. *Mathematical tradition in the north of England*. Durham.

Garland, M.M. 1980. *Cambridge before Darwin, the ideal of a liberal education, 1800–1860*. Cambridge.

Gregory, D.F. [1839] 1865. On the solution of linear differential equations with constant coefficients. *Cambridge Mathematical Journal* 1: 22–33. Reprinted in *Mathematical writings*, Cambridge. (page references are to reprint).

[Hamilton, W.]. 1831. On the state of the English universities. *Edinburgh Review* 53 (106): 384–427 (June 1831) and 54 (108): 478–504 (Dec. 1831). Reprinted in *Discussions on philosophy and literature, education and university reform, chiefly from the Edinburgh Review*, London 1852, 386–463.

———. 1833. Review: 'Recent publications on logical science'. *Edinburgh Review* 58 (115): 194–238 (Apr. 1833). Reprinted in *Discussions on philosophy and literature, education and university reform, chiefly from the Edinburgh Review*, London 1852, 116–174.

———. 1836. Review: 'Thoughts on the study of mathematics as a part of a liberal education, from W. Whewell'. *Edinburgh Review* 62 (126): 409–55 (Jan. 1836). Reprinted as 'On the study of mathematics as an exercise of mind' in *Discussions on philosophy and literature, education and university reform, chiefly from the Edinburgh Review*, London 1852, 257–327.

Hamilton, W. 1860. *Lectures on metaphysics and logic*, vols III and IV (on logic). Edinburgh.

———. 1877. *Lectures on metaphysics and logic*, vols I and II (on metaphysics). Edinburgh.

Herschel, J.F.W. 1858. Obituary notice on G. Peacock. *Proceedings of the Royal Society* 9: 536–41.

Herschel Manuscripts. Royal Society Library, Royal Society of London.

Jenkins, J., and D.C. Jones. 1950. Social class of Cambridge University alumni of the 18th and 19th centuries. *British Journal of Sociology* 1: 93–116.

Kelland, P. 1840. On general differentiation, I and II. *Transactions of the Royal Society of Edinburgh* 14: 567–603 and 604–18.

Lardner, D. 1834. Babbage's calculating engine. *Edinburgh Review* 59: 263–327. Reprinted in Babbage 1989, vol. 2, 118–86.

Locke, J. 1694. *Essay on human understanding*. 2d ed. London.

MacHale, D. 1985. *George Boole, his life and work*. Dublin.

Mill, John Stuart 1843. *System of logic ratiocinative and inductive*. London.

Morrell, J., and A. Thackray. 1981. *Gentlemen of science, early years of the British Association for the Advancement of Science*. Oxford.

Panteki, M. 1987. William Wallace and the introduction of Continental calculus to Britain: A letter to George Peacock. *Historia Mathematica* 14: 119–32.

Peacock, G. 1820. *A collection of examples on the calculus of the applications of the differential and integral calculus*. Cambridge.

———. 1830. *Treatise of algebra*. Cambridge.

———. 1833. A report on the recent progress and actual state of certain branches of analysis. *Proceedings of the British Association for the Advancement of Science*, London, pp. 185–351. Reprinted as monograph, Cambridge 1834.

———. 1841. *Observations on the statutes of the University of Cambridge*. Cambridge.

———. 1845. Arithmetic. *Encyclopædia Metropolitana*, vol. 1, 369–523.

Pedersen, O. 1963. The 'philomaths' of XVIIIth century England, a study in amateur science. *Centaurus* 8: 238–62.

Playfair, J. 1778. On the arithmetic of impossible quantities. *Philosophical Transactions of the Royal Society* 68: 318–43.

[Playfair, J.] 1808. Review: 'Traité de mécanique céleste, par P.S. La Place, membre de l'Institut National de France, et du Bureau des Longitudes. Paris'. *Edinburgh Review* 11 (22): 249–84 (Jan. 1808).

———. 1810. Review: 'A reply to the calumnies of the Edinburgh Review against Oxford; containing an account of studies pursued in that University. Oxford'. *Edinburgh Review* 16 (31): 158–87 (Apr. 1810).

Ponteil, F. 1968. L'éveil des nationalités et le mouvement libéral. *Peuples et Civilisations*, vol. 15. Paris.

Richards, J. 1980. The art and the science of British algebra: a study in the perception of mathematical truth. *Historia Mathematica* 7: 343–65.

Rogers, L. 1981. A survey of factors affecting the teaching of mathematics outside the universities in Britain in the nineteenth century. In *Social history of nineteenth century mathematics*, edited by H. Mehrtens, H. Bos, and I. Schneider. Boston.

Rouse Ball, W.W. 1889. *A history of the study of mathematics in Cambridge*. Cambridge.

Stephen, L. 1890. Hamilton, Sir William. *Dictionary of national biography*. London.

Stewart, D. 1814. *Elements of philosophy of human mind,* vol. 2. London.
Tillyard, A.I. 1913. *A history of university reform from 1800 to the present time, with suggestions towards a complete scheme for the University of Cambridge.* Cambridge.
Wallis, P.J. 1972. British philomaths: mid-eighteenth century and earlier. *Centaurus* 17: 301–10.
Whately, R. 1826. *Elements of logic.* London. Reprinted in *Encyclopædia Metropolitana,* London 1849.
Whewell, W. 1836. *Thoughts on the study of mathematics as part of a liberal education.* Cambridge.
———. 1837. *History of the inductive sciences, from the earliest to the present time.* London.
———. 1840. *The philosophy of the inductive sciences founded upon their history.* London.
———. 1845. *On a liberal education in general, and with particular reference to the leading studies in the University of Cambridge.* Cambridge.
Wilkes, M.V. 1990. Herschel, Peacock, Babbage and the development of the Cambridge curriculum. *Notes Rec. of the Royal Society of London* 44: 205–19.
Winstanley, D.A. 1940. *Early Victorian Cambridge.* Cambridge.
Woodhouse, R. 1801. On the necessary truth of certain conclusions obtained by means of imaginary quantities. *Philosophical Transactions* 91: 89–120.
———. 1802. On the independence of the analytical and geometrical methods of investigation, and on the advantages to be derived from their separation. *Philosophical Transactions* 92: 85–125.
Yeo, R. 1993. *William Whewell, natural knowledge and public debate in early Victorian Britain.* Cambridge.

Université de Paris 8 and REHSEIS (Recherches épistémologiques et historiques sur les sciences exactes et les institutions scientifiques), France

MARIA PANTEKI

THE MATHEMATICAL BACKGROUND OF GEORGE BOOLE'S *MATHEMATICAL ANALYSIS OF LOGIC* (1847)

This paper examines the mathematical stimuli that induced George Boole to conceive of a general calculus of symbols, and eventually of his first algebra of logic, as in his *Mathematical Analysis of Logic* (1847). We focus our attention particularly on Boole's first mathematical masterpiece 'On a General Method in Analysis' (1844), a work grounded on the laws of combination of non-commutative symbols, which apparently convinced Boole of the immense power afforded by symbolical methods, a power largely due to their main property of not depending upon conditions of their interpretation. Our aim is twofold. Firstly we argue that Boole's discovery of a *general method* was motivated by R.L. Ellis's study of the differential equation involved in the theory of the earth's shape. Secondly, we pick up indicative instances from his first work on logic which manifest the implicit or explicit import that his *general method* had upon the genesis of his general method in logic.

1. INTRODUCTION

1.1. Scope of the Paper

George Boole (1815–1864) was awarded the Royal Medal of the Royal Society of London in 1844 for his memoir 'On a General Method in Analysis' (Boole 1844), hereafter cited as GMA.[1] Boole's new general symbolical method for the solution of differential equations of importance to physics not only surpassed in generality and applicability the methods of his predecessors, but also challenged an amazing overgrowth of symbolical methods in mid-19th-century Britain. Today, however, Boole's name is mainly known in connection with the foundation of algebraic logic, as in the little-read *Mathematical Analysis of Logic* (Boole 1847a) and his mature work *Laws of Thought* (Boole 1854)—hereafter MAL and LT respectively. With a view to unveiling some significant links between MAL and GMA we study GMA and its background, claiming that this first mathematical masterpiece of Boole's would never have been composed had there not been the motivation of seeking improvements in the theory of the earth, and that thus MAL might not have been created either.

J. Gasser (ed.), A Boole Anthology, 167–212.

The import of the earth-figure equation as a par-excellence motivation for Boole's GMA was raised by Philip Kelland in 1858. On the occasion of the award of the Keith Medal to Boole for his memoir on probabilities, Kelland (1858, 85) referred to GMA as follows:

> Many problems of no very great apparent complication had baffled the ingenuity of mathematicians. Solutions were, it is true, obtained, but the processes were so indirect and unsatisfactory, that they were something like excrescences on the smooth face of science. Of this class of problems is an equation which occurs in the theory of the figure of the earth. Mr Airy, in his 'Tracts', gives simply the result, without the slightest indication of a process. Mr Gaskin and Mr Leslie Ellis had attacked this individual problem with partial success. But Mr Boole's 'New Method' not only set the logical question of dealing with separation of symbols in a clear light, but completely effected the solution of all that class of problems, of which this was a particular example.

Despite its diverse consequences, Boole's 'New Method' was seldom studied by historians until recently. Boole's GMA was briefly acknowledged in Koppelman (1971) but with the role of the earth-figure equation completely omitted. The exposition in Laita (1977), despite its promising title, 'The Influence of Boole's Search for a Universal Method in Analysis on the Creation of His Logic', is even more disappointing as it omits nearly all the mathematical elements. These inadequate studies of Boole's memoir did, however, provide the main motivation for an extensive study of Boole's symbolical methods and their consequences in Panteki's Dissertation, 'Relationships between Algebra, Differential Equations and Logic in England, 1800–1860' (Panteki 1992), which our survey draws upon.

1.2. Plan of the Paper and Its Limits

The paper is divided into three sections, covering respectively the stimuli that led to Boole s GMA (§2), the paper GMA as such (§3), and finally the connections between GMA and MAL (§4). A significant source for our study in §4 is the Grattan-Guinness-Bornet edition of Boole's *Selected Manuscripts on Logic and Its Philosophy*, cited as Boole (1997). Our survey ends in §5 with a brief over-view of Boole's aspirations, according to which logic forms part of universal reasoning, whose laws are best revealed by symbolical means. We take for granted all recent historical studies of Boole's contributions to logic; Boole (1997) is referred to for biographical and bibliographical information. Moreover, as GMA is lengthy and complicated, we omit detailed demonstrations, and refer the reader instead to original sources as well as to Panteki (1992).

2. SYMBOLICAL METHODS IN ENGLAND BEFORE 1844; THE EARTH-FIGURE EQUATION

2.1. The French Influence: Cambridge, 1800–1820

The method of separation of symbols of operation, such as $D = d/dx$ or Δ, from symbols of quantity, such as $u(x)$, was introduced by L.F. Arbogast in 1800. In writing Lagrange's theorem,

$$\Delta^n u = \left(\exp\left(h\frac{du}{dx}\right) - 1\right)^n \tag{2.1.1}$$

in the symbolic form

$$\Delta^n u = \left(\exp\left(h\frac{d}{dx}\right) - 1\right)^n u \tag{2.1.2}$$

he separated the symbol of operation d/dx from du/dx, a technique that gave rise to operator methods still in use. After Arbogast, B. Brisson and J.F. Français deployed this method to differential equations in 1807 and 1813 respectively, but no systematic study of the laws governing operators was given until after F.J. Servois introduced the terms 'commutative' and 'distributive' as properties of operations in 1814.[2]

At Cambridge, Lagrangian methods were diffused and developed in the 1810s by C. Babbage and F.J. Herschel who, together with G. Peacock, founded the Analytical Society in 1812–1813 (Enros 1983). After translating Lacroix (1816), they produced a sequel of *Examples* (Babbage et al. 1820), which incorporated Herschel's elaboration of D-operator methods and Babbage's study of functional equations. Relying upon induction and analogy, Babbage and Herschel showed no concern for the foundation of their methods. Interestingly, Herschel applied separation of symbols to the general linear differential equation with constant coefficients in 1814, but his pioneering effort became known too late to have any impact on the revival of the calculus of operations in the 1830s. However, his calculus of finite differences, treated symbolically, would endure the test of time.[3]

Analytical methods were founded on Lagrange's assumption that any function afforded an expansion in Taylor series. Algebraic in character, by not relying on geometrical procedures, limits or physical notions, these methods offered a vast capacity for uniformity and generalization, unlike the Newtonian 'synthetic' or 'mixed' mathematics which prevailed at Cambridge at that time (Becher 1995). R. Woodhouse had seen the need for a change in the curriculum in the early 1800s, but his plans failed (Becher 1980). Following on his lines, Babbage and Herschel were more successful, but their algebras were too abstract to attract attention, and their project to establish a reform of the curriculum had been only partially fulfilled by the time W. Whewell entered the scene in the late 1810s.

2.2. Whewell's Influence and the Earth-figure Equation, 1820–1839

After Peacock's reform in 1817, Newton's fluxional notation was once and for all replaced by the differential notation, while the Tripos papers were enriched by the Continental calculus (Becher 1995). When Whewell took over, his main concern was with the updating of mechanics, not mathematics. Realizing that a minimum of analytical methods were indispensable in mechanics, he nevertheless placed all emphasis on mixed mathematics, believing firmly that geometrical methods were the best vehicle for the training of the student's mind. Thus, contrary to Peacock's move towards pure mathematics, Whewell enriched the Tripos with more questions in mixed mathematics, encouraging only the best students to study the textbooks furnished by the Analytical Society.[4]

In the early 1820s, Whewell was convinced that advanced students should be led gradually from Newtonian mechanics to P.S. Laplace's *Mécanique céleste* (1799–1825). However, instead of supporting Woodhouse's *Physical Astronomy* (1818), which drew on Laplace's work, Whewell set off to produce 'elementary' textbooks, in order to initiate the student in the true spirit of Newtonian mechanics. Whewell decided to introduce the theory of the earth's shape into the curriculum, and proposed to his disciple B. Airy—wrangler in 1823—to prepare a textbook on the subject. The outcome was Airy's *Tracts* (1826), a blend of analytical and geometrical procedures, lacking explanations of the physical assumptions as well as indications of how he arrived at mathematical results, as for instance in the case of the earth-figure equation.

The first reactions to Airy's *Tracts* took place in the mid-1830s. By that time, the curriculum had been expanded so as to include elements from Laplace's potential theory (Becher 1980). J.H. Pratt, wrangler in 1833, published his *Mechanical Philosophy* (1836), simplifying and clarifying Laplace's theory and shedding light on Airy's obscure treatment of the earth's shape. Nonetheless, the question of the earth-figure equation remained open, first tackled by a frustrated student, A.J. Ellis in 1836, and subsequently by T. Gaskin in 1839.

The differential equation involving the earth's ellipticity, hereafter cited as EFE, became standard from 1826 in the form

$$\frac{d^2y}{dx^2} + n^2y = \frac{6y}{x^2}, \tag{2.2.1}$$

while its solution became known in the form

$$y = c_1\left[\left(1-\frac{3}{n^2x^2}\right)\sin(nx+c_2)+\frac{3}{nx}\cos(nx+c_2)\right]. \tag{2.2.2}$$

It was in the fifth volume of Laplace's *Mécanique* that equivalents of (2.2.1–2.2.2) first officially appeared. While (2.2.1) was peculiar to the theory of the earth, the generalized form

$$\frac{d^2y}{dx^2}+n^2y=\frac{m(m+1)}{x^2} \tag{2.2.3}$$

was initially connected with the theory of heat.[5] Solutions to many problems of physics would turn out to depend upon (2.2.3), being particular cases of the class

$$\frac{d^2y}{dx^2}+n^2y+P=0 \tag{2.2.4}$$

whose treatment by Euler in 1739 opened the way for the analytization of Newton's astronomy.[6]

Airy opened his *Tracts* with equation (2.2.4), confining its study only to cases where P is a function of x and constants. Indicating that in this case the general solution is derived after multiplication of (2.2.4) with $\cos nx$ and integration by parts, he deduced from it the solution of equation (2.2.4) for $P = 0$,

$$\frac{d^2y}{dx^2}+n^2y=0, \tag{2.2.5}$$

in two arbitrary constants, as

$$y = a\cos nx + b\sin nx \tag{2.2.6}$$

(Airy 1826, 1–3). Considering the earth as a heterogeneous fluid body, Airy followed Clairaut's method and was eventually led to the EFE (2.2.1) of which he claimed, without explanation, that its solution is (2.2.2) (1826, 107). The first solution of (2.2.1) was given by A.J. Ellis in his own handwriting on the back of his copy of Airy's *Tracts* as follows.[7]

Dispensing of the factor n^2 by a simple transform, Ellis considered (2.2.1) initially as devoid of its last term, as its solution is known to be (2.2.6) in two arbitrary constants a, b, with $n = 1$. Deploying the method of variation of constants, by differentiation and substitution of (2.2.6) in (2.2.1), he arrived at the condition

$$A = B = 0, \text{ where } A = \frac{d^2b}{dx^2}-2\frac{da}{dx}-6\frac{b}{x^2}, \tag{2.2.7}$$

B being of the same form as A, with a and b interchanged. To determine the functions a and b he assumed them, in analogy with the third term in (2.2.1), to be of the polynomial form $k+q/x+t/x^2$. Combining this assumption with (2.2.7), and by substitution of a and b in (2.2.6), he arrived at the known solution, (2.2.2).

While Ellis's procedure, apparently the simplest one possible, was never followed by physicists or mathematicians, Gaskin's study of (2.2.3) had an immediate and enduring influence on the growth of symbolical methods from 1839 onwards. Gaskin was the second moderator, after Peacock, to diverge from the standard curriculum, isolating the EFE from its physical context in a Tripos question in 1839. Gaskin presented the students with the solution of (2.2.3) in the symbolical form (2.2.14), and demanded that (2.2.2) be deduced from it. In what follows below we draw on Gaskin's procedure, as given in J. Hymers's *Differential Equations* (1839).[8]

Assuming (2.2.3), Gaskin let $y=u/x^m$, thus reducing it to the form

$$\frac{d^2u}{dx^2}+n^2u=\frac{2m}{x}\cdot\frac{du}{dx} \tag{2.2.8}$$

written more conveniently, by separating the symbol $D=d/dx$, as

$$x(D^2+n^2)u=2mDu. \tag{2.2.9}$$

Differentiating (2.2.9) twice we have:

$$D^3u+D^3u+xD^4u+n^2Du+n^2Du+xn^2D^2u=2mD^3u. \tag{2.2.10}$$

Multiplying (2.2.9) with n^2, and adding to it (2.2.10), we have

$$x(D^2+n^2)^2u=2(m-1)D(D^2+n^2)u. \tag{2.2.11}$$

Repeating the same process m times, we arrive at equation

$$(D^2+n^2)^{m+1}u=0. \tag{2.2.12}$$

Its solution is known to be

$$u=(a+a_1x+\ldots+a_mx^m)\cos nx+(b+b_1x+\ldots+b_mx^m)\sin nx. \tag{2.2.13}$$

For $m=2$, by differentiation of u and substitution in (2.2.9) we arrive at the solution of (2.2.8), and considering that $u=yx^2$, the solution (2.2.2) is obtained.

Without any explanation, Hymers claimed that the symbolical form

$$y = Cx^{-m}\left(\frac{d}{dr}\right)^{m}\left(\frac{\cos\left(x\sqrt{r}+a\right)}{\sqrt{r}}\right), \quad r = n^2, \tag{2.2.14}$$

where C, a, two arbitrary constants, and r to be put equal to n^2 after the performance of the differentiation, satisfies the general EFE (2.2.3) for any integer m (Hymers 1839, 83–5). Unfortunately we lack any clue as to how Gaskin arrived at this form, incorporated in the Tripos question. What we know is that Gaskin's result stimulated R.L. Ellis to deal with a more general class of differential equations in 1841 (§2.3), challenging British mathematicians to furnish a variety of symbolical methods in order to attain other elegant, symbolic forms, most of them incorporated in Glaisher (1881).[9]

2.3. Boole's Contemporaries, 1836–1843

In his paper 'On a General Method in Analysis' (1844, 225), Boole acknowledged D.F. Gregory for his foundational study of the calculus of operations (COO), his applications in the papers published in the *Cambridge Mathematical Journal*, and his book of *Examples* (1841). The names of R. Murphy and A. De Morgan were also mentioned, rather vaguely though, 'in connection with the history of this branch of analysis' (Boole 1844, 225). Lastly, R.L. Ellis was credited for his study of the EFE in 1841, and B. Brownin for his study of finite difference equations in 1843 (Boole 1844, 252, 278). What exactly was Boole's debt to the above-mentioned mathematicians?

A wrangler in 1829, Murphy gave a serious study of French analytics and a uniquely general development of Laplace's coefficients in his *Electricity* in 1833. Murphy also contributed to the 'direct' calculus of definite integrals, as established by Fourier, Poisson and Cauchy after Laplace, perceiving the necessity of an 'inverse' calculus as well, which 'would conduct us from the observed phenomenon, to the laws of the elementary actions' (Murphy 1833, 353–4). Murphy realized the inadequacy of the COO for dealing rigorously with inverse and distributive functions and dedicated his lengthy memoir 'On the Theory of Analytical Operations' (1837) to this subject.[10]

Murphy's apprehensions in connection with the shaky foundations of the calculi of functions and operations were partly anticipated in De Morgan's treatise 'Calculus of Functions' (1836). With a view to systematizing the forgotten work of Babbage and Herschel, De Morgan—wrangler in 1827—studied inverse and commutative operations, anticipating Murphy's interest in the transmutation of non-commutative operations.[11] Murphy's work was acknowledged in De Morgan's article 'Operation' (1840) and a detailed study of the close link between the EFE and the Riccati equation was furnished in his *Calculus* (1842). De Morgan also commented on the solutions of the EFE given by Gaskin and R.L. Ellis (see Glaisher 1881).

There can be no doubt that Gregory (1841) in conjunction with De Morgan (1842) formed the perfect background for Boole (1844), by offering access to all contributions then current in the field of the COO. In connection with partial differential equations, Gregory developed the ideas presented by S.S. Greatheed in 1837, the first Cambridge mathematician to make a free use of separation of symbols in applications. The only figure absent from the above treatises was Brownin, a Cambridge outsider, the most prolific of Boole's followers after 1844.[12] In what follows we display only those methods and formulae which had a direct influence on Boole's GMA.

Assuming the general linear differential equation (LDE) with constant coefficients,

$$\frac{d^n y}{dx^n} + A\frac{d^{n-1}y}{dx^{n-1}} + \ldots + R\frac{dy}{dx} + Sy = f(x) \tag{2.3.1}$$

where $f(x)$ is a known function, Gregory wrote it in the symbolic form

$$F(D)y = f(x), \tag{2.3.2}$$

letting D stand for d/dx and $F(D) = D^n + AD^{n-1} + \ldots + RD + S$. Operating on both sides with $(F(D))^{-1}$, he claimed that a 'complementary' function should be added to the result which will vanish when the operation $F(D)$ is performed on it; the solution is thus given as

$$y = (F(D))^{-1} f(x) + (F(D))^{-1} \cdot 0. \tag{2.3.3}$$

Gregory considered the operation $F(D)$ to be made up of the combination of n binomial operations of the form $(D - a)$ or $(D + a)$, and he handled these operations purely algebraically, upon the following grounds.[13]

Many theorems of algebra, held Gregory, admit of extended applications, since they depend only on the laws of combination to which the symbols are subject, and are therefore true for all symbols, which are subject to the same laws of combination. For instance, in the binomial theorem, $(a + b)^n$, it suffices for a and b to obey the rules of commutativity, distributivity and the index law in multiplication. Similarly, it is legitimate to consider $(D - a)^n$ as an operation, since the following properties evidently hold true:

$$\left.\begin{array}{lll} \text{(i)} \quad aDx = Dax \quad \text{and} & \text{(i)}' \quad \left(\frac{d}{dx}\right)\left(\frac{d}{dy}\right)(z) = \left(\frac{d}{dy}\right)\left(\frac{d}{dx}\right)(z) \\ \text{(ii)} \quad D(x+y) = Dx + Dy & \\ \text{(iii)} \quad D^n D^m x = D^{n+m} x, & \end{array}\right\} \tag{2.3.4}$$

as D or d/dx commute not only with constants, but also with other symbols of partial differentiation (Gregory 1839a, 24–5).

Claiming that for any functions u and v of x

$$\left(\frac{d}{dx}\right)(uv) = u\frac{dv}{dx} + v\frac{du}{dx}, \tag{2.3.5}$$

Gregory let d'/dx be an operation acting only on v, and wrote (2.3.5) as

$$\frac{d}{dx}(uv) = \left(\frac{d'}{dx} + \frac{d}{dx}\right)uv. \tag{2.3.6}$$

Since these operations are distributive, and commute with each other, we may take the nth differential on both sides of (2.3.6), expand the compound in the right-hand side according to the binomial theorem, and thus arrive at Leibniz's theorem,

$$\left(\frac{d}{dx}\right)^n (uv) = u\frac{d^n v}{dx^n} + n\frac{du}{dx}\frac{d^{n-1}v}{dx^{n-1}} + \frac{n(n-1)}{1\cdot 2}\frac{d^2 u}{dx^2}\frac{d^{n-2}v}{dx^{n-2}} + \dots \tag{2.3.7}$$

Supposing now $v = e^{ax}$, we have $dv/dx = av$, or $d'/dx = a$, and hence the formula

$$(D+a)^n u = e^{-ax}(D)^n e^{ax} u, \tag{2.3.8}$$

which holds, after a change of signs, for $(D - a)$ too, as well as for negative powers, with D^{-1} standing for integration (Gregory 1839b, 108–12).

This theorem, known to Cauchy and Murphy, had useful consequences for the COO. Indeed, if we let $D = d/d\theta$, and if $F(D)$ be any integral polynomial of powers of D, we can deduce from the basic 'translation' theorem (2.3.8) the following formulae:[14]

$$F(D)e^{m\theta} = F(m)e^{m\theta}, \quad F(D)e^{m\theta}u = e^{m\theta}F(D+m)u \tag{2.3.9}$$

and

$$D(D-1)\dots(D-n+1)u = x^n\left(\frac{d}{dx}\right)^n u \tag{2.3.10}$$

Formula (2.3.8) was deployed by Cauchy and consequently by Gregory for the solution of (2.3.2). If $r_1, \dots, r_n$ are the distinct roots of $F(z) = 0$, then $F(D)$ is resolved in factors of the form $(D - r)$, and successive applications of their inverse to the right-hand side of (2.3.2) lead to what became known as the 'Heaviside expansion theorem':

$$y=\sum_{k=1}^{n}\frac{f(x)}{F'(r_k)\cdot(D-r_k)}. \tag{2.3.11}$$

Gregory overlooked the fact that (2.3.11) fails for multiple roots. This was amended in Boole (1841). But Gregory seldom used this theorem in applications.[15] His favourite method was to develop the binomials in powers of D as follows.

Assume the equation in the form $D^2y - 4Dy + 4y = x^2$, where $F(D) = (D-2)^2$, and the solution according to (2.3.3) is $y = (D-2)^{-2}x^2 + (D-2)^{-2}\cdot 0$. To deal with the first term, he developed $(D-2)^{-2}$ in ascending powers of D, as $(2^{-2} + 2\cdot 2^{-3}D + 3\cdot 2^{-4}D^2 + ...)$ and performing these operations on x^2, the result was $x^2/2^2 + 4x/2^3 + 6/2^4$. However, for the second term he wrote: $(D-2)^{-2}\cdot 0 = D^{-2}(1-2D^{-1})^{-2}\cdot 0 = (1-2D^{-1})^{-2}(cx + c')$. He next developed $(1-2D^{-1})^{-2} = 1 + 2\cdot 2D^{-1} + 3\cdot 2^2D^{-2} + ...$, and performed these multiple integrations on cx and c' separately, arriving at the final result $(C' + (c + 2c')x)e^{2x}$, by letting the infinite series $1 + 2x + 2^2x^2/1\cdot 2 + ... = e^{2x}$ (Gregory 1839a, 22).

While Gregory avoided extending his method beyond equations with constant coefficients, Boole did so, motivated by R.L. Ellis's two papers 'On the Integration of Certain Differential Equations' (1841a, b). Claiming that the lacuna left by Laplace 'is not supplied in the works on the subject generally made use of in Cambridge' (Ellis 1841a, 169), Ellis opened his first paper with equation (2.2.1), or

$$\frac{d^2y}{dx^2}+n^2y=\frac{6y}{x^2} \tag{2.3.12}$$

'both because from its physical application it has an interest for some who care but little for pure analysis, and because it will exemplify the general method' (ibid.).

Assuming the solution of (2.3.12) to be of the form $y=\Sigma a_m x^m$, by differentiation and substitution he obtained the so-called 'scale' of the equation,[16] that is the relation between the coefficients of the series y, as

$$(m-3)(m+2)a_m+n^2a_{m-2}=0 \tag{2.3.13}$$

a finite difference equation. His basic idea was to effect a suitable transformation of the scale, so that the reduced one would be the scale of a known equation, namely of (2.2.5). To this end, Ellis got rid of the factor $(m-3)$ in (2.3.13) by assuming a new series $z=\Sigma b_m x^m$, which depended upon the former series by means of the relation

$$(m+2)a_m=(m-1)b_m. \tag{2.3.14}$$

By substitution of the new value of a_m in (2.3.13) he obtained the relation

$$m(m-1)b_m+n^2b_{m-2}=0. \tag{2.3.15}$$

It was readily seen that (2.3.15) is the scale of the equation (2.2.5), or

$$\frac{d^2 z}{dx^2} + n^2 z = 0, \tag{2.3.16}$$

whose solution is (2.2.6). Given z, the value of y was deduced from (2.3.14) and (2.3.15) by summation, as $y = z + 3/n^2 \cdot (z/x)'$, which after the calculations leads to the known form (2.2.2) (Ellis 1841a, 169–70).

This procedure was applied p times to the general EFE

$$\frac{d^2 y}{dx^2} + n^2 y = \frac{p(p-1)y}{x^2} \tag{2.3.17}$$

and lastly to the general class

$$\frac{d^m y}{dx^m} + n^m y = p(p-1) \cdot \frac{1}{x^2} \cdot \frac{d^{m-2} y}{dx^{m-2}}, \tag{2.3.18}$$

showing that the latter class depended on the solution of

$$\frac{d^m y}{dx^m} + n^m y = 0, \tag{2.3.19}$$

(Ellis 1841a, 170–1, 174–5).

The above process was slightly altered in his second paper; the results, however, were exhibited 'in a very different form' (Ellis 1841b, 193). While the former paper was motivated by Gaskin's consideration of the general EFE (2.2.3) or (2.3.17), the latter was motivated by his symbolical solution of that equation, and his transform of the dependent variable. Indeed, now Ellis studied equations of the class

$$\frac{d^p u}{dx^p} + n^2 u = \frac{pm}{x} \cdot \frac{d^{p-1} u}{dx^{p-1}} \tag{2.3.20}$$

which is equivalent to the class (2.3.18). For instance, for $p = 2$, (2.3.20) yields

$$\frac{d^2 u}{dx^2} + n^2 u = \frac{2m}{x} \cdot \frac{du}{dx}, \tag{2.3.21}$$

which is the EFE in Gaskin's form (2.2.8), after effecting the transform $u = yx^m$ to (2.2.3) or (2.3.17) for $p = m$. Ellis first tackled the general case (2.3.20) and then for $p = 2$ he obtained Gaskin's result (2.2.14), or

$$y = C\left(\frac{d}{dr}\right)^m \frac{\sin\left(x\sqrt{r} + a\right)}{\sqrt{r}}, \tag{2.3.22}$$

(Ellis 1841b, 193–5). At the end of his paper (1841b, 201), Ellis furnished another symbolic solution, namely

$$u = Cx^{2m}\left(\frac{d}{dx}\cdot\frac{1}{x}\right)^m \sin(nx + a). \tag{2.3.23}$$

The equivalence between the two forms was proved by De Morgan (1842, 702–3).

What Ellis managed to show was that the series-method, considered clumsy by the advocates of symbolical methods, offered generalization, in contrast with its common application by Lagrange and Laplace, and the ability to lead to elegant, finite forms. Interestingly, Ellis anticipated the impact of his work, claiming that (1841b, 201):

> Many applications and modifications of the method we have employed, will readily present themselves but the subject is not of sufficient importance to deserve a fuller discussion. It is not difficult to multiply artifices, by means of which particular equations may be solved, but the results will, generally speaking, be of little value.

3. BOOLE'S 'GENERAL METHOD IN ANALYSIS', 1844

3.1 The Scope and Limits of Boole's Paper

The scope of Boole's paper 'On a General Method in Analysis' (1844)—GMA—was summarized by Philip Kelland as follows:[17]

> The object which the author proposes in this paper is to illustrate the importance of conducting processes of reduction through the medium of separation of symbols by the assumption of an exponential form for the principal quantity operated on. The method, therefore, consists in the transformation of given forms, from x to θ, by means of the relation $x = e^{\theta}$, and operating on the results by certain refined processes. Although the facilities which the exponential form affords, under certain circumstances, have long been known to analysts, in no investigation does the importance of this form appear in the manner in which it is exhibited in this paper. Here the author has shown a method which classifies and comprehends operations and which has the advantage (in some cases at least) of showing their limits.
>
> The author applies his method to the solution of differential equations and of equations of differences. In the finite solution of differential equations, the beauty and utility of the method are strikingly exhibited. Certain forms of equations, the solutions of which have been effected by transformations or by a happy assumption, are made, by the author's method, to exhibit their solutions in dependence on a law. Both the processes by which the transformations are effected and the results obtained by them are of the highest interest and importance; and this part of the paper is so valuable that, alone, it would entitle the author to honorary distinction.

By 'this part of the paper', Kelland alluded to Boole's finite solution of a certain class of differential equations of physical import, including the EFE.[18] In fact, covering 13 out of the 57 pages of the whole paper, this specific part not only reflected the 'beauty and utility' of Boole's GMA; it actually contained the very germ of the whole method. Let us examine Boole's introductory statements as to the principal novelties of his work, and his aspirations expressed in the end of the paper.

Acknowledging his debt to Gregory, Boole displayed the three basic laws (2.3.4) that govern the combination of the symbols of operation D and Δ (Boole 1844, 225), further taking for granted the theory of LDE with constant coefficients by referring to theorem (2.3.11). Boole noted that by being restricted to equations involving commutative operations only, the 'calculus of operations has tended rather to simplify the processes of analysis than to extend its power' (1844, 226), and claimed that:

> The object of this paper is to develope [sic] a method in analysis, which, while it operates with symbols apart from their subjects, and may thus be considered as a branch of the calculus of operations, is nevertheless free from the restrictions to which we have alluded.

Boole held that he had established 'a perfectly general method' for the solution of LDE total and partial in series, and for the calculus of generating functions, by basing it on 'reductions, transformations and developments', the expression of which 'is independent of the forms' of the variable coefficients of the equations studied (Boole 1844, 226–7). Moreover, this analysis 'affords facilities, which are believed to be peculiar to itself, for the finite integration of linear equations, and for the classification of integrable forms' (1844, 227).

He concluded by commenting on the lacuna left by Euler and Laplace in connection with the solution of differential equations by the series method and with the theory of generating functions. Many of the difficulties which characterized the processes of these analysts would be found resolved in Boole's paper. In particular, Boole was happy to announce that 'The law of the series being known, or the equation of differences given, the differential equation is known by inspection. The rule is absolutely general, whatever may be the forms of the coefficients' (Boole 1844, 228). Secondly, 'The form under which the differential equation is presented offers great facilities for its integration. Those facilities are chiefly owing to the circumstance, that the form of the equation permits us [...] to effect the requisite transformations by general theorems' (ibid.).

Boole was very proud of his method, but he could not include the study and presentation of all its diverse applications in one paper. As he wrote in the 'Postscript' (Boole 1844, 282):

> Fearful of extending this paper beyond its due limits, I have abstained from introducing any researches not essential to the development of that general method in analysis which it was proposed to exhibit. It may however be remarked that the principles on which the method is founded have a much wider range. They may be applied to the solution of functional equations, to the theory of expansions, and, to a certain extent, to the integration of non-linear differential equations. The position which I am most anxious to establish is, that any great advance in the higher analysis must be sought for by an increased attention to the laws of the combinations of symbols. The

value of this principle can scarcely be overrated; and I only regret that in the absence of books, and under circumstances unfavourable for mathematical investigation, I have not been able to do that justice to it in this essay which its importance demands.

Further applications of his method in the theory of expansions and the finite solution of partial differential equations, such as the Laplace equation, would be given between 1845 and 1847 (§3.4). Of great value to our research are his textbooks on *Finite Differences* (1860) and *Differential Equations* (1859)—especially the latter, where the presentation of the GMA, apparently for educational purposes, followed a different order from Boole (1844), much closer to the order of discovery. A striking omission in both paper and textbook, though, is a detailed, rigorous study of inverse operations, as already given in Murphy (1837).[19]

3.2 Structure and Content of Boole's Paper

For convenience we divide Boole (1844) into eight distinct stages, strictly following his order of presentation and in most cases borrowing his section titles.

Part 1: 'Preliminary Theorems' (pp. 228–32)

Boole furnished the 'General Theorem of Development' (GTD) which determined the expansion of the function $f(\pi+\rho)$ in ascending powers of ρ, where the distributive operations π and ρ combined according to the law $\rho f(\pi)u = f(\phi(\pi))\rho u$. By letting $\phi(\pi) = \pi + \Delta\pi$, Boole deduced from the general formula of expansion (3.4.5) certain novel symbolical formulae, Taylor's theorem, and lastly the known formulae (2.3.8–2.3.10). He admitted that the latter are known relations, but claimed that 'With a view to the maintenance of an unbroken analogy, it has, however, been thought better to deduce them from the properties of the more general system in π and ρ, than to assume them as already proved' (p. 232).

Part 2: 'Fundamental Theorem of Development' (pp. 232–3)

Letting $x = e^{\theta}$, and $D = d/d\theta$, Boole first cast the LDE with polynomial coefficients in powers of x to the convenient symbolical form (3.3.10). With the aid of this peculiar symbolical representation, the 'Fundamental Theorem of Development' (FTD) (3.3.12) followed readily, being a general relation connecting any LDE or system of LDE with a corresponding equation or system of difference equations.

Part 3: 'On the Solution of LDE by Series' (pp. 234–44)

Boole provided a systematic theory based on the FTD for the solution of LDE in series. He was proud to have improved on Euler's method, which failed in certain cases.

Part 4: 'On the Integration of LDE in Finite Terms' (pp. 244–57)

In this section of the paper, Boole focused specially on a full study of the binomial equation $u + \phi(D)e^{r\theta}u = V$, or (3.3.16), furnishing two propositions for its convenient transformation to cases where the solution was known. In particular, Boole studied the symbolical solution of those equations whose solution depends on that of $d^n v/dx^n \pm q^n v = X$, where both X and v are functions of x. Among his illustrations were both forms of the EFE, corresponding to the above class for $n = 2$. Boole observed that 'the first conceptions of the theory developed' in this section were 'aided' by the study of Ellis's researches (p. 252).

Part 5: 'Some Illustrations of a More General Method' (pp. 257–61)

Continuing his investigation as in Part 4, Boole used some new consequences of the GTD to establish 'a more general method of resolution'. The result was a far too abstract, general and complex theory to lead to any useful applications. However, as he was happy to ascertain that this method 'enables us to integrate almost every class' of LDE that admits of integration in finite terms, Boole was apparently convinced that he was near to a genuinely general method, where 'almost every class' would soon turn into 'every single class'.

Part 6: 'Theory of Series and of Generating Functions' (pp. 261–70)

This portion of Boole's theory exemplifies his virtuosity in exploiting instances from the theory of transformations introduced in Part 4, as well as the potential of his method for applications. Conversely to what was involved in Part 3 of the paper, here, given a series, or a relation between the coefficients u_m, the problem of summation of the series under study, or of the determination of the generating function of u_m, was reduced to that of the solution of the corresponding LDE via the FTD.

Part 7: 'Evaluation of Certain Definite Integrals' (pp. 270–5)

Boole had already studied the theory of definite integrals in papers published in 1843. His novel approach, though, is more sophisticated, drawing on the theory introduced in Parts 2, 3, 4 and 6 of his paper. Many of the integrals under study had been collected in Gregory's *Examples.*

Part 8: 'On the Solution of Difference Equations in Series' (pp. 276–82)

This was the last section of Boole's paper and the only part, besides Part 5, to draw on the GTD. As with Part 5, the result is too general to summarize in a few words. The germs of the theory of difference equations would be fully developed in chapter 13 of Boole (1860). Here Boole expressed his immense debt to Gregory, along with mentioning Brownin's relevant work (1860, 278–9).

Consulting the sequels to his GMA until 1847, his textbook (1859), his letters to De Morgan prior to 1844, and most importantly, comparing Boole's techniques with those furnished by Murphy, Gregory and Ellis, it follows that the most plausible order of invention is the following: the FTD (Part 2), the study of binomial equations (Part 4—the EFE), Parts 3, 6, 7 and lastly the GTD (Part 1) with the accompanying Parts 5 and 8. This order is largely tentative. However, wherever Boole mentioned the basic ingredients of his method—focusing on LDE—he isolated the FTD and the theory exposed in Part 4, the most significant part of the paper, according to Kelland (§3.1). Based on the above-mentioned sources, we present our sketch for the genesis of Boole's GMA, illustrating the theory with the example of the EFE, an example which apparently motivated the theory itself.[20]

3.3. On the Genesis of Boole's Method

In the early 1840s, Boole was aware of Gregory's study of LDE with constant coefficients, and of Murphy's inquiry into the transmutation of non-commutative operations (§2.3). As he became acquainted with Gaskin's and Ellis's studies of the EFE, Boole felt challenged to extend the realms of the calculus of operations, so as to make it apply to a wide class of LDE with variable coefficients, and to establish a method free from the restriction of commutativity. We contend that the main stimuli and ideas stemmed from Ellis (1841a), and wish to record once again Ellis's basic technique.

Assume the EFE in the standard form (2.2.1), or

$$\frac{d^2y}{dx^2} + q^2y = \frac{6y}{x^2}, \tag{3.3.1}$$

with its corresponding scale, when $u = \sum a_m x^m$,

$$(m-3)(m+2)a_m + q^2 a_{m-2} = 0. \tag{3.3.2}$$

Choosing a new series $v = \sum b_m x^m$, connected with u by means of the relation,

$$(m+2)a_m = (m-1)b_m, \tag{3.3.3}$$

Ellis arrived at the new scale

$$m(m-1)b_m + q^2 b_{m-2} = 0, \tag{3.3.4}$$

which is the scale of equation (2.2.5), that is of

$$\frac{d^2v}{dx^2}+q^2v=0, \tag{3.3.5}$$

whose solution is known to be $v=a\cos(qx+b)$ (§2.2–3).

We recall that this trick was not peculiar to the simple case of (3.3.1), but could be given successive applications to a wide class of differential equations, such as (2.3.18). An obvious question for Boole to consider would have been the following: could Ellis's first attempt to establish a general method for the finite solution of a wide class of LDE with variable coefficients yield any insight towards the establishment of a general symbolical method serving the same purpose and perhaps prone to further generality in applications?

With his background in symbolical methods and formulae, it was fairly easy for Boole to write (3.3.1) as $x^2d^2u/dx^2+q^2x^2u=6u$ and cast it to a convenient symbolic form. Based on the known formula (2.3.10), or

$$D(D-1)(D-2)\ldots(D-n+1)u=x^nd^n/dx^nu, \tag{3.3.6}$$

he could write $x^2d^2/dx^2u = D(D-1)u$, where $D = d/d\theta$. Further letting $x = e^\theta$, the EFE would attain the form $D(D-1)u-6u+q^2e^{2\theta}u=0$, or

$$(D-3)(D+2)u+q^2e^{2\theta}u=0, \tag{3.3.7}$$

an equation strictly analogous in form to the scale of the EFE (3.3.2). In analogy with Ellis's technique, Boole sought a way to reduce (3.3.7) to

$$D(D-1)v+q^2e^{2\theta}v=0, \tag{3.3.8}$$

which is the symbolical form of (3.3.5).

Upon these considerations the FTD would follow readily. Drawing on Boole (1844), assume any LDE with polynomial coefficients, such as

$$(a+bx+cx^2+\ldots)\frac{d^nu}{dx^n}+(a'+b'x+c'x^2+\ldots)\frac{d^{n-1}u}{dx^{n-1}}+\ldots=X \tag{3.3.9}$$

where $X = X(x)$. Multiplying the equation on both sides with x^n, all factors of the form x^nd^n/dx^n, for all values of n, can be replaced by means of (3.3.6) in terms of $D = d/d\theta$, putting $x = e^\theta$. For instance, the first term of (3.3.9), $(a + bx +\ldots)x^nd^n/dx^n$, will become $aD(D-1)\ldots(D-n+1)u + b(D-1)\ldots(D-n)e^\theta u +\ldots$. The aggregate of such results will produce an equation of the form

$$f_0(D)u+f_1(D)e^\theta u+f_2(D)e^{2\theta}u+\ldots=U, \tag{3.3.10}$$

where $U = U(e^\theta)$ (Boole 1844, 232–3).

Assuming further that $u = \sum u_m e^{m\theta}$, and observing that $f_0(D)e^{m\theta} = f_0(m)e^{m\theta}$, etc., according to (2.3.9) or

$$f(D)e^{m\theta}u = e^{m\theta}f(D+m)u, \quad f(D)e^{m\theta} = f(m)e^{m\theta}, \tag{3.3.11}$$

we have readily from (3.3.10) an instance of the FTD,

$$f_0(D)u + f_1(D)e^\theta u + \ldots = \left((f_0(m))u_m + \ldots)e^{m\theta}\right) \tag{3.3.12}$$

This equation provides an equivalence between differential and difference equations, an automatic equivalence, unavailable by Ellis's reduction from (3.3.1) to (3.3.2) (Boole 1844, 233).

Having fully established the desired analogy between the symbolical form of the differential equation in question and its scale, Boole proceeded to inquire into the possibility of transforming equation (3.3.7) to (3.3.8). Switching from the coefficients a_m, b_m to the variables u and v, the transform sought for would be in analogy with (3.3.3):

$$u = \frac{D-1}{D+2}v \tag{3.3.13}$$

Before justifying its validity, let us see how it worked, quoting directly from Boole's paper.[21] Given that $v = a\cos(qx+b)$ or $c\sin(qx+c')$, we have from (3.3.13), $u = (D-1)/(D+2)v = (1-3(D+2)^{-1})c\sin(qx+c') = c(1-3e^{-2\theta}(D)^{-1}e^{2\theta})\sin(qx+c') = c(1-3/x^2(xd/dx)^{-1}x^2)\sin(qx+c') = c(\sin(qx+c') - 3/x^2\int x\sin(qx+c')dx$, or the solution of the EFE (3.3.1) is

$$u = c(1-3/q^2x^2)\sin(qx+c') + 3/qx\cos(qx+c'), \tag{3.3.14}$$

that is (2.2.2), the known solution of the EFE.

Part 4 of Boole's paper (§3.2) concerned the generalization, justification and further illustration of this procedure, which as already noted, he attributed in part to Ellis. Considering the general form of a LDE (3.3.10), Boole operated on both sides of the equation by $(f_0(D))^{-1}$, and for convenience worked thereafter with the form

$$u + \phi_1(D)e^\theta u + \ldots + \phi_n(D)e^{n\theta}u = U, \tag{3.3.15}$$

by writing U for $(f_0(D))^{-1}U$, and replacing $f_1(D)/f_0(D)$ by $\phi_1(D)$ etc. (ibid., 224). Boole then claimed that the theory of equations (3.3.15) is deducible from that of equation

$$u+\phi(D)e^{r\theta}u=U, \tag{3.3.16}$$

furnishing a complete study of (3.3.16), based on the following two propositions.

According to *Proposition 1,* (3.3.16) can at once be converted to

$$v+\phi(D+n)e^{r\theta}v=V, \tag{3.3.17}$$

by means of the relations $u=e^{n\theta}v$ and $U=e^{n\theta}V$. Indeed, by assuming $u=e^{n\theta}v$, and substituting in (3.3.16), equation (3.3.17) follows readily according to the known formula (3.3.11) (Boole 1844, 247).

Proposition 2, which is the much stronger of the two propositions, enables the transformation of (3.3.16) to the form

$$v+\psi(D)e^{r\theta}v=V, \tag{3.3.18}$$

by means of the relations

$$u=P_r\frac{\phi(D)}{\psi(D)}v,\quad U=P_r\frac{\phi(D)}{\psi(D)}V, \tag{3.3.19}$$

where P_r stands for the 'infinite product':

$$P_r\frac{\phi(D)}{\psi(D)}=\frac{\phi(D)\phi(D-r)\phi(D-2r)\ldots}{\psi(D)\psi(D-r)\psi(D-2r)\ldots} \tag{3.3.20}$$

Proof

Let $u=f(D)v$. Substituting in (3.3.16), dividing by $f(D)$ and taking under consideration that $e^{r\theta}f(D)v=f(D-r)e^{r\theta}v$, we have

$$v+\frac{\phi(D)f(D-r)}{f(D)}e^{r\theta}v=\left(f(D)\right)^{-1}U \tag{3.3.21}$$

Comparing now (3.3.21) with (3.3.18), we reach a recursive definition of $f(D)$, given by

$$f(D)=\frac{\phi(D)}{\psi(D)}f(D-r). \tag{3.3.22}$$

Then, by iteration from (3.3.22) it follows that $f(D)$ indeed coincides with the right-hand side of (3.3.20) (Boole 1844, 247–8).

Boole went on to inquire into the necessary conditions which render the product (3.3.20) finite, thus indicating which cases of (3.3.16) are susceptible to integration in finite terms (ibid., 248–9). He then determined the general characteristic of those equations whose solution depends upon that of

$$d^n x / dx^n \pm q^n v = X \tag{3.3.23}$$

(ibid., 249–50). The EFE followed in the standard and general form, as a special case of (3.3.23) for $n = 2$ (ibid., 250–2). It now remains to prove the validity of the transform (3.3.13) which was initially conceived by analogy with Ellis's relation (3.3.3) and which led to the invention of the useful propositions mentioned above.

Assume the EFE in the symbolical form (3.3.7), and apply *Proposition 2* for (3.3.7) in place of (3.3.16) and (3.3.8) in place of (3.3.18) respectively. We notice that $U = V = 0$, and that the coefficients $\phi(D)$ and $\psi(D)$ stand respectively for $\phi(D) = q^2 /(D-3)(D+2)$ and $\psi(D) = q^2 / D(D-1)$. Substituting these values in (3.3.20), we have

$$P_r \frac{\phi(D)}{\psi(D)} = \frac{D-1}{D+2}, \tag{3.3.24}$$

in other words the transform (3.3.13).

Boole was referring to equation (3.3.15) when he wrote to De Morgan in June 1843: 'It is on this equation that the reductions are effected and not on the scale as in Mr Ellis's mode—the reductions are effected at a single step and not successively and in general by symbols of differentiation and not of integration, some few cases excepted. This section also contains an extension of the theory of equations with constant coefficients to a large class of equations with variable coefficients'.[22]

3.4. *On the Theory of Developments and the Laplace Equation, 1844–1847*

We deviate slightly from our strict chronological order to draw on chapter 17 of Boole's *Differential Equations* (1859), in order to hint at the reasoning that apparently led him to the GTD (§3.2) in 1844. In this chapter, Boole furnished first the FTD and the theory for the finite solution of binomial equations (Boole 1859, 412–36), proceeding to the solution of LDE by series with applications in the field of generating functions and definite integrals (ibid., 437–46). The GTD was mentioned last in the section entitled 'Generalizations of the Foregoing Theory' (ibid., 446–50) as follows.

Claiming first that all the propositions used so far were based on the formulae (3.3.6) and (3.3.11), Boole repeated the latter

$$f(D)e^{m\theta}u = e^{m\theta}f(D+m)u, \tag{3.4.1}$$

observing that these propositions would remain true 'for any symbols π and ρ, whatever their interpretation, which combine according to the same formal law' (ibid., 446),

$$f(\pi)\rho^{m}u = \rho^{m}f(\pi+m)u. \tag{3.4.2}$$

If we replace D and e^{θ} by π and ρ respectively in (3.3.18), *Proposition 2* could still be applied, after replacing D by π in (3.3.20). He observed further that the law (3.4.2) holds also for combinations of the 'pure symbols' D and e^{θ}, namely for

$$\pi = D - n\phi(D)e^{\theta} \text{ and } \rho = \phi(D)e^{\theta}. \tag{3.4.3}$$

Letting $m = 1$ in (3.4.2) we can see, claimed Boole, that ρ can be transferred from the right to the left of $f(\pi)$, 'if we, so to speak, add to π the constant increment 1' (ibid., 447).

It was this mode of reasoning, absent in Boole (1844) and its sequels, which readily suggested the more general law

$$f(\pi)\rho u = \rho f(\pi+\Delta\pi)u, \tag{3.4.4}$$

and challenged the determination of the development of $f(\pi + \rho)$ in ascending powers of ρ, for π and ρ distributive symbols combining according to (3.4.4). The main result, whose lengthy and complicated proof we omit, was the formula

$$f(\pi+\rho) = f(\pi) + \frac{\Delta}{\Delta\pi}f(\pi)\rho + \frac{\Delta^2}{\Delta\pi^2}f(\pi)\frac{\rho^2}{1\cdot 2} + \ldots, \tag{3.4.5}$$

where

$$\frac{\Delta}{\Delta\pi}f(\pi) = \frac{f(\pi+\Delta\pi) - f(\pi)}{\Delta\pi}. \tag{3.4.6}$$

When $\Delta\pi$ vanishes, the symbols π and ρ become commutative, and (3.4.5) takes the form of Taylor's theorem. Suitably replacing π and ρ by compounds of x, d/dx and $\exp(d/dx)$, Boole was led to new results, and eventually to the known formulae (3.3.6), (3.3.11) (Boole 1844, 228–32; 1845, 215–7; 1859, 447–9).

These three instances were the only ones in which Boole's GTD was presented. While it was of limited utility in Boole (1844), serving at this stage as a reinforcement of the generality of the method, it showed its full import in the paper 'On the Theory of Developments' (1845). Boole's peculiar approach in conceiving and proving this theorem, apparently independently of the study of non-commutative operations as in Murphy (1837),[23] was not followed, but individual ideas and results led to the development of the COO around 1848–1850. One of the consequences of (3.4.5) was the formula

$$f\left(\frac{d}{dx}+\phi'(x)\right)u = e^{-\phi(x)} f\left(\frac{d}{dx}\right)e^{\phi(x)}u, \tag{3.4.7}$$

a generalization of the known formula (3.3.11). By suitably specifying $\phi(x)$ as $\log x$, an outcome of (3.4.7) was the expansion

$$f\left(\frac{d}{dx}+\frac{1}{x}\right)u = f\left(\frac{d}{dx}\right)u+\frac{1}{x}f'\left(\frac{d}{dx}\right)u, \tag{3.4.8}$$

while letting $\phi'(x) = x$ we get from (3.4.7) the formula

$$f\left(\frac{d}{dx}+x\right)u = \left(f_0\left(\frac{d}{dx}\right)+f_0'\left(\frac{d}{dx}\right)x+\frac{1}{1\cdot 2}f_0''\left(\frac{d}{dx}\right)x^2+\ldots\right)u \tag{3.4.9}$$

where

$$f_0(x) = \exp\left(\frac{1}{2}\frac{d^2}{dx^2}\right)f(x) \tag{3.4.10}$$

(Boole 1845, 217–20).

All these formulae were accompanied by their duals (i.e. the expansion of $f(x + d/dx)$), but most of the procedures were omitted, as he found them quite 'trouble-some'. The paper ended with some applications of (3.4.9) to equations of the form

$$xf(d/dx)u + f'(d/dx)u = 0, \tag{3.4.11}$$

a theory further developed by Hargreave later on (§5). Boole was particularly proud of certain 'remarkable' analogies he noted between commutative and non-commutative functions. He was impressed by the fact that the development of (3.4.8) will terminate at the second term 'whatever may be the form of the function denoted by f' (ibid., 219). Moreover, the coefficients of (3.4.9) follow, after the first term, the law of Taylor's theorem, a 'remarkable circumstance, seeing that the symbols x and d/dx are not commutative' (ibid., 220–1). Interestingly, the former remark would be incorporated in MAL (§4.4), while the latter was communicated to De Morgan prior to publication of the paper 'because you are interested in the Calculus of operations of which it appears to me to constitute one of the most singular results'.[24]

Boole's last contributions in the realms of the COO at that time involved a general study of the Laplace Equation (LE), first in his paper 'On the Equation of Laplace's Functions' (1846) and consequently in the far too abstract query 'On a Certain Symbolical Equation' (1847b). Considering it in the form

$$\frac{d}{d\mu}(1-\mu^2)\frac{du}{d\mu}+\frac{1}{1-\mu^2}\frac{d^2u}{d\phi^2}+n(n+1)u=0, \tag{3.4.12}$$

Boole introduced suitable transformations, replacing $(d/d\phi)\sqrt{-1}$ by a—which he regarded as a constant—so as to reduce it to an equation in one variable, susceptible to the theory of the binomial equation (3.3.16).[25] As R.L. Ellis shrewdly perceived in 1841, 'It is not difficult to multiply artifices, by means of which particular equations may be solved, but the results will, generally speaking, be of little value' (§2.3). Indeed, with Boole's papers on the COO, not only were particular artifices multiplied, but new, general methods were put forward and a far wider class of equations were found to be integrable in finite terms, with new ardent followers from 1848 onwards (§5). However, hardly any of the particular results obtained ever proved to be of any value other than as a proof of the potential power of the COO. Neither the EFE nor the LE would benefit, physically or mathematically speaking, from the numerous variations of the symbolical solutions provided by Boole and his followers.[26] Boole himself foresaw the limited practical import of his sophisticated researches, which he, nevertheless, considered as fruitful illustrations of the doctrines of symbolical algebra.

We conclude the present survey with two quotes from Boole (1847b, 8, 12):

> This equation is a symbolical one, and it admits of two conjugate solutions [...] which are also purely symbolical; i.e. their validity does not depend on the significance of the symbols which they involve, but only on the truth of the laws of their combination [...]. Practically, the solutions of Laplace's equation, which we are thus made acquainted with, are of little utility [...]; but they throw an interesting light on the subject of Symbolical Algebra, and serve to illustrate some general doctrines in Analysis [...].

> The investigation we have entered upon is chiefly valuable, as presenting to us what will be thought a very curious chapter in symbolical algebra, and introducing us to the family of which Laplace's equation is a member. But it must be confessed that they are an interesting rather than an amiable group.

4. TOWARDS A GENERAL METHOD IN LOGIC, 1847

4.1. Anticipations of Boole's Algebraization of Logic, 1807–1847

Boole's MAL came at a time when the discipline of logic was experiencing a remarkable revival. The first to defend Aristotelian logic against the criticisms it had suffered since the time of Descartes and Locke[27] were R. Kirwan with his *Logick*

(1807) and R. Whately with his *Elements of Logic* (1826). The subject attracted the attention of a diversity of writers, such as G. Bentham, A. De Morgan, T. Solly and W. Hamilton, the Scottish philosopher. This revival culminated with Boole's MAL and De Morgan's *Formal Logic,* two books published in October of 1847, with vividly differing approaches in their extension and mathematization of logic.[28]

Innovations introduced by Bentham, Solly and De Morgan in treatises published in the late 1830s unfortunately went unnoticed and were acknowledged only after 1847, that is only after being overshadowed by the new contributions of Boole and De Morgan. However, new textbooks formed only one component of the revival of logic in the 1830s. Critical reviews of Whately's popular book, as well as disputes between Hamilton, Whewell and De Morgan on the nature and educational import of logic and mathematics, were also significant factors. The provocative tone of the disputes, which culminated in the De Morgan-Hamilton quarrel over the 'quantification of the predicate' (QPR), had even more impact than the treatises themselves, as we can see from the opening of Boole's MAL:[29]

> In presenting this Work to public notice, I deem it not irrelevant to observe, that speculations similar to those which it records have, at different periods, occupied my thoughts. In the spring of the present year my attention was directed to the question then moved between Sir W. Hamilton and Professor De Morgan; and I was induced by the interest which it inspired, to resume the almost-forgotten thread of former inquiries.

Despite the stimulation that resulted from such disputes, Boole's own algebra of logic reveals no signs of being profoundly influenced by the recent developments. Still, he had one predecessor in his peculiar adaptation of symbolical methods to logic, Thomas Solly, author of *A Syllabus of Logic* (1839) which became fairly well known late in 1847. In what follows, we provide a brief outline of the developments that took place within formal logic in England from 1807 until 1847 and sustain the view that when Boole entered the scene in the mid-1840s the climate was already favourable to a fruitful application of mathematical methods to logic. We claim that the period under discussion, and especially the transitional period of the 1830s, deserves further research.[30] We do no more than hint at the possible importance of the disputes alluded to above.

Kirwan was the precursor of Whately who first drew attention to the scientific role of logic and to the plausibility of connections between logic and mathematics. In his *Logick* (1807, 1–3) he claimed that logic is 'both a science and an art', and also that 'in no other science have the rules of logic been so well observed, as in that of mathematics' (ibid., iii–v). Kirwan recognized that logic is not immune from fallacies, but declared nevertheless that logical reasoning is superior to mathematical reasoning and emphasized the usefulness of logic in 'detecting the errors of a mathematician' (ibid.).[31]

Kirwan was succeeded by Whately, an Oxford intellectual, author of treatises on rhetoric, political economy and logic. His *Elements* (1826), which originally appeared in 1823 under the entry 'Logic' in the *Encyclopedia metropolitana,* formed the introduction to logic for Bentham, De Morgan, Solly and Boole. Strictly confining his account to traditional logic, Whately offered no technical innovations and deployed no mathematical methods or symbols. Moreover, his views on the role of logic were found shaky by his critics, none of whom would deny, though, Whately's contribution to the revival of the study of logic.[32]

Whately enriched his textbook with a lucid account of fallacies and argued for the utility of logic in botany, ethics, law and mathematics. Furthermore he shared Kirwan's concern for grammar and language. He called logic 'the Grammar of Reasoning', claiming that 'Syllogism' should be regarded as 'the form to which *all* correct reasoning may be ultimately reduced' (Whately 1826, 11). To reinforce logic's formal character, Whately drew interestingly on algebra, arguing as follows. As in algebra we proceed with 'arbitrary signs representing numbers in the abstract', so does logic 'pronounce on the validity of a regularly-constructed argument, equally well, though arbitrary symbols may have been substituted for the terms; and consequently, without any regard to the things signified by those terms' (ibid., 11–4).

Bentham took the next step, with his *Outline of a New System of Logic* (1827), the main part of which consists of a critical commentary on Whately's *Elements.* As a botanist, Bentham realized the necessity of extending the limited realms of traditional logic. To this end, he introduced the QPR, by splitting each of the four categorical propositions in two, depending on whether the predicate was taken over the entire quantity or not. For instance, the proposition 'All X is Y' was divided into 'All X are some Y' and 'All X are all Y'. In fact, he proposed a quasi-algebraic notation, writing for example 'All X are some Y' in the equational form '$tX = tY$', with t standing for 'toto', meaning 'all'; he also used p for 'parte', meaning 'some', = for identity and || for negation (Bentham 1827, ch. 8). Bentham did not make use of all eight symmetrical forms obtained, but reduced them instead to a system of five forms which, as he realized, was sufficient for purposes of classification in botany.[33] Bentham's novel scheme, strikingly absent in Hamilton's review 'Recent Publications on Logical Studies' (1833), was first acknowledged by De Morgan (1850, 32–4).

Although his review referred to ten treatises of logic published between 1826 and 1832, Hamilton focused exclusively on Whately's book, commenting only en passant on Bentham's *Outline.* The latter work was viewed as a work 'in which Dr Whately is alternatively the object of censure and encomium' (Hamilton 1833, 199). Well versed in the history of logic, Hamilton acknowledged Kirwan as a precursor of both Whately and Bentham, but overlooked the new ideas presented in chapter 8 of Bentham's *Outline,* just as, a few years later, he would neglect Solly's innovations (Panteki 1993, 142, 153). Hamilton's own contribution to logic became fairly well known after 1846, when he accused De Morgan of plagiarism in connection with the invention of the QPR. The dispute endured for a long time and not only motivated Boole to resume his 'former inquiries', but also induced De Morgan to reexamine his own syllogistic scheme.[34] By 1847, Boole was also aware of Hamilton's polemic review of Whewell's *Thoughts on the Study of Mathematics as a Part of a Liberal Education* (1835), published in 1836.[35]

Whewell was especially opposed to Whately's logic, which formed the core not only of the Oxford curriculum (along with classics) but also, according to Whewell, of Whately's political economy—which he abhorred.[36] Arguing for geometrically and physically grounded mathematics, and against symbolical algebra, Whewell entered into a comparative study of mathematics and logic in his *Thoughts,* holding that mathematics properly taught was the best teaching tool, whereas logic was considered as utterly redundant in the curriculum (Whewell 1835, 5–8, 44). This 'invalid' contrast made Hamilton (1836, 413) furious, and he in turn claimed that mathematics is the most dangerous tool for educating men in reasoning. Boole would accordingly devote his 'Introduction' to MAL to the refutation of Hamilton's views. What is less known, though, is that beyond Whewell's provocative comparison lay a review of Peacock's *Algebra* published by De Morgan (1835), a few months before the *Thoughts.* For in this review, De Morgan attacked the educational system at Cambridge University, arguing first for the indispensability of logic, and second for that of algebra, as subjects with which a student should be acquainted prior to his encounter with the higher calculus. It was for this reason that Whewell apparently had fears about the stability of his system, and put forward his 'implicit' answer to De Morgan—or to any potential moderator under his influence—in the *Thoughts.*[37]

We have no evidence of how widespread De Morgan's, Whewell's and Hamilton's opposing views were at that time within Cambridge University. We know, however, that Solly was a student there in the mid-1830s, publishing his *Syllabus* in 1839 with the intention of establishing the scientific character of logic upon new, solid grounds. To this end, Solly combined Kant's philosophy with mathematical notions and methods in an original, indeed pioneering way. His breakthrough went unnoticed, however, and Solly, who moved to Germany in 1843, was not immediately aware of the limited circulation of his book in England. He sent a copy of it to De Morgan in 1847, begging him to take notice of his contributions. The copy was received before *Formal Logic* went to press, and De Morgan summarized the essence of Solly's symbolical calculus in the end of his *Formal Logic,* as follows:[38]

> The symbolical expression here given is of a peculiar character, the algebraic signs are adopted in a sense which preserves the rules of sign, while the symbols represent the terms of the syllogism, or else the notions of particular and universal. Thus, if p stands for particular, u for universal, and m for one of the terms of a syllogism, $m = u$ or $m - u = 0$ implies that m is a universal term, and $(m - u)(n - p) = 0$ implies the alternative that either m is universal, or n is particular. By means of such alternative relations, the conditions of validity of the various figures are expressed.

By presenting the known conditions of validity of the figures, first in the traditional way, and consequently by symbolical methods, Solly aimed to 'add a theoretical completeness to the science, which it can never obtain when the symmetry of its results is arrived at by empirical methods alone' (Solly 1839, iii). Aware of the boldness of his endeavour, he first explained the function of his method, borrowing terms from Gregory's study of the nature of symbolical algebra.[39] For instance, if any set of symbols, like $(m - u)(n - p)$, be juxtaposed and equated to zero, the equation will indicate that at least one of the factors may be equated to zero. In that case, claimed Solly, 'all its constituent symbols may receive whatever interpretation may arise from the transpositions of terms, or any other operations, that are formally analogous to those of arithmetical algebra, and have been admitted in our particular application of the science' (Solly 1839, 112–3).

Besides the QPR, introduced but discarded as redundant, the symbolical expression of syllogistic laws was Solly's unique novelty within traditional logic. Deviating, however, from the main object of his book, Solly also delved into Kant's transcendental logic, his ultimate aim being to clarify philosophical notions with the aid of mathematical ones. Within a rather eccentric blend of algebra, analytic geometry, operator methods and metaphysics, Solly raised certain interesting analogies, as for instance between abstract conceptions and enveloping curves. Deploying alternatively the method of separation of symbols and the theory of infinitesimals, Solly was happy to show the manifestation of properties and laws of philosophical notions unattainable otherwise. This aspect of Solly's work would not flourish with his successors. Nevertheless, Solly anticipated many of Boole's later speculations; Solly drew on the theory of singular solutions of differential equations so as to define philosophical conceptions, whereas Boole, late in life, was concerned with the philosophical interpretation of the properties of singular solutions.[40]

As early as 1839, Solly deployed algebraic symbolical methods in the wider realms of logic and philosophy, not only as a tool for symmetry, rigour, brevity, systematization and generalization of known theories, but also as a means of discovery of hidden laws and properties. In this respect, Solly's work echoes the epistemological approach of his predecessors and contemporaries, such as Babbage, Herschel, Murphy, De Morgan and Gregory who developed ardently the calculi of operations and functions. It is beyond the scope of this section to inquire further into a joint study of the revival of the COO and formal logic in the 1830s. Nevertheless, Solly's case reveals the early multidimensional import of symbolical methods in the late 1830s. Given further the disputes already mentioned between leading intellectuals on the connection between logic and mathematics, we hold that prior to Boole's first works on the subject, there already existed a promising environment for the application of mathematics to logic.

4.2. Scope and Structure of MAL

Boole opened his 'Introduction' to MAL stating: 'They who are acquainted with the present state of the theory of Symbolical Algebra, are aware, that the validity of the processes of analysis does not depend upon the interpretation of the symbols which are employed, but solely upon the laws of their combination' (Boole 1847a, 3). It was upon this doctrine that Boole founded his own symbolical calculus of logic, in so doing making ample use, as we shall see, of analogies with the calculus of differential operators.

The elements of Boole's logical calculus, which constitute what we may call for convenience 'basic logic', were presented in the first two chapters of MAL (see Van Evra (1977)). Here Boole defined the laws of combination of his symbols (ch. 1), giving next the symbolical expression of the four standard forms of the categorical propositions of traditional logic (ch. 2). There followed applications of this basic method to issues of 'conversion' (ch. 3), 'syllogism' (ch. 4) and 'hypotheticals' (ch. 5), while the last two chapters, largely deviating from the realms of traditional logic, suggested a 'general method in logic' (GML), nearly as general and sophisticated as his GMA.

The order of his presentation appears to be very close to his order of discovery. Interestingly, Boole had not anticipated the existence of a GML. As he admitted in the 'Introduction', his object was initially confined solely 'to the expression of the received logic, and to the forms of the Aristotelian arrangement' (Boole 1847a, 7). Boole was thus orientated towards an algebraization of traditional syllogistic, and made no 'special effort to arrive at novel results' (ibid., 8). He was soon compelled, however, to 'interrogate the method itself for an expression of the just limits of its applications' (ibid.). Eventually these latter inquiries led to general theorems in logic, which Boole was happy to accept 'not only as a proof that the laws I had investigated were really the laws of thought, but also as a means of giving to the process of the calculus an analytical generality and simplicity, which they could not otherwise have had'.[41]

The object of MAL, in his own words, was as follows (ibid., 6):

> Every logical proposition, whether categorical or hypothetical, will be found to be capable of exact and rigorous expression, and not only will the laws of conversion and of syllogism be thence deducible, but the resolution of the most complex systems of propositions, the separation of any proposed element, and the expression of its value in terms of the remaining elements, with every subsidiary relation involved. Every process will represent deduction, every mathematical consequence will express a logical inference. The generality of the method will even permit us to express arbitrary operations of the intellect, and thus lead to the demonstration of general theorems in logic analogous, in no slight degree, to the general theorems of ordinary mathematics.

In §4.3 we present the basic elements of Boole's logical calculus, then in §4.4 we investigate the plausible connections between his GMA and GML. We draw on material from MAL and also on manuscripts written around 1847–1848 to argue that GMA was a constant source of motivation and justification for Boole's symbolical procedures in MAL, but much less evidently so in LT.[42]

4.3. The Elements of Boole's Logical Calculus

The variables of Boole's calculus, x, y, z, stood for mental acts by means of which we select objects of a specific property from a given class. If X, Y, Z stand for individual members of classes, Boole introduced the elective symbols x, y, z as possessing the following character: 'The symbol x operating upon any subject comprehending individuals or classes, shall be supposed to select from that subject all the Xs which it contains' (Boole 1847a, 15). Accordingly, 'the product xy will represent, in succession, the selection of the class Y, and the selection from the class Y of such individuals of the class X as are contained in it, the result being the class whose members are both Xs and Ys' (ibid., 16).

Boole argued that 'The result of an act of election is independent of the grouping or classification of the subject' (ibid.), as well as that 'It is indifferent in what order two successive acts of election are performed' (ibid., 17). These two properties of the elective symbols were expressed respectively as

$$x(y+z)=xy+xz \tag{4.3.1}$$

and

$$xy=yx. \tag{4.3.2}$$

Boole's principal innovation for logic was the third law, according to which 'The result of a given act of election performed twice, or any number of times in succession, is the result of the same act performed once' (Boole 1847a, 17), or

$$x^{n}=x. \tag{4.3.3}$$

As he wrote in his 'Notes' dating from circa 1848, 'The laws $x(y + z) = xy + xz$ and $xy = yx$ are the only laws of algebraic symbols devoid from the conception of quantity' (Boole 1997, 42).[43] In fact, distributivity and commutativity characterized the combination of differential operations with themselves and constants (§2.3). In that same passage, Boole also claimed that:

> [...] the law $x^m x^n = x^{m+n}$ is a law quite independent of the nature of the operation denoted by x. It is for example true in the Differential Calculus that $(d/dx)^m(d/dx)^n = (d/dx)^{m+n}$. To the above [meaning the laws (4.3.1)–(4.3.3)] it would seem that we ought to add the law whose expression is $x+y=y+x$.[44]

Boole's allusion to Gregory's three laws that governed the foundations of the COO at its first stage, (2.3.4), is evident here. Called the 'index' law, (4.3.3) was the only law peculiar to logic, and the key to all interpretations. As consequences of this law, Boole formed the equations

$$x(1-x)=0, \quad \text{and} \quad x+(1-x)=1, \tag{4.3.4}$$

which for him expressed the law of contradiction and the law of excluded middle respectively. The symbol 1 stood for the universe, 0 denoted 'nothing' and $(1 - x)$ denoted the class of not-Xs—there being no symbol for negation in his account. The main distinction between the basic logic and the general method was that the latter involved numerical constants other than 0 and 1. A. Cayley was the first reader of MAL to be perplexed by the appearance of numerical constants in Boole's GML, and queried him on this issue on the 2nd of December 1847.[45]

Cayley's attention was first captured by Boole's rather too general statement made towards the end of chapter 1 of MAL. There he asserted that in virtue of the properties of distributivity and commutativity 'all the direct processes of common algebra are applicable to the present system' (Boole 1847a, 18). As he clarified in his reply to Cayley's first letter, 'When I speak of the operations of common algebra being applicable to my system, I mean of course the symbolical operations—those which depend upon laws of combination, not upon interpretation' (Boole 1997, 191–2). In fact, as Cayley noticed, Boole had remarked in this text that certain algebraic processes do affect the interpretation of an elective equation, by limiting the expression of a truth. For instance,

$$\text{if } y = z, \text{ then it follows that } xy = xz, \tag{4.3.5}$$

involves a perfectly 'legitimate' inference, 'but the fact which it declares is a less general one than was asserted in the original proposition' (Boole 1847a, 19). In other words, the converse of (4.3.5) does not in general hold true. Furthermore, division was excluded from the fundamental operations of basic logic, and was only implicitly effected through the GML.[46]

In chapter 2 of MAL, Boole quoted Whately (Boole 1847a, 20), and displayed the four standard forms of categorical propositions, A, E, I and O, for which he supplied corresponding symbolical expression. His list was enriched with several alternative forms, deploying a vaguely defined new elective symbol v. We conclude our preliminary account of Boole's logical calculus by commenting on the formulation of the symbolical expressions of A and I.

Table of the Standard Categorical Forms

A	All *X*s are *Y*s	$x(1-y)=0$	(1)	$x = vy$	(1)′	(4.3.6)
E	No *X*s are *Y*s	$xy = 0$	(2)	$x = v(1-y)$	(2)′	
I	Some *X*s are *Y*s	$v = xy$	(3)	$vx = vy$	(3)′	
O	Some *X*s are not *Y*s	$v = x(1-y)$	(4)	$vx = v(1-y)$	(4)′	

Considering proposition A, Boole reasoned as follows: 'As all the *X*s which exist are found in the class *Y*, it is obvious that to select out of the Universe all *Y*s, and from these to select all *X*s, is the same as to select at once from the Universe all *X*s' (ibid., 21), writing thus $xy = x$. Then based on the property of distributivity and the law of transposition (missing in MAL), he wrote $xy = x$ as $x(1-y) = 0$, form (1) of (4.3.6). To express I he claimed: 'If some *X*s are *Y*s, there are some terms common to the classes *X* and *Y*. Let those terms constitute a separate class *V*, to which there shall correspond a separate symbol v, then $v = xy$. And as v includes all terms common to the classes *X* and *Y*, we can indifferently interpret it, as Some *X*s, or Some *Y*s' (ibid., 21).

Having thus formulated the forms (1)–(4), Boole claimed that equations expressing particular truths can be viewed as necessary deductions from one general proposition,

$$x = y. \tag{4.3.7}$$

Multiplying (4.3.7) with x we have $x^2 = x = xy$ or $x(1-y) = 0$, that is form (1) of (4.3.6). He next regarded the latter as an equation in which x is to be determined in terms of y, claiming that its 'general solution' is $x = vy$, form (1)′ of (4.3.6), suggesting verification merely by substitution in (1). Finally, multiplying (1)′ with v, implicitly assumed as non-empty, he arrived at form (3)′ which implies that 'Some *Y*s are *X*s and some *X*s are *Y*s' (ibid., 22–4).

Boole applied his basic method to issues of traditional logic in chapters 3–5 of MAL, often replacing the standard forms (1)–(4) by their alternatives (1)′–(4)′. In fact, his procedure has certain weak points. Firstly, the actual mode proposed for the solution of elective equations was not furnished until the last chapter of MAL (§4.4). Secondly, it has been proved that to replace a proposition, (1) for example, by its 'solution' (1)′, altered the existential import of the proposition. Moreover, $x = vy$ is not deducible from $x(1 - y) = 0$ 'using the rules that Boole explicitly states' (Corcoran and Wood 1980, 624). Finally, the indefinite symbol v, originally defined via formula (3), read as 'The class *V* is not empty', perplexed all of Boole's readers.[47] Apparently, in certain instances such as form (1)′, symbol v had a role analogous to that of arbitrary constants in the solution of differential equations, just as symbols x, y, z imitated the function of differential operators (§4.4).

4.4. Connections between Boole's GML and GMA

The basic component of Boole's GML was Maclaurin's theorem,

$$f(x) = f(0) + f'(0)x + \frac{f''(0)x^2}{1\cdot 2} + \frac{f'''(0)x^3}{1\cdot 2\cdot 3} + \ldots, \qquad (4.4.1)$$

deployed for the expansion of an arbitrary elective function $f(x)$ in terms of x, upon the grounds that 'elective symbols combine according to the laws of quantity' (Boole 1847a, 60). Evidently, his justification is largely erroneous; he should have said instead that (4.4.1) could be applied to elective symbols, since it demands only the properties of commutativity and distributivity, which indeed are obeyed by these symbols, the only laws 'devoid from the conception of quantity' (§4.3), as he wrote later in the 'Notes'.

Letting $x^2 = x^3 = \ldots = x$ according to the index law, Boole replaced x by 1 in (4.4.1), arriving by simple algebraic procedures at the formula

$$f(x) = f(1)x + f(0)(1 - x), \qquad (4.4.2)$$

the so-called 'General Theorem of Development' in logic, hereafter cited as GTDL. Calling $f(0)$ and $f(1)$ the 'moduli' of $f(x)$, Boole showed that the necessary condition for $f(x)$ to obey the index law was that its moduli obey this law (ibid., 60–2). If $f(xy)$ is an elective function of two variables, by expanding it first in terms of x, by (4.4.2), and consequently expressing each of its moduli in terms of y, there resulted a symmetrical expansion for $f(xy)$, namely (ibid., 62–3):

$$f(xy) = f(00)(1 - x)(1 - y) + f(01)y(1 - x) + f(10)x(1 - y) + f(11)xy. \qquad (4.4.3)$$

Though much simpler than the GTD (3.4.5) in Boole's GMA, inasmuch as it required the property of commutativity, the GTDL (4.4.2–4.4.3) played a role fairly close to that of the abstract and general theorem of development in the COO. In fact, there is an interesting observation to be made in this context: Boole was pleased to note that the expansion formula (4.4.2) is not 'the only case in which an expansion stops at the second term', appending in a footnote formula (3.4.8), or

$$f\left(\frac{d}{dx}+\frac{1}{x}\right)u = f\left(\frac{d}{dx}\right)u + \frac{1}{x}f'\left(\frac{d}{dx}\right)u \tag{4.4.4}$$

(ibid., 61). What might have been Boole's point?

In fact, that was the only instance in MAL where a logical formula was compared to a mathematical one involving non-commutative operations. The whole of Boole's GMA concerned combinations of non-commutative operators, like x and d/dx, hardly serving as a model for the GML which dealt exclusively with commutative symbols. Nevertheless, Boole's allusion to (4.4.4) is striking, inducing the conjecture that at that stage he was strongly under the influence of his GMA, instances of which, like (4.4.4), served as a means of confirmation that he was working on the right track. This influence faded away from the late 1840s ownards, and even Maclaurin's theorem was found to be redundant as a component of Boole's GML in the LT.[48]

Assuming an elective equation

$$f(xy\ldots)=0, \tag{4.4.5}$$

Boole claimed that according to the GTDL (4.4.3) $f(xy\ldots)$ can always be considered in the form

$$f(xy\ldots)=a_1t_1+a_2t_2+\ldots+a_nt_n, \tag{4.4.6}$$

where t_i are compounds of x, y, $(1-x)$, etc., satisfying the index law, and a_i are numerical constants. Boole proved that if 'at' is a random term of $f(xy\ldots)$, where 'a' is a modulus of the given function which does not vanish, then for each such term we have $at = 0$, or, since $a \neq 0$, $t = 0$, and 'the combined interpretation of these several equations will express the full significance of the original equation' (ibid., 64). The proof was straight forward, demanding only multiplication of (4.4.6) with t, while another similar proposition was proved next concerning the interpretation of equations of the form $f(xy\ldots)=w$ (ibid., 64–6).

When Cayley wrote to Boole about numerical coefficients allowed in his system, Boole replied by alluding to the above-mentioned proposition:[49]

> And now for the numerical constants. They never appear in the expression of a proposition and all the details can be accomplished without them. [...] In the chapter on the General Properties of Elective Functions I proposed a different object [than in the previous ones] viz. what are the properties of a function $\varphi(xyz)$ in which xyz... are elective

symbols but the function unrestricted so as to allow of its involving numerical constants? You will allow that this was a thing which I was at liberty to do. The general result is that any equation $\varphi(xyz) = 0$ is reducible to the form $a_1t_1 + a_2t_2 \ldots + a_nt_n = 0$ in which $a_1, a_2 \ldots a_n$ are numerical constants or 0 and $t_1, t_2 \ldots t_n$ constituent elective functions which *are interpretable in logic*—and further that this equation is resolvable into a series of equations of the form $t = 0$, *all interpretable in logic*. Prop. 2nd, p. 64f. that the numerical elements disappear altogether from the final result. I did not anticipate this, I thought it exceedingly unlikely that every equation $\varphi(xyz) = 0$ should be interpretable but when I found that this was the case and that it gave us the power of reducing the solution of equations to *general theorems* I accepted it not only as proof that the laws I had investigated were really the laws of thought but also as a means of giving to the process of the calculus an analytical generality and simplicity which they could not otherwise have had.

This passage is amply revealing of Boole's reasoning and induces us to draw a heuristic analogy between Gregory's COO, as based on commutativity, and Boole's algebraization of traditional logic, as in the first five chapters of MAL. Just as in the opening of GMA (§3.1) Boole claimed to have established a general method in analysis 'free from the restrictions', which had characterized the COO so far, similarly he felt induced to investigate the true limits of his basic logic, and consider general elective functions 'unrestricted so as to allow of [their] involving numerical constants'. Thus, the role played by the consideration of non-commutative constants within the COO was now the role played by general elective functions, with coefficients other than 0 and 1. And here comes the rhetorical question: 'You will allow that this was a thing which I was at liberty to do'. Boole's confidence at this stage was the confidence of an analyst, who had already extended the limited realms of Gregory's COO, so as to encompass LDE with variable coefficients, largely motivated by the EFF (§3). In fact, he felt he had discovered a genuine, general method, trying to prove its utility in diverse branches of the integral calculus. We also know that he did so gradually, first contemplating particular cases and then letting his operators assume the most general forms possible (§3.4). Had Boole not gone beyond LDE with constant coefficients, it is unlikely that he would have gone beyond his basic logic in MAL.

We conclude this survey with one more instance which manifests a close link between Boole's mathematical background and MAL, namely the connection between the solution of differential and elective equations. The first mode presented for the solution of elective equations was based on the assumption that the most general expression of an elective symbol, y, in terms of another, x, shall always be

$$y = vx + v'(1 - x), \qquad (4.4.7)$$

where v, v' were arbitrary elective symbols. The method put forward by Boole by replacing the symbol to be determined by (4.4.7), and then determining the moduli v, v' by comparison of the coefficients, largely recalls the common method of undetermined coefficients in differential equations. Boole took (4.4.7) for granted in MAL, calling it in LT the 'dichotomy theorem', deployed there for the proof of the GTDL (4.4.2).[50]

We confine our account to the first example of an elective equation solved by Boole in the opening of chapter 7 of MAL, namely of equation

$$(1-x)y=0, \tag{4.4.8}$$

where y is to be determined in terms of x. Using (4.4.7), he arrived after substitution of y in (4.4.8) at $(1-x)(vx+v'(1-x))=0$, or $vx(1-x)+v'(1-x)^2=0$, or according to the index law and its consequences at $v'(1-x)=0$. Now, if we substitute the latter in (4.4.7), we obtain the desired solution $y=vx$. Interestingly, Boole reasoned as follows: Since $v'(1-x)=0$, 'may be generally true, without imposing any restriction upon x, we must assume $v'=0$, and there being no condition to limit v, we have:

$$y=vx, \tag{4.4.9}$$

[...] the complete solution of the equation' (ibid., 70).

Boole overlooked the fact that xy can be zero without necessarily one of the classes x or y being empty. Secondly, by reading vx as 'some Xs', it is implied that the algebraic solution $y=0$ of (4.4.8) is omitted. Let us assume the same equation, writing its solution algebraically as

$$y=(1-x)^{-1}0. \tag{4.4.10}$$

If x is a symbol of quantity, then $y=0$. However, let x stand for d/dx, then

$$y=\left(1-\frac{d}{dx}\right)^{-1}0=ce^{x}, \tag{4.4.11}$$

where c is an arbitrary constant, closely resembling v in (4.4.9). The hint for this illustration, absent in MAL, was provided by Boole in a letter to Cayley, quoted below:[51]

> When I speak of the operations of common algebra being applicable to my system, I mean of course the symbolical operations—those which depend upon laws of combination, not upon interpretation. Thus if $(z^2-a^2)u=0$ and z and a follow the laws of quantity, we have: $(z+a)u=(z-a)^{-1}0$. If z and a are symbols of quantity the second member is 0 but not unless. If z were d/dx it would give a term involving an arbitrary constant—but so far as operations alone are concerned we proceed with reference only to the laws of combination. The equation $x(u-v)=0$ gives $u-v=x^{-1}0$ whether the symbols are quantitative or elective but in the former case the result is equivalent to $u-v=0$—in the latter to $u-v=w(1-x)$, w being an arbitrary elective symbol.

Upon these grounds, it is clear that Boole viewed his elective symbols largely in analogy with differential operators. Such clarifications are missing in MAL, where Boole confined himself to more general remarks on a posteriori analogies noted between mathematical and logical procedures or results. For instance, he claimed at the opening of the last chapter: 'It is to be observed of such equations, that from the very nature of elective symbols, they are necessarily linear, and that their solutions have a very close analogy with those of linear differential equations, arbitrary elective symbols in the one, occupying the place of arbitrary constants in other' (Boole 1847a, 70). In fact, after giving as an example the solution of (4.4.8), he proceeded to the solution of equation $(1 - x)zy = 0$, where y was sought in terms of x and z. This latter example illustrated the analogy between the solution of elective equations with more than two variables, and partial differential equations. He deliberately chose not to pursue this comparison any further in MAL, as it 'would minister to curiosity rather than to utility', but preferred 'to contemplate the problem of the solution of elective equations under its most general aspect' (ibid., 72).

Considering the equation

$$f(xy) = 0, \tag{4.4.12}$$

Boole expanded $f(xy)$ according to the GTDL (4.4.3), replacing y by the expression (4.4.7). Equating next the coefficients of x and $1 - x$ to zero, 'in order that this equation may be satisfied without any way restricting the generality of x', he readily determined v and v', which replaced in (4.4.7) gave as the general solution of (4.4.12)

$$y = \frac{f(10)}{f(10) - f(11)} x + \frac{f(00)}{f(00) - f(01)} (1 - x), \tag{4.4.13}$$

(ibid., 72–3). He then suggested another method, which led to the same result, expanding $f(xy)$ first with respect to y, deducing thus

$$y = \frac{f(x0)}{f(x0) - f(x1)}, \tag{4.4.14}$$

a result which 'might have startled those who are unaccustomed to the processes of Symbolical Algebra [...] because of the apparently meaningless character of the second member' of (4.4.14). 'Such a result', he went on, 'would however have been perfectly lawful, and the expansion of the second member would have given us the solution above obtained' (ibid., 73).

Boole concluded his account by deploying formulae such as (4.4.13) for the solution of elective equations, claiming that his method, in combination with Lagrange's method of indeterminate multipliers, 'may be very elegantly applied to the treatment of simultaneous equations' (ibid., 76). As he wrote in his 'Notes' of 1848, 'In applying the method of indeterminate multipliers in the calculus of Logic we pass through uninterpretable forms to forms finally interpretable through development'.[52] The existence of such uninterpretable terms, like the second member of (4.4.14), evidently did not impede Boole's work, given his rich experience with dealing with uninterpretable terms within his advanced COO. His biggest discovery within logic was that every equation was ultimately interpretable, and we have serious doubts he would ever have endeavoured to seek a GML had he not been so confident working with symbolical methods in the course of formulating his GMA.

5. BOOLE'S ASPIRATIONS; AN EPILOGUE

Boole's most ardent and prolific successor was B. Brownin, a clergyman of Denby, Yorkshire.[53] By 1847, Brownin, stimulated by Boole (1845), furnished papers to the *Cambridge and Dublin Mathematical Journal,* which included direct proofs, generalizations and applications of Boole's theorem (3.4.7) and its dual. W. Thomson, editor of the journal, asked Boole to be a referee for Brownin's recent paper and received the following answer: 'Although I do not think Mr Brownin's paper contains any thing very useful it ought to be printed because it is desirable to have records of our own progress even in directions in which nothing is to be hoped for' (Smith 1984, 23). Boole's attitude at that time is highly interesting, for on one hand he was sceptical about the utility of new elaborate symbolical theorems—a concern already anticipated by R.L. Ellis in 1841 (§2.3)—but on the other hand was more anxious than ever about achieving a universal symbolical method in the realm of differential equations. In that same letter sent to Thomson in April 1847, Boole claimed that:

> It is very easy by means of the known properties of x and d/dx to construct equations involving these symbols under functional signs and admitting of complete solutions. In general such labour is useless because it is impossible to foresee whether an equation presented under the ordinary form can be reduced to one of those functional forms or not [...]. What we want is a *genuine* and *complete* method.[54]

Regarding his own method as 'in its present state *incomplete*', Boole placed many hopes on Brownin's promising work and stated that 'If Mr Brownin would undertake this, for which he appears to have both the leisure and ability, he would render a service to analysis' (Smith 1984, 23). Brownin did indeed exploit both 'leisure and ability' in the service of analysis, but in his third paper published in 1848, entitled 'On the Solution of a Particular Differential Equation' (Brownin 1848), he had the audacity to perceive 'errors' in Boole (1847b), the latter's too abstract study of the LE (§3.4). Boole, furious with this unexpected insult, hurried to clarify the dispute in his 'Remarks on a Paper by the Rev. Brice Brownin' (1848). Having proved that Brownin's accusations were groundless, involving 'both a false analogy and an erroneous principle' (Boole 1848, 414), Boole concluded with the following remarks, worthy of noting (ibid., 418):

> It is indeed with sincere pleasure that I quit a topic, upon which, had my reputation alone been concerned, I would willingly have been spared the necessity of writing, and acknowledge that upon some other points Mr. Brownin has added to the completeness of the investigation [...]. It is further but just to Mr. Brownin to remark, that in questions like that which he had undertaken to examine, far more than the ordinary caution which the mathematician is bound to exercise appears to be demanded [...].
>
> Dismissing, however, the more immediate subject of these comments, I cannot but observe upon the fact as somewhat remarkable, that in pure mathematics, controversy

> and misunderstanding should be so rife, as for some years they appear to have been. It would, I conceive, be interesting to inquire into the causes of a state of things which, judging *à priori,* we should so little expect. One reason is undoubtedly to be found in the unmeasured capabilities of the modern analysis for the expression of general theorems, and in the practically frequent employment of analogy and induction, especially as suggestive aids, in contradistinction to the purely deductive processes and more limited conclusions of the ancient geometry. But it may be doubted whether this is a sufficient explanation of the fact in question. A far more influential cause is, I believe, to be found in the almost entire absence of any direct study of the laws of correct reasoning in connexion with the practical discipline of modern science. But this is a topic which I do not venture upon the present occasion more fully to discuss.

Another amateur mathematician to enter the scene in 1848 was C. Hargreave, a lawyer by profession. Stimulated on one hand by Boole (1845), and on the other by the EFE, Hargreave devoted his first masterpiece 'On the Solution of Linear Differential Equations' (1848) to a study of equations of the form (3.4.11). This paper was followed by a far more general and abstract study, 'General Methods in Analysis for the Resolution of Linear Equations in Finite Differences and Linear Differential Equations' (1850). Hargreave (1850) and Brownin's elaborate paper 'On the Solution of Linear Differential Equations' (1851) manifested the peak of abstraction and generalization in the search after a 'genuine' and 'complete' method. Dispensing as far as possible with artificial techniques peculiar to the COO, Hargreave followed very closely along the lines of Boole (1844), without, however, achieving the desired goal. On the other hand, Brownin was very dependent upon Boole's techniques, but had serious doubts as to the existence of a true, unique general method. Having devoted a decade and about forty papers to the COO, Brownin stated emphatically in 1851:

> If we consider the very different forms which the solutions of Differential Equations differing very little from each other frequently take, and the very different processes often required in each particular case to obtain the solution, we shall be led to conclude that the discovery of any universal or general method of solving them must be a hopeless case. We cannot therefore regard particular methods, especially when applicable to a large number of cases, as useless speculations.[55]

The period of the 1850s saw, besides Hargreave and Brownin, many new ambitious authors who tried to elaborate on certain aspects of Boole's work, or to offer slightly different solutions of the EFE and the LE, the par-excellence source of inspiration and motivation for the development of the COO. What is indeed amazing is the fact that most of them, just like Hargreave and Brownin, were not mathematicians by profession but devoted their leisure to putting forward new 'beautiful' and 'elegant' theorems or solutions for the two above-mentioned equations. The astronomer W. Donkin, for instance, was quite reasonably drawn to the study of the EFE and the LE. However, he was particularly attracted also by the discovery of the analogy between commutative and non-commutative symbols and carried out research that was unlikely to render any service to the science of astronomy.[56]

The leading analyst of this period in connection with the COO was R. Carmichael, the author of the first treatise wholly devoted to the COO, which appeared in 1855. Carmichael's refusal to follow Boole's sophisticated method was striking but not unreasonable. Greatly favouring Gregory's methods, Carmichael claimed in a paper of 1854 that:

> It is proposed, in the following paper, to offer some practical remarks upon the subject of integration, more especially in reference to three capital defects under which many of the methods in common use appear to me to labour. The first defect in the methods [...] is their extremely artificial character, which occasions much embarrassment to the student at first, and considerable difficulty in his effort to retain them. The second great defect in these methods is, that they seem wholly unsusceptible of useful generalization. The third defect is less common, and consists in this, that some of the processes employed are circuitous, terms being introduced which subsequent operations cause to disappear. The method here put forward appears to be free from these defects, and, in so far as it is calculated to reduce and simplify the labours of the student, a practical good.[57]

Carmichael's arguments, in conjunction with Brownin's own three years earlier, amply refute the search for a universal symbolical method. With a view to producing a treatise where the laws of correct reasoning in the COO would be presented in a coherent way, Carmichael was the first author on symbolical methods to appeal to 'the student'. Unfortunately, we lack Boole's reactions to the epistemological conclusions arrived at in the writings of Brownin and Carmichael in the early 1850s. Apparently, the search for a 'genuine' and 'complete' symbolical method had ceased to be among his projects, as simplicity won out over artificial methods and techniques. In fact, Boole switched from the study of symbolical methods as such to a study of the foundations and philosophy of symbolical methods. Among his new projects was to produce a sequel to MAL and LT where the theory of logic, hidden in these works beneath excessive symbolism, would be stripped of its mathematical veil. His manuscripts on this effort, now edited as Boole (1997), prove the difficulties he was confronted with in carrying out his task. It is beyond the scope of this epilogue to enter into any discussion of Boole's manuscripts on the philosophy of logic in the late 1850s. We wish to claim, though, that this endeavour must not have been totally unrelated to results that were recent at that time in the development of symbolical methods, whose defects were readily pointed out.[58]

Many of Boole's novel speculations, partly diffused in the LT and above all in the manuscripts on his book on philosophy of logic, were nicely encapsulated in his textbook on *Differential Equations* (1859). As already revealed in the 'Postscript' of GMA (§3.1), Boole had meant since 1844 to delve further into the principles governing symbolical methods. Educational concerns induced him to furnish some of his mature epistemological views in the 'Introduction' to his (1859) which crowns his queries in the field of the COO. Boole no longer felt that symbolical methods were indispensable, and students were advised to follow the historical sequence of discovery and deal with symbolical methods only when ready to understand their nature. For 'Of the many forms of false culture, a premature converse with abstractions is perhaps the most likely to prove fatal to the growth of a masculine vigour of intellect' (1859, vi).

Referring to his early views on the close connections he perceived between logic and language (views which had formed part of his introductions to both MAL and LT), Boole stated in his 'Introduction' (1859, vii–viii):

> The question of the true value and proper place of symbolical methods is undoubtedly of great importance. Their convenient simplicity—their condensed power—must ever constitute their first claim upon attention. I believe however that, in order to form a just estimate, we must consider them in another aspect, viz. as in some sort the visible manifestation of truths relating to the intimate and vital connexion of language with thought—truths of which it may be presumed that we do not yet see the entire scheme and connexion. But, while this consideration vindicates to them a high position, it seems to me clearly to define that position. As discussion about words can never remove the difficulties that exist in things, so no skill in the use of those aids to thought which language furnishes can relieve us from the necessity of a prior and more direct study of the things which are the subjects of our reasonings. And the more exact, and the more complete, that study of things has been, the more likely shall we be to employ with advantage all instrumental aids and appliances.

Moreover, as clearly hinted at in both GMA and MAL, Boole did not view the basic principle governing the COO as peculiar to mathematics (§3.1, §4.2). The same belief was once more expressed in the context of his discussion of 'Forms purely symbolical' in chapter 16 of his (1859). Wondering whether one is allowed to treat expressions such as $f(d/dx)$ 'as if d/dx were a symbol of quantity', Boole claimed that (1859, 399):

> Now all special instances point to the conclusion that this is permissible, and seem to indicate, as a general principle, that the mere processes of symbolical reasoning are independent of the conditions of their interpretation. In the few instances we may have occasion to employ, verification will be easy. We take occasion to notice that, whatever view may be taken of this principle, whether it be contemplated as belonging to the realm of *a priori* truth, or whether it be regarded as a generalization from experience, it would be an error to regard it as in any peculiar sense a mathematical principle. It claims a place among the *general* relations of Thought and Language.

As Sophie Bryant pointed out in 1901:[59]

> The symbolism of the Calculus of Operations, is, indeed, the natural mean term, in the development of mathematical invention, between common mathematics and that general formal science which Boole called Mathematics in the wider sense of the word. We may—perhaps we must—call it General Formal Logic; but General Mathematics would be better. The name does not, however, affect Boole's contention that mathematical form is a true type of logical form in general, and that mathematical language as already established and highly organized is the natural language of symbol generally.

More recently, Laita (1977) and Grattan-Guinness (1982) have argued with Bryant that Boole viewed logic and mathematics as particular cases of a 'universal calculus of symbols'. Their research draws not only on MAL but also on Boole's earlier mathematical papers, as well as works by his wife on his religious influences. Indeed, the background of Boole's MAL was multidimensional, enriched with speculations on religion, philosophy and language. However, mathematics, and particularly the COO, formed the main key to his logical investigations. Summing up, we firmly believe that Boole would never have entered into such a sophisticated study of symbolical methods and their underlying philosophy, let alone have tried to apply them to logic, had he not witnessed through his own work in 1844 the immense power of these methods in generality and range of applications. As we claimed in §4.4, his GML can largely be seen as an analogue to his GMA. And we wish to stress that the latter was largely due to Ellis's valuable hint for the solution of the EFE (§2.3, §3.3).

ACKNOWLEDGEMENTS

I wish to express my deepest gratitude to the organizers of the George Boole Conference on the 150 years since his *Mathematical Analysis of Logic* at the University of Lausanne, and especially Dr James Gasser for his kind invitation and constant support for over a year. I was motivated by the theme of the Conference to revise my earlier research on British symbolical methods and the genesis of algebraic logic. This paper is an extended version of my talk at the Conference in September 1997, which I gave in conjunction with that of my ex-supervisor and dear friend, Professor Ivor Grattan-Guinness whom I thank for his enormous help and support even after my Ph.D. was over! I also wish to thank my friend Tassos Tokmakidis who helped me to print this paper, and who went through endless corrections of my draft, as well as Ivor and Harvey Becher for commenting on portions of this survey, and above all James for his thorough linguistic corrections.

NOTES

[1] The following abbreviations are used in this article: COO, Calculus of operations; EFE, Earth-figure equation; FTD, Fundamental theorem of development; GMA, General method in analysis; GML, General method in logic; GTD, General theorem of development; GTDL, General theorem of development in logic; LDE, Linear differential equation(s); LE, Laplace equation; LT, *Laws of Thought;* MAL, *Mathematical Analysis of Logic;* QPR, Quantification of the predicate.

[2] On the history of the calculus of operations in France, see Panteki (1992, ch. 1), and for a brief account Grattan-Guinness (1994) and Koppelman (1971, 155–75).

[3] On Herschel's and Babbage's early contributions, see Panteki (1992, ch.2) and Grattan-Guinness (1992a, b). Particularly on Herschel's theorem in the calculus of finite differences, see De Morgan (1842, 307) and Panteki (1992, §§2.3 and 4.8). The work of these analysts had no impact on Boole (1844).

[4] On the transitional period under study here, see Panteki (1992, §3.2). On the Tripos tradition and Whewell's educational views, see also Becher (1980, 1992).

[5] In the original version of (2.2.1), Laplace had a instead of x, and $h\int r a d a$ instead of y, where h stood for ellipticity, r for density, $r = (A/a)\sin na$, $n^2=2\pi/k$, A and k being constants, and a standing for the radius of the earth considered as a sphere. Laplace omitted to explain how he arrived at (2.2.2), and treated (2.2.3) clumsily by his series method (Laplace 1799–1825, bk XI, arts 6 and 9). See also Panteki (1992, §1.3).

[6] According to the complexity of the function P, equation (2.2.4) was involved in the theory of the two- or the three-body problem: see Wilson (1980, 98) and Woodhouse (1818, 25, 96–100). See also Panteki (1992, §3.2).

[7] Ellis's solution was traced accidentally on the cover of his 1831 edition of Airy's *Tracts*, bearing the date 1836, that is one year before Ellis's graduation (Panteki 1992, §3.2).

[8] On the Tripos question, originally posed by Gaskin in 1839, see Glaisher (1881, 810) and Panteki (1992, §4.2).

[9] Glaisher (1881) is devoted partly to Glaisher's original study of the EFE equation, and the closely related Riccati equation, and partly to a historical survey of the various symbolical forms produced by French and British mathematicians. Gaskin's attempt was saved from total oblivion thanks to Forsyth's textbook on differential equations. Interestingly, Glaisher was De Morgan's student, while A.R. Forsyth was Glaisher's student and biographer. For details, see references in footnote 8, and Forsyth (1914, 200).

[10] On Murphy's contributions, see Creedon (1992); Panteki (1992, §§3.2–3). Murphy's original procedures in the COO were not followed, being too general and complicated, and lacking applications. Boole was very vague as to his debt to Murphy. On Murphy's plausible influence upon Boole's GMA, see §3.4 below.

[11] On De Morgan (1836), see Panteki (1992, §§3.4–9). Regarded nowadays as a 'satellite' of the Analytical Society (Richards 1987, 10), De Morgan was the first mathematician to be deeply influenced by Peacock (1830) (cf. Pycior 1983), followed by Gregory (Panteki 1993, 138–9).

[12] Brownin became engaged with the EFE in 1843, deploying the series method. As his study had no influence on Boole, we omit a discussion of his contributions. On Greatheed's and Brownin's work on the COO, see Koppelman (1971, 188–9, 200–2) and Panteki (1992, §§4.2–3, 5.2), and §5 below.

[13] On the notion of the complementary function, anticipated by Liouville in 1832 and by Murphy (1833, 407; 1837, 188), see Creedon (1992, 66), Koppelman (1971, 195–6) and Panteki (1992, §3.3). It is worth stressing that Gregory (1839a) lacks Murphy's systematic study of inverse, distributive functions, the same defect that characterized Boole (1859); see Panteki (1992, §§4.4, 8.8).

[14] On the history of the formulae (2.3.9–2.3.10), diffused in England mainly through Gregory's work, see De Morgan (1842, 751–8); Gregory (1841, 29–32); Murphy (1837, 193–202); Koppelman (1971, 189–98) and Panteki (1992, §§3.3, 4.4).

[15] On the history of theorem (2.3.11), discovered by Cauchy in 1825, see Petrova (1987). On Gregory's peculiar procedures and their elaboration by Boole (1841), see Panteki (1992, §§4.4–5) and Petrova (1987, 13–4).

[16] On Ellis's treatment, see Gregory (1841, ch. 5); De Morgan (1842, 700–4); Forsyth (1914, 201, 204) and Panteki (1992, §4.3).

[17] On the reception of Boole (1844) by the Royal Society, and Kelland's testimonial, see MacHale (1985, 60–6).

[18] The import of the EFE in Boole (1844) would be explicitly stressed by Kelland (1858, 85). See also quotation in §1.1 above and reference in footnote 17 above.

[19] The absence of the rigorous foundation of inverse, distributive operations leads to the impression that Boole did not base his GMA upon Murphy's studies; see Panteki (1992, §§4.4, 4.5, 8.8), and §3.4 below. The only treatise to incorporate Murphy's study of inverse operations was Carmichael (1855), where Boole's general symbolical method was omitted (ibid., §5.7).

[20] Boole (1844), briefly mentioned by Koppelman and Laita (§1.1), was first studied in Panteki (1992, ch. 4); see §4.5 on the structure and contents of Boole's paper, an account enriched with examples. Exceptionally, an example from Boole's theory for the summation of series was displayed in MacHale (1985, 64–5). The reader may profit by reading Boole's own summary of his paper in his letter to De Morgan, 19 June 1843, in Smith (1982, 11–2).

[21] On Boole's procedure, see Boole (1844, 251; 1859, 422–3) and Panteki (1992, §§4.6–7). The latter investigation is more detailed than the present one. However, we are happy to claim that due to a thorough reassessment of Boole's work, motivated by the Boole Conference held at Lausanne in 1997, we have shed further light on the genesis of the GMA, and on Ellis's crucial impact in this respect.

[22] On Boole's letter to De Morgan on 19 June 1843, see Smith (1982, 12). This quote is indicative of the inspiration Boole obtained from Ellis (1841a) (see also §3.2 above). For details on calculations, missing in Boole's paper, and applications of *Propositions 1* and *2* to the general EFE (2.2.3), see Panteki (1992, §4.6).

[23] Both MacHale (1985, 63) and Creedon (1992, 59, 70) insist that Boole copied from Murphy's papers while working on his GMA. Our study so far has shown that at least the basic ingredients of Boole's GMA were based almost exclusively on formulae and ideas provided by Gregory and Ellis. Any influence of Murphy (1837) would have been in the realm of the GTD (3.4.5) in so far as both Murphy and Boole tried to establish the proof of Taylor's theorem upon a rigorous basis. Murphy anticipated formula (3.4.7), but Boole's procedures differ substantially from those of Murphy to whom he would surely have given credit—as he did to Ellis—if he had been significantly influenced by him. This explains why we have given no account of Murphy (1837) in this survey. See also footnotes 10 and 19 above.

[24] See Boole's letter of 24 February 1845 in Smith (1982, 15). On Boole's procedures, see Panteki (1992, §4.7), and Boole (1859, 455). Interestingly, in some manuscript notes with code number W21, Boole aimed at deducing Lagrange's theorem (2.1.1) from his GTD (on Boole's manuscripts and their classification, see Boole (1997, 201–8)).

[25] On the import of Boole's GMA, see Panteki (1992, ch. 5). Especially on the LE, Boole's procedures are recorded in §5.8 of the above work.
[26] See Panteki (1992), chs. 5 and 9, and especially §9.4.
[27] On the stagnation of research in logic in 17th and 18th century England and the alternative theories that prevailed at that time, see Dessi (1988, xvi–ii) and Van Evra (1984, 3–7).
[28] Boole and De Morgan corresponded on logical issues beginning in 1842 (Smith 1982). However, a study of their letters shows that the correspondence had hardly any influence on the shaping of their distinct approaches (Corcoran 1986). Upon these grounds we will not comment on De Morgan's work; see also Boole (1997, xlv).
[29] Boole (1847a, 1). The De Morgan-Hamilton dispute and its influence upon Boole's MAL is amply dealt with in Laita (1979). The issue of the QPR is briefly explained below in connection with Bentham. As to other stimuli that may have induced Boole to delve into logic, see references in Boole (1997, Part 1) and Panteki (1992, §7.1).
[30] Laita (1979) and Durand (1996) touch upon the Whewell-Hamilton dispute on the educational import of mathematics and logic, but this issue has been only fragmentarily dealt with so far, and De Morgan's views at the time have not been taken into account at all. Solly's *Syllabus* has been discussed recently by Panteki (1993), but a comprehensive study of all the tendencies during the crucial period of the 1830s is still missing.
[31] First mentioned by Van Evra (1984, 10, fn 21), Kirwan's contributions are discussed in Panteki (1992, §6.2); see also Panteki (1993, 140–1). Kirwan's fresh attention to the scientific role of logic was noted by Hamilton (1833, 204), apparently the only logician to have read his work.
[32] On Whately's contributions and their reception by Bentham, Hamilton, De Morgan and J.S. Mill, see Van Evra (1984). On Solly's critical response, see Panteki (1993, 145–6). In what follows we quote from Whately (1826), as reprinted in Dessi (1988), including interesting comments on Whately's role in the revival of logic in England.
[33] On Bentham's syllogistic scheme, see Panteki (1992, §6.3; 1993, 141–2) and Styazhkin (1969, 148–50). On the anticipation of the QPR by Continental logicians, see Lewis (1918, 33–6).
[34] On Hamilton's logical contributions, see Passmore (1968, 120–1) and Styazhkin (1969, 150–6). On the consequences of this dispute on De Morgan's work, see Panteki (1992, §§6.6–7). Ironically, as argued in Laita (1979), by supporting the QPR, Hamilton suggested that the theory of equations is the foundation of logic, despite his notorious hostility to mathematics. See also references in footnote 30 above.
[35] On Boole's reaction to Hamilton's views on the limited nature of mathematics and their danger in assisting the faculty of reasoning, see MAL, pp. 10–4, and Laita (1979, 60–1). On the Whewell-Hamilton dispute, see also Durand (1996, 456–7).
[36] On Whewell's disrespect for Whately's theory of political economy and logic, see Yeo (1993, 55, 93, 102–6, 193–8). For more on Whewell's role at Cambridge University, see Becher (1980; 1992) and §2.2 above.
[37] Our link between De Morgan (1835) and Whewell (1835) is tentative. Even if the latter was written without knowledge of the former, De Morgan's views on algebra and logic in 1835 still deserve attention (see Panteki (1992, §3.4)).
[38] See De Morgan (1847, 335–6), Merrill (1996) and Panteki (1993, 143).
[39] We lack exact evidence on the sources of Solly's inspiration in connection with mathematics. His plausible background and details on the procedures followed by him are presented amply in Panteki (1993), especially pp. 153–7.
[40] See Panteki (1993, 140, 150 and 165). This survey hints at other similarities between the views of these two logicians.
[41] This quote is from Boole's letter to Cayley of 10 December 1847. The Boole-Cayley correspondence of that period is appended at the end of Boole (1997, 191–7). The above-mentioned letter is fully quoted in §4.4 below. These letters, which have been useful in sustaining our arguments in this survey, were first used in Panteki (1992, chs. 7 and 8), devoted to Boole's MAL and LT respectively.
[42] As we stressed in §2.1 above, we take all recent studies of Boole's logic for granted, dispensing with details concerning logical issues. As it is beyond the scope of our survey to deal with LT, we refer the reader to Boole (1997) and Panteki (1992), especially §§8.1–4.
[43] The first of these laws is transcribed in Boole (1997) as '$x(y + z) = xy + yz$', but this is clearly a misprint.

[44] These 'Notes', fragmentarily presented in Boole (1997, 42–5), are fully discussed in Panteki (1992, §8.2). On the operation of addition of two classes, vaguely defined in MAL, see Grattan-Guinness (1991). Commutativity of addition was missing in MAL, like the property of associativity and other properties as well. On such crucial omissions, see Corcoran and Wood (1980).
[45] On Cayley's first letter to Boole, see Boole (1997, 191) and also Panteki (1992, §8.2). On the elements of Boole's basic logic, see ibid., §§ 7.2–4 and Van Evra (1977).
[46] The 'Notes' referred to in footnote 44 are the only work in which Boole considered division in logic to be the inverse of multiplication (Panteki 1992, §8.3). On Boole's approach to division, as in LT, which seriously puzzled all his readers, see Hailperin (1981) and Hooley (1966). This issue is only tangentially dealt with in §4.4 below. See more references and details in Panteki (1992, ch. 8).
[47] The indefinite symbol v was apparently influenced by Hamilton's scheme which was based on the definite and indefinite quantities (see Laita 1977). For a full discussion of Table (4.3.6), see Panteki (1992, §7.4).
[48] See footnote 42 above.
[49] See Boole (1997, 194) and Panteki (1992, §§7.7 and 8.2).
[50] See Panteki (1992, §§7.5, 7.7 and 8.4).
[51] This quote is from Boole's first reply to Cayley's letter, dated 6 December 1847; see Boole (1997, 191–2) and Panteki (1992, §7.5). We give only one example of a solved elective equation due to the limited length of this survey. For a detailed study, including more examples, see Panteki (1992, §7.5).
[52] See Boole (1997, 43) and Panteki (1992, §8.2).
[53] As we mentioned in §2.3 above, Brownin started working on differential equations in 1843. On his early work, see Panteki (1992, §4.3). Unfortunately, there exists no biography of this Cambridge outsider.
[54] Smith (1984, 23). On Brownin's papers published in 1847–1848 and Boole's remarks on them, see details in Panteki (1992, §5.2).
[55] Brownin (1851, 461). On the peak of the COO around 1850, see Panteki (1992, §5.4).
[56] On Donkin's simplification and generalization of laws furnished by Boole and Brownin, see Koppelman (1971, 203) and Panteki (1992, §5.5). On the whole, over twenty figures are mentioned in chapter 5 of the latter dissertation for their contributions to the COO from the late 1840s to the late 1860s. Among them feature, besides those referred to in the text, C. Graves, A. Curtis, W.H.L. Russell, W. Spottiswoode, H. Jellett, W.R. Hamilton, R. Greer and others. By 1860 and thereafter, interest in the COO distinctly faded away. Explanations for this phenomenon are given in chapter 9 of Panteki (1992).
[57] Carmichael (1854, 227). Carmichael's views were also repeated in his textbook (Carmichael 1855, 3–4). Unfortunately, we have hardly any biographical information on Carmichael. Historically, he is fairly well known for his textbooks (see comments in footnote 19 above and Koppelman (1971, 213)). For a detailed study of his research in the COO, see Panteki (1992, §5.7).
[58] On Boole's philosophical speculations from 1847 onwards, see Boole (1997) and Panteki (1992, ch. 8).
[59] Bryant (1901, 106). See also Panteki (1992, §7.2).

REFERENCES

Airy, G.B. 1826. *Mathematical tracts on the lunar and planetary theories, the figure of the earth, precession, nutation and the calculus of variations*. Cambridge: Deighton.

Babbage, C., J.F.W. Herschel, and G. Peacock. 1820. *Collection of examples*. Cambridge: Deighton.

Becher, H. 1980. William Whewell and Cambridge mathematics. *Historical Studies of the Physical Sciences* 11: 1–48.

———. 1992. The Whewell story: Essay review. *Annals of Science* 49: 377–84.

———. 1995. Radicals, Whigs and conservatives: the middle and lower classes in the analytical revolution at Cambridge in the age of aristocracy. *British Journal of the History of Science* 28: 405–26.

Bentham, G. 1827. *An outline of a new system of logic*. London: Hunt and Clark.

Boole, G. 1841. On the integration of linear differential equations with constant coefficients. *Cambridge Mathematical Journal* 2: 114–9.

———. 1844. On a general method in analysis. *Philosophical Transactions of the Royal Society of London* 134: 225–82.

———. 1845. On the theory of developments. Part I. *Cambridge Mathematical Journal* 4: 20–8.
———. 1846. On the equation of Laplace's functions. *Cambridge and Dublin Mathematical Journal* 1: 10–22.
———. 1847a. *The mathematical analysis of logic, being an essay towards a calculus of deductive reasoning.* Cambridge: Macmillan.
———. 1847b. On a certain symbolical equation. *Cambridge and Dublin Mathematical Journal* 2: 7–12.
———. 1848. Remarks on a paper by the Rev. Brice Brownin, On the solution of a particular differential equation. *Philosophical Magazine* 32: 413–8.
———. 1854. *An investigation of the laws of thought, on which are founded the mathematical theories of logic and probabilities.* London: Walton and Maberley.
———. 1859. *A treatise on differential equations.* London: Macmillan. Pagination referred to from a reprint of the 4th edition (1877) published by G.E. Sterchert (1931), New York.
———. 1860. *A treatise on the calculus of finite differences.* London: Macmillan. References from the 2nd edition by J.F. Moulton (1872). London: Dover.
———. 1997. *George Boole. Selected manuscripts on logic and its philosophy.* Edited by I. Grattan-Guinness and G. Bornet. Basel: Birkhäuser.
Brownin, B. 1848. On the solution of a particular differential equation. *Philosophical Magazine* 32: 256–62.
———. 1851. On the solution of linear differential equations. *Philosophical Transactions of the Royal Society of London* 141: 461–82.
Bryant, S. 1901. The relation of mathematics to general formal logic. *Proceedings of the Aristotelian Society,* n.ser. 2: 105–34.
Carmichael, R. 1854. Remarks on integration. *Cambridge and Dublin Mathematical Journal* 9: 227–33.
———. 1855. *Treatise on the calculus of operations.* London: Longman.
Corcoran, J. 1986. Correspondence without communication. *History and Philosophy of Logic* 7: 65–75. (Review of Smith (1982).)
Corcoran, J. and S. Wood. 1980. Boole's criteria for validity and invalidity. *Notre Dame Journal of Formal Logic* 21: 609–38. [Reprinted in this volume, pp. 101–128.]
Creedon, L. 1992. *The life and works of Robert Murphy 1806–1843.* (Msc Dissertation, National University of Ireland.)
De Morgan, A. 1835. A treatise on algebra. *Quarterly Journal of Education* 9: 91–110; 239–311. (Review of Peacock (1830).)
———. 1836. Calculus of functions. *Encyclopedia Metropolitana* 2: 305–92. (Date of original offprint publication.)
———. 1840. Operation. *Penny Cyclopedia* 16: 442–6.
———. 1842. *The differential and integral calculus.* London: Baldwin and Cradock. (Dated as of completion; published in numbers from 1836.)
———. 1847. *Formal logic; or, the calculus of inference, necessary and probable.* London: Taylor and Walton.
———. 1850. On the syllogism: II. On the symbols of logic, the theory of the syllogism, and in particular of the copula. *Transactions of the Cambridge Philosophical Society* 9: 79–127. (Read 1850; also in De Morgan (1966), 22–68.)
———. 1966. *On the syllogism, and other logical writings.* Edited by P. Heath. London: Routledge.
Dessi, P., ed. 1988. Introduction to Whately (1826). Bologna: Instrumenta Rationis.
Durand, M.J. 1996. L'Ecole algébrique anglaise: les conditions conceptuelles et institutionnelles d'un calcul symbolique comme fondement de la connaissance. In *L'Europe mathématique: Mythes, histoires, identités,* edited by C. Goldstein, J. Gray and J. Ritter. Paris: Editions de la Maison des sciences de l'homme.
Ellis, R.L. 1841a. On the integration of certain differential equations. *Cambridge Mathematical Journal* 2: 169–77.
———. 1841b. On the integration of certain differential equations. *Cambridge Mathematical Journal* 2: 193–201.
Enros, P.C. 1983. The analytical society (1812–1813): precursor of the renewal of Cambridge mathematics. *Historia Mathematica* 10: 24–47.

Forsyth, A.R. 1914: *A treatise on differential equations*. 4th ed. London: Macmillan. (1st edition dated 1885.)

Glaisher, J.W.L. 1881. On Riccati's equation and its transformations, and on some definite integrals which satisfy them. *Philosophical Transactions of the Royal Society of London.* 172: 759–828.

Grattan-Guinness, I. 1982. Psychology in the foundations of logic and mathematics: the cases of Boole, Cantor and Brouwer. *History and Philosophy of Logic* 3: 33–53.

———. 1991. The correspondence between George Boole and Stanley Jevons, 1863–1864. *History and Philosophy of Logic* 12: 15–35.

———. 1992a. Charles Babbage as an algorithmic thinker. *Annals of the History of Computing* 14: 34–48.

———. 1992b. The young mathematician. In *John Herschel 1792–1871: a bicentennial commemoration,* edited by D.G. King-Hele. London: Royal Society.

———. 1994. Operator methods. In *Companion encyclopedia of the history and philosophy of the mathematical sciences,* edited by I. Grattan-Guinness. London and New York: Routledge.

Gregory, D.F. 1839a. On the solution of differential equations with constant coefficients. *Cambridge Mathematical Journal* 1: 22–31. (Also in Gregory (1865), 14–27.)

———. 1839b. Demonstrations of theorems in the differential calculus and the calculus of finite differences. *Cambridge Mathematical Journal* 1: 212–22. (Also in Gregory (1865), 108–23.)

———. 1841. *Examples of the processes of the differential and integral calculus.* London: Deighton.

———. 1865. *The mathematical writings of D.F. Gregory.* Edited by Walton. Cambridge: Deighton.

Hailperin, T. 1981. Boole's algebra isn't Boolean algebra. *Mathematical Magazine* 54: 173–84. [Reprinted in this volume, pp. 61–77.]

Hamilton, W. 1833. Recent publications on logical sciences. *Edinburgh Review* 57: 194–238.

———. 1836. Thoughts on the study of mathematics as a part of a liberal education. *Edinburgh Review* 62: 409–55. (Review of Whewell (1835).)

Hargreave, C. 1848. On the solution of linear differential equations. *Philosophical Transactions of the Royal Society of London* 138: 31–54.

———. 1850. General method in analysis for the resolution of linear equations in finite differences and linear differential equations. *Philosophical Transactions of the Royal Society of London* 140: 262–86.

Hooley, J. 1966. Boole's method for solving boolean equations. *Mathematical Gazette* 50 114–8.

Hymers, J. 1839. *A treatise on differential equations and the calculus of finite differences.* Cambridge.

Kelland, P. 1858. Address on the award of the Keith Medal. *Proceedings of the Royal Society of Edinburgh* 84–7.

Kirwan, R. 1807. *Logick.* London.

Koppelman, E. 1971. The calculus of operations and the rise of abstract algebra, *Archive for History of Exact Sciences* 8: 155–242.

Lacroix, S.F. 1816. *An elementary treatise on the differential and integral calculus.* Translated by C. Babbage, J.W.F. Herschel and G. Peacock. Cambridge: Deighton.

Laita, L.M. 1977. The influence of Boole's search for a universal method in analysis on the creation of his logic. *Annals of Science* 34: 163–76. [Reprinted in this volume, pp. 45–59.]

———. 1979. Influences of Boole's logic: the controversy between William Hamilton and A. De Morgan. *Annals of Science* 36: 45–65.

Laplace, P.S. 1799–1825. *Traité de mécanique céleste*. Paris.

Lewis, C. 1918. *A survey of symbolic logic.* Berkeley: University of California Press.

MacHale, D. 1985. *George Boole. His life and work.* Dublin: Boole Press.

Merrill, D.D. 1996. Making sense of Solly's syllogistic symbolism. *History and Philosophy of Logic* 17: 199–207

Murphy, R. 1833. On the inverse method of definite integrals, with physical applications. *Transactions of the Cambridge Philosophical Society* 4: 352–408.

———. 1837. On the theory of analytical operations. *Philosophical Transactions of the Royal Society of London* 127: 179–210.

Panteki, M. 1992. Relationships between algebra, differential equations and logic in England: 1800–1860. Ph.D. Dissertation, London (Council of National Academic Awards).

———. 1993. Thomas Solly (1816–1875): an unknown pioneer of the mathematization of logic in England, 1839. *History and Philosophy of Logic* 14: 133–69.

Passmore, J. 1968. *A hundred years of philosophy.* London: Pelican.

Peacock, G. 1830. *A treatise on algebra.* Cambridge.

Petrova, S. 1987. Heaviside and the development of the symbolic calculus. *Archive for History of Exact Sciences* 37: 1–23.

Pratt, J.H. 1836. *The mathematical principles of mechanical philosophy.* Cambridge: Deighton.

Pycior, H.M. 1983. Augustus De Morgan's algebraic work: the three stages. *Isis* 74: 211–26.

Richards, J. 1987. Augustus De Morgan, the history of mathematics, and the foundations of algebra. *Isis* 78: 7–30.

Smith, G.C., ed. 1982. *The Boole-De Morgan correspondence 1842–1864.* Oxford: Clarendon Press.

———. 1984. *The Boole-Thomson letters.* Monash University, Clayton (Australia), Paper no 32.

Solly, T. 1839. *A syllabus of logic in which the views of Kant are generally adopted, and the laws of syllogism symbolically expressed,* Cambridge: Deighton.

Styazhkin, L. 1969. *History of mathematical logic from Leibniz to Peano.* London: M.I.T. Press.

Van Evra, J. 1984. Richard Whately and the rise of modern logic. *History and Philosophy of Logic* 5: 1–18.

———. 1977. A reassessment of George Boole's theory of logic. *Notre Dame Journal of Formal Logic* 17: 363–77. [Reprinted in this volume, pp. 87–99.]

Whately, R. 1826. *Elements of logic.* London. Pagination referred to from the reprint of this first edition by Dessi (1988).

Whewell, W. 1835. *Thoughts on the study of mathematics as a part of a liberal education.* Cambridge: Deighton.

Wilson, C.A. 1980. Perturbations and solar tables from Lacaille to Delambre: the rapprochement of observation and theory, Part I. *Archive for History of Exact Sciences* 22: 53–188.

Woodhouse, R. 1818. *An elementary treatise on astronomy, Vol. II containing physical astronomy.* Cambridge: Deighton.

Yeo, R. 1993. *Defining science. William Whewell, natural knowledge and public debate in early Victorian Britain.* Cambridge: Cambridge University Press.

Department of Mathematics, Aristotle University of Thessaloniki,
Thessaloniki, 54006, Greece

IVOR GRATTAN-GUINNESS

ON BOOLE'S ALGEBRAIC LOGIC AFTER *THE MATHEMATICAL ANALYSIS OF LOGIC*

THINGS THE SAME

Boole did not fundamentally change the lines of his research after publishing *The Mathematical Analysis of Logic* (Boole 1847; hereafter '*MAL*'). His logic was still mathematics applied to logic; consequences were to be drawn from premises rendered in algebraic form by means of expansion theorems which in effect solved class equations; logical expressions might (not) be interpretable; the role of signs remained prominent; language was still focused upon nouns and adjectives, although its role in general was still emphasised; classes were analysed in terms of part and whole; disjunction was defined only for disjoint classes; and so on.

Boole also remained blind to the singular solutions to which his equations were sometimes susceptible (Corcoran and Wood 1980). This was a strange oversight in a mathematician for whom such solutions to differential equations were a passion, with their positive and negative properties (a solution, but not one of the particular ones), in imitation of his logical universe 1 and its two complementary parts x and $(1 - x)$ (Grattan-Guinness 1982).

However, Boole also made various modifications and extensions. The main result was the second book, *The Laws of Thought* (Boole 1854; hereafter '*LT*'), the more famous now but less well received then; in addition, he put out a few papers. He also wrote a variety of manuscripts of which Gérard Bornet and I recently published a selection, prefaced by an extensive editorial introduction (Boole 1997, cited below as '*SM*').

To popularise his theory of logic, Boole began to write a third book, entitled 'The Philosophy of Logic', but he never completed it; the surviving identifiable material is published in (*SM*, chs. 8–11). Maintaining the strong relationship to thinking, he also wrote on the connections between this logic and arithmetic, and the role of signs; and he went back to contrast it with 'the Aristotelian logic'. In addition, he attempted a book on the philosophy of mathematics; the surviving chapter treats 'geometry as a science', largely within a Euclidean framework, including the role of syllogisms (*SM*, ch. 15).

J. Gasser (ed.), A Boole Anthology, 213–216.

THINGS DIFFERENT

The main changes after *MAL* will be summarised here. Some are of content, others of balance. The chronological details of the changes are not always noted, and indeed in several cases not known.

1. Boole reduced his basic law to $x^2 = x$ from the $x^n = x$ for any integer $n \geq 2$ of *MAL* when he showed that the second version involved the uninterpretable factor $(1 + x)$ (*LT*, 50 for $n = 3$). He was also more explicit about specifying the universe pertaining to a given logical exercise (*LT*, 32–4).

2. The technical means were graced by the property $f(0)f(1) = 0$ for any logical function f, and analogous versions for functions of several variables, which improved the methods of elimination (*LT*, 101). Boole proved the expansion theorems by evaluation of coefficients, the proof by Maclaurin's theorem being relegated to a footnote (*LT*, 72–3). In addition, he made more clear the division coefficients, especially '0/0' as denoting none, some or all of the members of the class for which it was the coefficient (*LT*, 89–92).

3. Boole clarified the relationship between the algebras of his logic and of common algebra, and the status of number was made more clear (*SM*, ch. 2); he noted the link between integers and iterations of x to produce $x, x^2, x^3, \ldots$ (*SM*, ch. 3), without being sure how to exploit it. '0' and '1' possessed only certain features in common with their arithmetical brothers.

4. *MAL* shows an element of model theory, in that Boole interpreted the theory in terms both of classes and 'elective symbols' denoting mental acts. The latter reading soon started to dim down, and classes became the standard version. However, the aim to prescribe thinking remained, as the title of *LT* shows; in its last chapter he even attempted to describe the 'constitution of the intellect'.

5. The role of syllogistic logic was also greatly diminished. In *MAL* it is a frequent point of reference for the new theory, whereas in *LT* Boole treated it only in the 15th of the 15 chapters concerned with logic proper. Seemingly the revolutionary character of his algebra did not dawn upon him at first (although he may have used syllogistic in *MAL* as a known source for the convenience of readers). The manuscripts written between the two books (*SM*, chs. 1–3) show the decline in interest very markedly.

6. By contrast, Boole came up with another reading of his laws (more model theory, as it were): probability theory, where compound events were construed as Boolean combinations of simple ones. The insight seems to have come to him early in 1849 (*SM*, 198–200). Most of the rest of *LT* after ch. 15 is taken up with this topic. Various novelties came out of this endeavour, most not well appreciated for a long time (Hailperin 1986).

7. *LT* also contains a new distinction of propositions: instead of the categorical and hypothetical categories, Boole now worked in ch. 4 with 'primary or concrete'

and 'secondary or abstract' types. The second names clarified the distinction; the former type of proposition related 'to *things*' and the latter 'to *propositions*'. Primary propositions were categorical, but once again they were not treated as 'atomic' entities; instead, they were components of secondary ones, which included hypothetical propositions such as 'If the sun shines the earth is warmed' (*LT*, 53). This was the closest that he came to the propositional calculus as an algebra in his sense; the reading to which we are now accustomed is due to (MacColl 1877).

8. Boole also invoked time in *LT*, ch. 11, in a surprising, and non-durable way: '$X = 0$' means that proposition X was never true, '$X = 1$' always so. This idea was to gain some currency among successors, especially Ernst Schröder, though not happily: '$= 0$' became multiply interpretable, as 'is false', 'is equivalent to a contradiction', and/or 'equals the empty class'. Logical compounds such as '$(1 = 0) = 0$' become a combinatorial nightmare of interpretations! (Grattan-Guinness 1975).

9. Perhaps at a late stage, Boole sought an epistemological basis for his logic. The topic was to be treated in the third book (*SM*, ch. 10), and some other manuscripts related to it (*SM*, chs. 4–5, and also the one of the 1860s printed in (Boole 1952, ch. 6)). He followed a standard approach of his time, though modified or oriented to fit his new logic. The act of Conception by the mind produced a Concept, such as 'trees' (or the class of them); then the act of Judgement of the copresence of Concepts produced a Proposition, such as 'this is an oak tree'; finally, the act of Reasoning produced a Conclusion, perhaps 'oak trees are ...'.

10. There is an important religious line in Boole, for he linked his logical Universe to ecumenism. In his later years he became very enthusiastic for the presentation of logic by the French theologian A.J.A. Gratry, in which claims such as God as the truth gained prominence, and also topics such as nullity versus unity, universal laws of thought, and the ultimate source of exercise of the human mind (Gratry 1855). With his wife, he became around that time a profound admirer of Frederick Denison Maurice, who suffered professionally for advocating such a brand of Christianity; Boole had his portrait placed by his death-bed. He never mentioned religiosity in *MAL* or in his technical papers; but some traces are evident in *LT*. He devoted ch. 13 of *LT* to logical analyses of passages from the writings of Samuel Clarke and Benedict Spinoza which surely did *not* just happen to deal with properties of the one and only God. In addition, there is a striking allusion to it at the end of the book, where he saw the 'juster conceptions of the unity, the vital connexion, and the subordination to a moral purpose, of the different parts of Truth, among those who [...] profess an intellectual allegiance to the Father of Lights'. This last phrase was then a code term for his kind of Dissenter to refer to the unifying Godhead. Apart from his widow, Boole's followers did not follow him here.

BIBLIOGRAPHY

In addition to the items cited in the text, others are included which bear upon the points made; comments are added to a few of the latter.

Boole, G. 1847. *The mathematical analysis of logic* ..., Cambridge: Macmillan, Barclay and Macmillan. [Also in (1952), 45–124 (with additions). Repr. 1948, Oxford: Blackwell.]

———. 1854. *The laws of thought* ..., London: Walton and Maberley. [Repr. 1958, New York: Dover.]

———. 1952. *Studies in logic and probability*. Edited by R. Rhees. London: Watts.

———. 1997. *Selected manuscripts on logic and its philosophy*. Edited by I. Grattan-Guinness and G. Bornet. Basel: Birkhäuser.

Corcoran, J., and S. Wood. 1980. Boole's criteria for validity and invalidity. *Notre Dame Journal of Formal Logic* 21: 609–38. [Reprinted in this volume, pp. 101–28.]

Gratry, A.J.A. 1855. *Philosophie—logique*. Paris: Douniot. [English trans. and introd. by H. and M. Singer of the 5th ed. (1868): *Logic*. La Salle; Open Court (1944).]

Grattan-Guinness, I. 1975. Wiener on the logics of Russell and Schröder. An account of his doctoral thesis, and of his subsequent discussion of it with Russell. *Annals of Science* 32: 103–32.

———. 1982. Psychology in the foundations of logic and mathematics: the cases of Boole, Cantor and Brouwer. *History and Philosophy of Logic* 3: 33–53. Also in *Psicoanalisi e storia della scienza*. Florence: Olschki (1983, 93–121).

Hailperin, T. 1986. *Boole's logic and probability*. 2nd ed. Amsterdam: North Holland.

Laita, L. 1980. Boolean algebra and its extra-logical sources: the testimony of Mary Everest Boole. *History and Philosophy of Logic* 1: 37–60. [Argues well for her basic reliability.]

MacColl, H. 1877. The calculus of equivalent statements and integration limits. *Proceedings of the London Mathematical Society* 9 (1): 9–20.

MacHale, D. 1985. *George Boole—his life and work*. Dublin: Boole Press.

Neil, S. 1865. Modern logicians—the late George Boole. *The British Controversialist and Literary Magazine* n.s., no 80, 81–94; no 81, 161–74. [Fine, little-known obituary; reprinted in this volume, pp. 1–25.]

Panteki, M. 1992. Relationships between algebra, differential equations and logic in England: 1800–1860. C.N.A.A. (London) doctoral dissertation. [Strong on context and pre-history.]

Prior, A.N. 1949. Categoricals and hypotheticals in George Boole and his successors. *Australian Journal of Philosophy* 27: 171–96. [Insightful on the role of universes.]

Styazhkin, N.I. 1969. *History of mathematical logic from Leibniz to Peano*. Cambridge, Mass.: MIT Press. [Useful on the reception of Boole in the late 19th century.]

Young, G.P. 1865. Remarks on Professor Boole's mathematical theory of the laws of thought. *The Canadian Journal* 10: 161–82. [Interesting early, and little-known, study; reprinted in this volume, pp. 27–43.]

Middlesex University, Queensway, Enfield, Middlesex EN3 4SF, U.K.

SRIRAM NAMBIAR

THE INFLUENCE OF ARISTOTELIAN LOGIC ON BOOLE'S PHILOSOPHY OF LOGIC: THE REDUCTION OF HYPOTHETICALS TO CATEGORICALS

> I am then compelled to assert, that, according to this view of the nature of Philosophy, *Logic forms no part of it*. On the principle of a true classification, we ought no longer to associate Logic and Metaphysics, but Logic and Mathematics (Boole 1847, 13).

George Boole (1815–1864) is widely regarded as the founder of mathematical logic. Many influential logicians at the beginning of the twentieth century, a time of rapid growth in logic, credited him with originating the tradition in which they worked. Many influential historians of logic credit him with playing a pivotal role in transforming logic from the mold in which it was set from ancient times to the dramatically growing mathematical form it has assumed in more recent times.

The mathematician Peano, a pioneer in modern logic, acknowledges the influence of the Boolean tradition in his *Principles of Arithmetic.*

> In this paper I have used the research of others. The notations and propositions of logic ... represent the work of many, among them Boole especially (Peano 1889, 102).

By the time of his *Principles of Mathematics*, Russell (1903, 5) sees mathematical logic as having acquired a more central role in mathematics.

> The fact that all Mathematics is Symbolic Logic is one of the greatest discoveries of our age; and when this fact has been established, the remainder of the principles of mathematics consists in the analysis of Symbolic Logic itself.

Establishing the reduction of mathematics to logic was the project which he undertook for the next decade. It was only after he had written almost all of this book that Russell first read Frege and detected his famous paradox. He recognizes Boole's role in the founding of symbolic logic (Russell 1903, 10).

> It is from the recognition of asyllogistic inferences that modern Symbolic Logic, from Leibniz onward, has derived the motive to progress. Since the publication of Boole's *Laws of Thought* (1854), the subject has been pursued with a certain vigour, and has attained to a very considerable technical development.

In their book *Symbolic Logic*, Lewis and Langford (1932, 7) recognize Boole's contribution.

J. Gasser (ed.), A Boole Anthology, 217–239.

> The foundations from which symbolic logic has had a continuous development were laid in Great Britain between 1825 and 1850. The signally important contribution was that of the mathematician George Boole [...].

Church, Tarski and Quine who, to a significant extent, have shaped the present form of mathematical logic and philosophy of logic also see Boole as originating this tradition.

> While [Boole] made contributions to other branches of mathematics, he is now remembered primarily as the founder of the Nineteenth Century algebra of logic and through it of modern symbolic logic (Church 1983, 55).

> A continuous development of mathematical logic began only near the middle of the 19th century, at about the time when the logical system of the English mathematician G. BOOLE was published ... the new logic found its full expression in the epochal work of ... A.N. WHITEHEAD ... and B. RUSSELL ... (Tarski 1941, 18 note).

> [Mathematical logic's] crude beginnings are placed with George Boole, in the middle of the last century. Fragments foreshadowing mathematical logic date back much farther than Boole—as far back indeed as Leibniz; but it was from Boole onward through Peirce, Schröder, Frege, Peano, Whitehead, Russell and their successors that mathematical logic underwent continuous development and reached the estate of a reputable department of knowledge (Quine 1940, 1).

However, in spite of the seminal role he is believed to have played in the creation of modern logic, and the fine biography by Desmond MacHale in 1985, there is no close study of the development of his work and hence no basis for an assessment of the influence he has had on later logicians (Corcoran 1987). In this paper I give a brief description of Boole's philosophy of logic and compare it with a philosophy of traditional logic, as depicted in Richard Whately's (1787–1863) well-known exposition of logic of 1827. This comparison brings out reasons that Boole saw himself as a follower of the Aristotelian tradition. The philosophy of logic which gave rise to Boole's endeavors led him to see it as a completion of a project first undertaken by Aristotle two millennia earlier. With this brief sketch as background I examine, in some detail, the influence of a traditional conception of the relation between categorical and hypothetical propositions on Boole, in order to show how an appreciation of traditional logic can improve our evaluation of Boole's project. Not paying attention to this tradition has led historians of logic to crucial misunderstandings of the work which is regarded as founding modern logic.

1. A BRIEF DESCRIPTION OF BOOLE'S PHILOSOPHY OF LOGIC

That Boole saw himself as reformulating and refining the Aristotelian tradition seems plausible on examining the structure of *The Mathematical Analysis of Logic* (MAL), his first work on logic written in 1847. In fact, here he states that his initial goal was simply to reformulate traditional logic in equations before he realized that his approach could be developed further. On the cover page of MAL is a quote from Aristotle. This preliminary evidence for the influence of traditional logic on Boole is confirmed on doing a more detailed study. Some of the fundamental components of Boole's philosophy of logic were clearly inherited from the tradition. I touch on a few very briefly in this section and one in some detail in later sections.

Of course, there are many different traditions in logic before Boole and many different traditional philosophies of logic. For my purpose, I do not need to examine all the traditional sources that Boole may have studied. Those aspects of his philosophy of logic which I take to have a traditional origin are in Richard Whately's 1827 textbook *Elements of logic*—an exposition and defense of Aristotelian logic. Whately was seen as a representative of Aristotelian logic by his contemporaries even though his defense of its study may have been original. Comparing Whately and Boole is sufficient to establish that Boole was influenced by traditional logic; it is not necessary to find the original sources of Whately's views or the other sources that Boole may have got it from. Some historians credit Whately with revitalizing the study of logic in Britain and thus setting the stage for Boole (Jongsma 1982). Boole recommends Whately's book as an introduction to Aristotelian logic in *The Laws of Thought* (LT), his second work on logic, written in 1854. He mentions it several times in MAL and, in fact, he reproduces parts of it there, sometimes word for word [for example compare (Whately 1827, 57) and (MAL 20); (Whately 1827, 107–8) and (MAL 48)]. The present-day interpretations of Aristotle's texts on logic are significantly different from earlier interpretations. I will use 'Aristotle's logic' to refer to the modern ones and 'Aristotelian logic' to refer to the earlier ones.

Boole uses 'the received logic' and 'the traditional logic' without making explicit which tradition he is referring to, or even acknowledging that there are several different traditions. I use 'the traditional logic' for the exposition of logic in Whately's *Elements*.

> The aim of these investigations was in the first instance confined to the expression of the received logic, and to the forms of the Aristotelian arrangement, but it soon became apparent that restrictions were thus introduced, which were purely arbitrary and had no foundation in the nature of things (MAL 7–8).

In MAL he describes how categorical propositions are to be translated into equations and then spends the bulk of this eighty-page book demonstrating that traditional logic can be reformulated in his approach. It is mainly in the last twenty pages that he progresses beyond the tradition. However, he seems to have made no systematic attempt at studying the apparent advantages of his approach.

> Still, however, there was no special effort to arrive at novel results. But among those which at the time of their discovery appeared to be such, it may be proper to notice the following (MAL 8).

In the preface to LT, written seven years after MAL, he states that he has further organized and expanded his results. He also seems to have thought through the relationship between his work and the tradition from which it arose.

> The processes of Formal Logic [of the traditional logic] are of two kinds, viz., "Conversion" and "Syllogism". By Conversion is meant the expression of any proposition of the above kind in an equivalent form, but with a reversed order of terms. By Syllogism is meant the

> deduction from two [categorical] propositions having a common term, whether subject or predicate, of some third proposition inferentially involved in the two, and forming the "conclusion". It is maintained by most writers on Logic, that these processes, and according to some, the single process of Syllogism, furnish the universal types of reasoning, and that it is the business of the mind, in any train of demonstration, to conform itself, whether consciously or unconsciously, to the particular models of the processes which have been classified in the writings of logicians.
>
> [...] The course which I design to pursue is to show how these processes of Syllogism and Conversion may be conducted in the most general manner upon the principles of the present treatise, and, viewing them thus in relation to a system of Logic, the foundations of which, it is conceived, have been laid in the ultimate laws of thought, to seek to determine their true place and essential character (LT 227–8).

This passage reveals two features of Boole's philosophy of logic which originate in traditional logic. It also reveals how he believes he has improved on the tradition. The first concerns the main goal of logic. The tradition takes the main goal of logic to be the generation of conclusions, implicitly contained in given premises, by the elimination of terms in the premises, thus making explicit all the information contained in the premises about these terms. This goal is different from the main goal of logic as conceived today, but a more general version of it is clearly in Boole.

> As the conclusion must express a relation among the whole or among a part of the elements involved in the premises, it is requisite that we should possess the means of eliminating those elements which we desire not to appear in the conclusion, and of determining the whole amount of relation implied by the premises among the elements which we wish to retain (LT 8).

Boole calls this the 'final problem' or the 'general problem' of logic (LT 10; 140)

The common view today is that the main goal of logic is to find criteria for determining of a given conclusion whether it follows from given premises (Gasser 1989). However, according to Boole, the main goal of logic is to find a method which generates a conclusion from given premises—the conclusion being further specified as the one containing all the information in the premises about the "relation" between a given term (simple or complex) and other given terms contained in the premises. Boole believed that he had accomplished the goal he had set for himself. Seeing the main goal of logic as the generation of conclusions may be one reason for Boole committing *the fallacy of solutions,* namely, the confusion of the solution of an equation with a consequence of the equation (Corcoran and Wood 1980).

A second feature of Boole's philosophy of logic arising from the tradition is that logic is seen as the study of the process of reasoning. This is not psychologistic since deductive reasoning is taken to be a mental process which proceeds according to laws which do not depend on the will (LT 11). Logic is seen, by the tradition and by Boole, as a science with principles or axioms according to which reasoning is carried out. Boole differs from the tradition, however, in what he takes to be the fundamental laws. The tradition takes the "Dictum de Omni et Nullo" to be fundamental whereas Boole believes the laws that he presents are fundamental and he further believes that this can be shown by deriving the laws of the tradition from his: he undertakes this task in several chapters of MAL and in chapter fifteen of LT. He takes fundamental laws to be axioms from which the others can be somehow deduced.

> Let us define as fundamental those laws and principles from which all other general truths of science may be deduced, and into which they may all be again resolved (LT 5).

There are, of course, further similarities and differences between Boole's views and those of the tradition. For example, another feature of the traditional philosophy of logic which Boole incorporated into his own is the division of the propositions studied by logic into two mutually exclusive and jointly exhaustive kinds, namely, categorical propositions and hypothetical propositions. Moreover, he also incorporates the traditional reduction of hypothetical propositions to categorical ones. In sections 2 and 5 of this paper I examine this reduction in detail and in section 4 I point out that the ramifications of the reduction were not appreciated by later historians of logic, perhaps because they did not pay sufficient attention to the impact of traditional logic. There are many respects in which traditional logic and the philosophy of logic as given in Whately's *Elements* influenced Boole. I do not undertake here to describe this in detail. However, establishing that Whately did influence Boole in his treatment of hypothetical propositions as categoricals should suffice to show the need for paying attention to the influence of Aristotelian logic on the founding of mathematical logic. Arthur Prior (1948) is one of the very few articles that does this.

Another respect in which Boole believed his logic was different from the tradition was in it being somehow mathematical, though the precise grounds for regarding it as mathematical are not clear. Despite what Boole says in the epigraph to this essay and in several other passages, two of which I quote below, he is obscure on the relation between mathematics and logic and in what way logic is to be seen as associated with mathematics.

> ... the ultimate laws of Logic are mathematical in their form... (LT 11)

> That Logic, as a science, is susceptible of very wide applications is admitted; but it is equally certain that its ultimate forms and processes are mathematical (LT 12).

A connection between logic and mathematics, an admittedly diffuse one, is seen to be already nascent in Whately (Van Evra 1984).

2. WHATELY'S REDUCTION

In this section, I introduce Boole's treatment of propositions partly to underline that he was accepting and developing the Aristotelian approach to logic. The traditional classification divided all propositions into categoricals and hypotheticals. "Some snake is a parrot" and "Every star is a planet" are examples of categorical propositions. There are four kinds of categorical propositions: the universal affirmative, the existential affirmative, the universal negative and the existential negative. Categorical propositions are similar to what we nowadays would treat in a very restricted monadic first-order logic. Hypothetical propositions are treated today in propositional logic; they would be described as truth-functional combinations of categorical propositions. For example, "If some star is a planet then no star is a sun" and "Some star is a planet or no star is a planet". Boole and other followers of the Aristotelian tradition, unlike Aristotle himself, paid little attention to the logical form of the propositions they were studying; "Socrates is a man" is taken as a universal affirmative categorical proposition. Boole used the equational language ordinarily used in the algebra of his time. Nowhere does he specify it precisely. On the other hand the language that Aristotle used to express the categorical propositions studied in his logic was precisely specified and his study of it was systematic. In fact, as far as being precise and systematic is concerned, Boole's logic was in many respects a step backwards from Aristotle. This was just as true of other followers and expounders of the Aristotelian tradition, including Whately, who almost never matched Aristotle in this regard. Unwarranted assumptions about the sorts of propositions being considered by Boole have often led his interpreters to misunderstand Boole's project.

In MAL, Boole accepts the traditional classification according to which every proposition is either categorical or hypothetical. However, following a tradition which can be traced to ancient followers of Aristotle, he took hypothetical propositions to be reducible to categorical propositions.

> For this purpose we must consider every Conditional Proposition as a Universal-affirmative categorical Proposition, of which the Terms are entire Propositions, *viz.* the antecedent answering to the *Subject,* and the consequent to the *Predicate. E.G.* [...] to say, "if Louis is a good king, France is likely to prosper," is equivalent to saying, "The case of Louis being a good king, is a case of France being likely to prosper:" and if it be granted as a minor Premiss to the Conditional Syllogism, that "Louis is a good king," that is equivalent to saying, " the present case is the case of Louis being a good king;" from which you will draw a conclusion in *Barbara,* (*viz.* "the present case is a case of France being likely to prosper,") exactly equivalent to the original Conclusion of the Conditional Syllogism: *viz.* "France is likely to prosper" (Whately 1827, 118–9).

Though I use Whately's terminology 'equivalent to' and 'reduced to' when writing about his claims, I do not wish to give the impression that these are precise, well-defined notions. In fact, no criteria are given to determine either when a given proposition is reducible to a second or when it is not reducible to it. This is a cause of obscurity, since the equivalence of propositions may itself be a topic of study in logic. Neither Whately nor Boole were the first to carry out this sort of reduction and they were not its last practitioners. To some extent it persists in modern times. For example, Quine reduces "Socrates is a man" to something like "Everything which Socratizes is a thing which manlizes" (Quine 1961, 78).

Whately believes that all reasoning can be reduced to Aristotelian syllogisms. As a result he performs elaborate contortions to reduce conditional propositions to categorical propositions and Modus Ponens to Barbara. It is worthwhile to examine in some detail what Whately did from the point of view of the modern logician because it sheds light on the similar contortions which Boole performed.

Whately attempts to treat the propositions which are the antecedent and consequent of a conditional as Aristotelian terms. In the example quoted above this is achieved by taking the proposition "Louis is a good king" and reducing it to the term "The case of Louis being a good king", and the proposition "France is likely to prosper" to the term "The case of France being likely to prosper". The results of this reduction are taken to be Aristotelian terms and the proposition "The case of Louis being a good king, is a case of France being likely to prosper" is taken to be a categorical proposition. Reducing the second premise to a form which will enable it to be treated as a premise in a syllogism got by reducing "If Louis is a good king then France is likely to prosper" to a categorical, requires an additional interesting contortion since it is necessary to have two terms in every categorical proposition. "Louis is a good king" is not now reduced to the term "The case of Louis being a good king", as done before, but to the categorical proposition "The present case is a case of Louis being a good king". This categorical proposition seems to have been got by reducing a hypothetical proposition. In order to treat Modus Ponens as Barbara, "Louis is a good king", which is a categorical proposition, seems first to have been reduced to a hypothetical such as "If this is the present case then Louis is a good king" and this reduced back to a categorical, a move I call *Whately's double reduction.*

Modus Ponens
Louis is a good king
If Louis is a good king then France is likely to prosper
? France is likely to prosper

The above argument is reduced to what Whately takes to be an instance of Barbara:

Barbara (according to Whately)
The present case is a case of Louis being a good king
The case of Louis being a good king is a case of France being likely to prosper
? The present case is a case of France being likely to prosper

Whately would take it to be equivalent to:

Every case which is a present case is a case of Louis being a good king
Every case of Louis being a good king is a case of France being likely to prosper
? Every case which is a present case is a case of France being likely to prosper

Notice that the universe of discourse of these categorical propositions contains cases and that terms such as "case of Louis being a good king" pick out the class of cases where "Louis is a good king" is true. As I show in section 5, Boole uses similar notions but is unable to carry out Whately's double reduction.

Later in the text Whately considers the proposition "pain is an evil" and reduces it to "the present case is not the case of pain being no evil" and finding that this 'produces a circuitous awkwardness of expression' allows it to be reduced to the metalinguistic proposition expressed by 'that pain is no evil is not true' (Whately 1827, 120). These reductions seem ad hoc and inconvenient, to say the least, and Whately did not require that they should be carried out in normal reasoning. His purpose in carrying out the reductions was to support a central thesis of his philosophy of logic that all reasoning may be reduced to Aristotelian syllogisms.

> [...] it is not intended that [hypothetical propositions] should be reduced to Categoricals, in *ordinary practice*, as the *readiest way* of trying their validity, (their own rules being quite sufficient for that purpose;) but only *that we should be able*, if required, to subject any argument whatever to the test of Aristotle's Dictum, in order to show that all reasoning turns upon one simple principle (Whately 1827, 121).

Whately did not require that drawing inferences from hypothetical propositions be carried out only after reducing them to categoricals. However, Boole's treatment of hypotheticals does.

That all deductive reasoning may be reduced to Aristotelian syllogism was a central presupposition of Whately's philosophy of logic, and in order to defend this view he attempts to reduce hypothetical propositions to categoricals and Modus Ponens to Barbara. Boole incorporates Whately's presuppositions as an integral and unavoidable part of his approach to logic. Despite dissatisfaction with the original division of propositions into categorical and hypothetical in MAL and the metaphysics presupposed in reducing hypotheticals to categoricals, in LT Boole continues to maintain a similar division of propositions into primary and secondary propositions and he attempts to reduce secondary propositions to primary propositions in a similar way (LT 159–76).

3. BOOLE'S REDUCTION

Boole devises a way of translating categorical sentences into equations. He believes that it can be extended to translating hypothetical sentences since, following Whately, he takes hypotheticals as reducible to categoricals; in addition, he reduces other natural-language sentences to equations. Boole saw the equational language used in mathematics as better suited for reasoning and believed that he could improve Aristotelian logic by using it. He attempted to do this by a process of translating from the Aristotelian categorical sentences into his equational language and then, after carrying out a derivation using his equations, translating back into the categorical sentences. He called the process of translating from categorical sentences to equations 'expressing' and the process of translating back from equations to categorical sentences 'interpretation'. It is necessary to presuppose that hypotheticals may be reduced to categoricals in order to take both as expressible in equations the way Boole does.

Boole's translation of English categorical sentences into equations is a sort of reduction. Many different equations may be used to translate an English sentence. Boole often uses reasoning in his metalanguage, that is, in his writing, to justify his translation. This indicates that his process of translation involves a process of reasoning. Boole was aware that since the process of reduction carried out by the Aristotelian logicians itself presupposed reasoning it could not be seen as giving an account of reasoning. He criticized aspects of the traditional reduction in MAL (18, note).

> What are the elements involved in that process of reduction? Clearly they are as much a part of general reasoning as the dictum itself. [...] Any account of the process of reasoning is insufficient, which does not represent, as well the laws of the operation which the mind performs in that process, as the primary truths which it recognises and applies.

Boole was surprisingly oblivious, however, to the reduction involved in translating English sentences into his equations—and also to the reasoning presupposed by this process, and by the process of reducing hypothetical propositions into categoricals in order to translate them into equations.

The letters in his equations denote the classes corresponding to the common nouns in the categorical sentences. '0' denotes the empty class and '1' the universal class. (Quotes convention: I enclose an expression in single-quotes to form the name of the expression, for example: 'Two' denotes two; 'Two is even' is a sentence with three words. I enclose an expression in double-quotes to form the name of the concept or proposition expressed by the expression, for example: "Not", "or", and "every" are logical concepts, whereas "is an element of" and "successor of" are not; "Two is even" is a true proposition.) The concatenation of two letters denotes the intersection of the classes denoted by each of them. For example, if '*p*' denotes the class of primes and '*e*' the class of evens then '*pe*' denotes the class of even primes. The term formed by joining two letters by a plus-sign (+) denotes the union of the two classes if the two classes are disjoint and is nondenoting otherwise. For example, '$p + e$' is nondenoting if '*p*' and '*e*' denote the classes in the previous example. If '*p*' denotes the class of Presidents of the United States and '*w*' the class of women, '$p + w$' denotes the class containing the union of the class of women and the class of Presidents. Likewise, the term formed by joining two letters by a minus-sign (–) denotes the difference of the two classes if the class denoted by the first letter contains the class denoted by the second and is nondenoting otherwise. For example, the term '$1 - x$' is denoting for any class denoted by '*x*' since every set is a subset of the universe but '$x - (1 - x)$' is nondenoting since '$1 - x$' denotes the complement of the class denoted by '*x*' and therefore has no members in common with *x*.

Using these symbols Boole was able to translate categorical sentences such as 'Every snake is a parrot' into '$s(1 - p) = 0$' and 'No snake is a parrot' into '$sp = 0$'. "$s(1 - p) = 0$" amounts to "the intersection of the class of snakes with the complement of the class of parrots is the empty class" and "$sp = 0$" amounts to "the intersection of the class of snakes with the class of parrots is the empty class". Surprisingly 'Some snake is a parrot' is not translated as the negation of '$sp = 0$' but as '$sp = v$' or as '$vs = vp$'. He calls v an indeterminate class. There has been much discussion about how to interpret this but I will not get into that here. It is interesting to note that he always uses equations, never using inequations even when they seem convenient and natural.

The plus operation is used in equations which translate English sentences with 'or'. For example, 'No prime number is either even or odd' may be translated into '$p(e + d) = 0$' where 'p' denotes the class of prime numbers, 'e' denotes the class of even numbers and 'd' the class of odd numbers. '$e + d$' denotes the union of the two classes since the class of even numbers has no members in common with the class of odd numbers. "$p(e + d) = 0$" amounts to "the intersection of the class of prime numbers with the union of the class of even numbers with the class of odd numbers is the empty set". According to the commutative law for classes $xy = yx$ hence $p(e + d) = (e + d)p$ and hence $(e + d)p = 0$. The last equation may be translated back into English as 'No number which is either even or odd is prime'. This is the converse of the first English sentence. The process of converting English sentences into equations, carrying out a derivation using laws and then translating back into equations yields a consequence of the proposition expressed by the first English sentence.

There are several uses of the word 'or' in English. These divide into cases where the 'or' is used to compound words, either nouns or adjectives, as in: "A fireman or a policeman was on the scene" or "The victim was either dead or asleep"; or cases where the 'or' connects sentences expressing propositions as in "Abe is able or Ben is brave". There are other uses of 'or' in English but it is only in connection with the third usage mentioned that it makes sense to speak of exclusive or non-exclusive "or". Although it is true that the word 'or' is used between class names, for example, 'the class of natural numbers or the class of rational numbers may be taken as the universe of discourse of number theory', nevertheless the word 'or' in normal English is never used to indicate union. Therefore when Boole translates from English to his equational language he never used the plus-sign as a substitute for 'or'. It is a category mistake to say that plus is the logical connective "or".

In LT (32–3) Boole attempts to draw an analogy between the use of the words 'and' and 'or' in English and the plus-sign, '+', within his equational language. However, he does not make the analogy when he first introduces the operations plus, minus and concatenation in MAL. From this it appears that his initial purpose in introducing the class operations is not to provide substitutes for the English words 'and' and 'or'.

Examining examples of the use of compound nouns will help clarify the logical form of the propositions they are used to express.

A. Every fireman or policeman must be brave.
B. Every fireman and policeman must be brave.

Sentences A and B have compound nouns as subjects, the first formed with a disjunction and the second with a conjunction. Surprisingly, the propositions expressed by these two sentences are logically equivalent as an analysis reveals.

Sentence A is elliptical for 'Every person who is either a fireman or a policeman must be brave' and this sentence is elliptical for 'Every person who either is a fireman or is a policeman must be brave' and this in turn is elliptical for 'Every person who is a policeman or who is a fireman must be brave'. The occurrence of 'or' in this nonelliptic sentence expresses a propositional connective and not a connective joining concepts. Likewise, sentence B is elliptical for 'Every fireman must be brave and every policeman must be brave'. The occurrence of 'and' in this nonelliptic sentence clearly expresses a propositional connective. Understanding that 'or' and 'and' are propositional connectives enables us to see why the propositions expressed by the original sentences are logically equivalent.

Most sentences of ordinary English, unlike those of a logically perfect language, do not mirror the logical structure of the propositions they express. More precisely, the grammatical forms of ordinary English do not mirror the logical forms of the propositions expressed by the sentences of ordinary English. Ambiguity, especially structural ambiguity, and ellipsis are among the factors causing the mismatch. The grammatical form of the two nonelliptic sentences above mirrors the logical form of the propositions they express and a deduction of each from the other will show that they are logically equivalent.

In his second work LT, Boole subscribes to the common view that language can express reasoning and that studying the structures of language aids the study of reasoning (LT 24). However he believed that the language of equations best captures those aspects of the structure of propositions which are relevant for reasoning. Boole appears to be aware that the propositions expressed by the two sentences, A and B, are logically equivalent but made no attempt to understand the relation between the two propositions in terms of the structure of A and B. His project, which involved in part trying to reduce all reasoning in natural language to an algebraic form, seems to have blinded him to the necessity of studying natural language more closely. He would express the two equivalent propositions with the same equation thus reducing one to the other.

The necessity of studying the relationship between a sentence with a compound noun using 'or' and a similar one using 'and' is even more apparent when it is noticed that they do not express equivalent propositions when they occur in the predicate position of a universal affirmative proposition.

A′. Every number is either even or odd.
B′. Every number is even and odd.

The first expresses a true proposition the second a false one; they are obviously not logically equivalent. This should have led Boole to notice that "and" and "or" may not both be reduced to plus, if he had studied the relationship between natural language and his equations more closely.

4. KNEALE'S MISTAKE: THE EXCLUSIVE-"OR"

In this section, I give examples of some common mistakes by historians of logic and examine in some detail the confusion of taking plus to be the exclusive-"or" by the influential historian of logic William Kneale. I believe that this and similar confusions have led to our misunderstanding of Boole's project and his achievements, making it difficult to come to an accurate assessment of the philosophy of logic involved in the founding of mathematical logic. Seeing the influence of Aristotelian logic on Boole will help us to begin to interpret him more faithfully, as I try to show in the next section.

In his 1997 biography of the logician Kurt Gödel, John Dawson presents a synopsis of the history of logic in the third chapter. Here Boole is praised, in the manner common among historians, but for the wrong reasons.

> Real progress in developing a logical calculus was first achieved by George Boole in his *Mathematical Analysis of Logic*, published in 1847. Boole recognized the formal analogies between the use of the arithmetical operations "plus" and "times" and the logical connectives "or" and "and", and he showed how Aristotle's four basic statement forms could be expressed as equations within a symbolic algebra ... (Dawson 1997, 39–40).

Boole does attempt to translate English sentences with 'and' and 'or' into equations with the plus-sign and the times-sign (or concatenation). However, "and" and "or" are propositional connectives and Boole took plus and times as operations on classes. It is a misrepresentation to claim that Boole discovered a formal analogy between arithmetic operations and propositional connectives. In MAL Boole never made such a claim. This is a common mistake and I discuss it in more detail in what follows. It would be closer to the truth to say that Boole recognized some (not *the*) formal analogies between the arithmetic operations of addition and multiplication and the class operations of union and intersection. Later critics of Boole believe that this is not one of his achievements but one of his faults. As some have observed, using algebraic notation for logical operations interfered with the symbolization of the logic of algebra (Frege 1882, 88; Frege 1883, 93; Peano 1889, 103). Boole's apparent attempt to give an axiomatic treatment of logic was not only a bold innovation but one that misled many logicians about the nature of logic for the rest of the century. It also led to misinterpretations by influential interpreters, Łukasiewicz for example, of what Aristotle took logic to be (Lear 1980, 8–13).

Following the Aristotelian tradition, Boole takes hypothetical propositions to be reducible to categoricals. He employs the equational language, used to translate categorical propositions, to translate reduced hypothetical propositions. In order to do this he has to change the universe of the categorical propositions to what he calls 'the hypothetical universe'. The hypothetical universe according to him contains 'all conceivable cases and conjunctures of circumstances' (MAL 49). Historians of logic pay little attention to the reduction which Boole carries out or to the metaphysics Boole retrofits to try to make it work. As I show in the next section this in part explains why at times they misunderstand Boole.

The following is a quote from William Kneale, the historian of logic:

> In the *Mathematical Analysis of Logic* and again in the *Laws of Thought* Boole suggests a convention whereby "$x = 1$" may be taken to mean that the proposition X is true and "$x = 0$" that X is false. In accordance with this usage the truth values of more complicated propositions can be represented by functions of small letters, *e.g.* the truth value of the conjunction of X and Y by "xy" and the truth value of the exclusive disjunction of X and Y by "$x + y$". It will be seen that we have here all that is needed for an interpretation of Boole's system in terms of the truth values of propositions with "1" for truth and "0" for falsity. But this interpretation, like the numerical one, satisfies also the stricter system which involves [the principle according to which either $x = 1$ or $x = 0$] (Kneale 1948, 166 (cf. Kneale and Kneale 1962, 413)).

In the passage quoted here Boole is seen as suggesting a convention by translating the equation '$x = 1$' as the metalinguistic English sentence 'X is true' and '$x = 0$' as 'X is false'. However, in both MAL and LT Boole tries to provide an account of the universe of hypothetical propositions on the basis of which he tries to show that '$x = 0$' may be translated as 'X is false' and '$x = 1$' as 'X is true'. He takes the translations to need corroboration on the basis of what he had already said about the universe of hypothetical propositions. Rather than as conventions, he takes the translations as the sort of thing which may be false and something about which he may be wrong. Here Kneale reveals a misunderstanding of Boole's project, which shows itself again when he suggests that Boole has what amounts to the notion of the truth-value of a complex proposition and that the truth-value of the exclusive disjunction of propositions X and Y is denoted by '$x + y$'. Frege, in his 1879 *Begriffsschrift,* was the first to use the notion of truth value. It was Russell who first used the expression 'truth-value', in the appendix to his *Principles of Mathematics* of 1903 where he discussed Frege's notion which was not well known at the time.

The exclusive disjunction of two propositions X and Y is true if one is true and the other false and is false otherwise. The plus-sign does not stand for a logical connective but for a partial function defined on disjoint classes. The exclusive disjunction of two true propositions is meaningful but false; for example, the exclusive disjunction of "Zero is even" with "One is odd" is entirely meaningful, though false. However, for Boole, '$x + y$' is meaningful only if x and y have no members in common. The exclusive-"or" is always meaningful but the plus operation is not. Boole always uses the plus as an operation between classes, and the exclusive-"or" is never a class operation. The plus cannot be interpreted as a disjunction whether exclusive or non-exclusive. Though today the plus is often read as a disjunction it is anachronistic to read Boole as having it. Taking Boole's translation of '$x = 1$' to 'X is true' and '$x = 0$' to 'X is false' to be a convention, and '$x + y$' to stand for the exclusive disjunction of X and Y, indicates that Kneale did not take seriously Boole's own account of the hypothetical universe or his account of the nature of the operations expressed in his equations. It also indicates that Kneale did not appreciate the nature of the reduction that Boole was conducting when he translated English sentences to equations.

Kneale's confusion of "$x + y$" with exclusive disjunction is common in the literature. He is in the company of Frege, Lewis and Langford and many others (Frege 1882, 97; Lewis and Langford 1932, 14, 16–7]. A reason why this confusion is so prevalent may be that defining plus in English, as Kneale does, leads to believing that there are two senses of 'plus' corresponding to the two senses of 'or'.

Kneale writes:

> In the algebra of classes the complex sign $x + y$ may be used to signify the class of things which belong either to the class denoted by x or to the class denoted by y. But the word 'or' has two distinguishable meanings in ordinary English; it is used sometimes in an inclusive and sometimes in an exclusive sense (Kneale and Kneale 1962, 409).

That is, in order for the plus of two classes to be a given class it is necessary and sufficient that an object is in the given class if and only if it is in the first class or it is in the second class. The plus is supposed to have an exclusive or non-exclusive sense according as the occurrence of 'or' in the definition is exclusive or non-exclusive. However, while the definition with 'or' in the non-exclusive sense does define plus as the union of two classes in the usual way, the definition taken with 'or' in the exclusive sense has plus denoting the symmetric difference of two classes—defined on all classes. If plus is taken as a partial function defined only on disjoint classes, the same partial function is defined with 'or' taken in either the exclusive or the non-exclusive sense. Again, there is no way that plus can be taken as exclusive disjunction.

The passage just quoted reveals a plausible reason for why the confusion of plus for exclusive disjunction is common. However, there are places in Boole's writings where he uses the word 'exclusive' in two senses, which may have helped increase the confusion though it is clear that Boole himself saw these as distinct senses.

> Thus, if we take the Proposition of the preceding example, viz. Either X is true, or Y is true, and assume that the two members of this Proposition are not exclusive, insomuch that in the enumeration of possible cases, we must reckon that of the Propositions X and Y being both true, then the mutually exclusive cases which fill up the Universe of the Proposition, ... (MAL 52).

The first occurrence 'not exclusive' in this passage refers to the disjunction in the hypothetical proposition expressed in English. The second occurrence 'mutually exclusive' refers to classes of "cases" in which the propositions are true. These classes are mutually exclusive and jointly exhaustive of the universe of discourse of the hypothetical propositions. Interestingly, Boole was the first person to use the now prevalent expression 'universe of discourse' (LT 42). The notion of class of cases is obscure but it plays an important role in Boole's philosophy of logic, which we can better appreciate by seeing how he got it from the Aristotelian tradition, as I show in what follows. I should emphasize that the idea of reducing hypotheticals to categoricals and its accompanying metaphysics is not in Aristotle even though some of his later interpreters took him to be using it (Englebretsen 1980). Corcoran and Scanlan in their 1981 review of Englebretsen's paper point out that as far back as William of Ockham it was recognized that in order to infer "Some snake is a not a plant or some star is not a planet" from "It is not the case that both every snake is a plant and every star is a planet" it is necessary to have both hypothetical and categorical modes of analysis.

5. BOOLE'S HYPOTHETICAL UNIVERSE

In order to translate the hypotheticals into his equations Boole has first to reduce them to categoricals and then change the universe of discourse to what he calls the 'hypothetical universe'. The hypothetical universe seems to be retrofitted onto the reduced hypotheticals in order to translate them into Boole's equations. In this section I propose to interpret Boole's text in order to determine what he takes the constituents of the hypothetical universe to be. Boole claims that the hypothetical universe contains 'all conceivable cases and conjunctures of circumstances'. 'Conceivable', 'cases', 'circumstances', 'possible', etc., are words that occur several times in the pages where Boole describes the hypothetical universe and where he describes how the equations translating the reduced hypotheticals are to be read. On my account of these parts of Boole's text the hypothetical universe contains logical possibilities, or possible worlds.

> Thus, what we have to consider is not objects and classes of objects, but the truths of Propositions, namely, of those elementary Propositions which are embodied in the terms of our hypothetical premises.
>
> To the symbols X, Y, Z, representative of Propositions, we may appropriate the elective symbols x, y, z, in the following sense.
>
> The hypothetical Universe, 1, shall comprehend all conceivable cases and conjunctures of circumstances.
>
> The elective symbol x attached to any subject expressive of such cases shall select those cases in which the Proposition X is true, and similarly for Y and Z (MAL 48–9).

In the passage quoted above Boole is changing the interpretation of the equational language, which was developed to translate categorical sentences, in order to translate hypothetical propositions. The capital letters X, Y, Z each stand for propositions and the small letters x, y, z each for the class of cases where these propositions are true. As in the earlier categorical interpretation the small letters denote classes and concatenated letters denote the intersection of the classes denoted by the letters. The plus denotes a partial function defined on disjoint classes. The term formed by joining two letters with a plus denotes the union of the two classes if the classes are disjoint and is nondenoting if they are not disjoint. However, unlike in the earlier interpretation, the universe of discourse, denoted by '1', does not consist of objects nor are the classes denoted by the letters classes of objects. A close examination of the passage quoted above, along with other passages adjoining it in MAL, shows the universe to be taken as consisting of possible worlds and the classes denoted by the letters to contain possible worlds in which a proposition is true. It would be anachronistic to use 'possible worlds' in its present-day sense, as for example, in Bradley and Swartz (1979), or to believe that Boole had a clear conception of it. However, the expression was made well-known before Boole's time by Leibniz and it is useful in conveying the nature of the individuals in the hypothetical universe even though it is vague.

In the passage quoted Boole writes 'x attached to any subject expressive of [conceivable cases and conjunctures of circumstances] shall select those cases in which the Proposition X is true'. On my account this means that 1 contains all possible worlds and 'x' concatenated with '1' denotes the class of all possible worlds in which X is true. Likewise 'xy' denotes the class of possible worlds in which X and Y are both true.

Drawing the analogy between categoricals and reduced hypotheticals helps see why Boole does this. The conditional "If Abe is able then Ben is bold" is reduced to the universal "Every case of Abe being able is a case of Ben being bold" or equivalently "Every case of "Abe is able" being true is a case of "Ben is bold" being true". The motivations for such an approach are difficult to appreciate till the approach described in Whately and its motivations are appreciated. The influence of this approach on Boole is clear.

English sentence:
Categorical: Every star is a planet
Reduced hypothetical: Every possible world in which "Abe is able" is true is a possible world in which "Ben is bold" is true

Universe of discourse:
Categorical: Heavenly objects
Reduced hypothetical: Possible worlds

Equations translating the English sentences:
Categorical: $s(1 - p) = 0$
Reduced hypothetical: $a(1 - b) = 0$

Classes denoted by the letters:
Categorical: 's' denotes the class of stars; 'p' denotes the class of planets
Reduced hypothetical: 'a' denotes the class of possible worlds (cases) in which "Abe is able" is true; 'b' denotes the class of possible worlds in which "Ben is bold" is true

Other passages make this clearer, for example:

> If we confine ourselves to the contemplation of a given proposition X, and hold in abeyance every other consideration, then two cases only are conceivable, viz. first that the given Proposition is true, and secondly that it is false. As these cases together make up the Universe of the Proposition, and as the former is determined by the elective symbol x, the latter is determined by the symbol $1 - x$ (MAL 49).

This passage shows that when only one proposition is considered there are two possibilities—that it is true or that it is false—both of which are contained in the universe. The term 'x' denotes the class consisting of the first possibility and the term '$1 - x$' the class consisting of the second. When two propositions are considered there are four distinct possibilities: that both propositions are true, that the first is true and the second false, that the first is false and the second true, or that they are both false. From other passages it is clear that he does not take the hypothetical universe to change as the number of propositions considered increases. This is in exact analogy to the earlier interpretation in which as additional terms are considered the universe of objects is divided into smaller classes, mutually exclusive and jointly exhaustive of the universe, but with no change in the extent of the universe itself. For example, if the universe of discourse is the class of animals and we consider a single term, say 'snake', it divides the universe into the class of snakes and the class of nonsnakes. If taken together with a second term, say 'parrot', the universe is divided into four jointly exhaustive and mutually exclusive classes: the class of animals which are both snakes and parrots; the class of animals which are snakes but not parrots; the class of animals which are parrots but not snakes; the class of animals which are neither snakes nor parrots. More such terms generate more such classes but the universe of discourse itself does not change. Of course, strictly speaking the classes are not mutually exclusive in the degenerate cases of the empty set or the universal set.

In exact analogy as additional propositions are considered the additional terms which they generate subdivide the hypothetical universe further and do not increase its extent. Boole notes:

> But if other considerations are admitted, each of these cases will be resolvable into others, individually less extensive, the number of which will depend upon the number of foreign considerations admitted (MAL 49–50).

> Thus the extent of the hypothetical Universe does not at all depend upon the number of circumstances which are taken into account. And it is to be noted that however few or many those circumstances may be, the sum of the elective expressions representing every conceivable case will be unity (MAL 50).

It is clear that Boole treats hypotheticals as categoricals just as Whately did in his treatment of traditional logic and that in order to do this Boole changes the universe of discourse. However, he has not carefully thought through what the constituents of the hypothetical universe are and this becomes clear on a close examination of his use of the word 'case'. Following Whately, in order to treat conditionals as universal categoricals, Boole has to treat the hypothetical universe as containing possible worlds and he does so in his writing, for example, when he writes that the symbol 'x' selects the cases in which the proposition X is true. On the other hand, he uses the word 'case' for the class of possible worlds denoted by expressions such as 'x', 'xy', '$xy(1 - z)$', etc., for example when he writes 'each of these cases will be resolvable into others individually less extensive'. (An attempt at giving an account of Boole's propositional logic in terms of a universe which varies in extent with the number of propositions considered is given in Hailperin's 1984 paper.) This use of 'case' is close to but different from the use on page 50 of MAL just quoted, where the different possible combinations of truth and falsity of the propositions being considered are called cases. However, in each of these uses 'case' applies to the correlate, whatever it may be, in the hypothetical universe of false propositions as well as of true propositions. The account Boole gives of the hypothetical universe allows him to translate conditionals into equations by reducing them to universal categoricals, but it does not allow him, in any convenient way, to translate a categorical proposition after carrying out Whately's double reduction on it. Boole is not able to follow Whately all the way. In order to translate 'X is true' he uses the word 'case' in a way that is drastically opposed to his earlier use.

According to the account I have given of Boole's hypothetical universe, '$x = 1$' translates to 'X is true in all possible worlds' and '$x = 0$' to 'X is true in no possible world'. However, in order to carry out simple inferences such as Modus Ponens it is necessary to reduce nonhypothetical propositions to hypotheticals, just as Whately did with his double reduction. It is necessary to reduce the proposition "Abe is able" to a proposition with two terms of which the equivalent of "Abe is able" is one—for example, to the metalinguistic proposition ""Abe is able" is true"—and then to translate it into symbols. The account that Boole gives of the hypothetical universe does not allow him to do this in any obvious way. However, he tries to accomplish this by ignoring the description of the hypothetical universe he has given thus far and making a claim that conflicts with this description, namely, that there is no possible world in which a false proposition is true. This amounts to using the word 'case' in a way that conflicts with his earlier use (compare MAL page 49 down 14 lines and MAL page 51 down 4 lines) but it enables Boole to translate 'X is true' as '$x = 1$' and 'X is false' as '$x = 0$'.

> To express that a given Proposition X is true.
>
> The symbol $1 - x$ selects those cases in which the Proposition X is false. But if the Proposition is true, *there are no such cases* in its hypothetical Universe, therefore
>
> $1 - x = 0,$
>
> or $x = 1$, (25) (MAL 51 [my emphasis]).

In order to translate the sentence 'X is false' he does something similar.

> To express that a given Proposition X is false.
>
> The elective symbol x selects all those cases in which the Proposition is true, and therefore if the Proposition is false,
>
> $x = 0$, (26).

> And in every case, having determined the elective expression appropriate to a given Proposition, we assert the truth of that Proposition by equating the elective expression to unity, and its falsehood by equating the same expression to 0 (MAL 51).

The passages quoted above also make clear that Boole was not taking '$x = 1$' and '$x = 0$' to express 'X is true' and 'X is false' by convention but as something that required proof based on his metaphysics. This is a metaphysics of possible worlds in a hypothetical universe which he inherited from the tradition but which he revises at this crucial point in order to carry out the equivalent of Whately's double reduction. The revision that Boole here makes to his account of false cases in the hypothetical universe amounts to changing the translation of his equations from being about logical relations to being about material relations. I give examples below of the two ways of translating Boole's equations. The change also misleads interpreters into believing that Boole had the notion of truth-value.

Boole never uses the proposition "$x = 1$ or $x = 0$" and though he gives a different account of the universe of hypothetical propositions in LT he writes that they have 'an exact analogy ... to primary [i.e. categorical] propositions'. 'x' is taken as denoting that portion of the universe of discourse for which X is true and '$1 - x$' the portion for which it is false (LT 167–8).

Seeing Boole's universe as consisting of possible worlds, as he did before he revised it, leads to an interesting interpretation of his equations, though this would not allow 'X is true' to be expressed in his equations. The translations of his equations when the universe is taken as consisting of possible worlds and when it is taken as consisting of truth-values are logical and material analogs. Boole never seems to have become clear about the distinction between logical and material relations between propositions and it is this confusion which seems to persist among later logicians working in the Boolean tradition. Lewis and Langford, who worked in the tradition originating from Boole largely independent of Frege, seem to recognize this to some extent.

> The fact that almost all developments of symbolic logic are based upon this relation of extensional or material implication—a relation of truth-values, not of content or logical significance—is due to their having been built up, gradually, on the foundation laid by Boole; that is, upon a calculus originally devised to deal with the relations of classes. There is nothing more esoteric, behind these peculiar properties of material implication, than this somewhat unfortunate historical accident (Lewis and Langford 1932, 89).

They are right to consider the prominence of material implication a historical accident and perhaps right to trace it to Boole. However, Boole did not have the notion of truth-value and that Boole's calculus deals with the relations of classes is not sufficient to explain his having the notion of material implication. It is necessary in addition to understand Boole's incorporation of both the traditional reduction of hypotheticals to categoricals and his conflicting modification to it. The equations, when seen as about possible worlds, can be translated into English sentences about logical relations. Examining equations from both perspectives should make the contrast obvious.

Here are some examples of equations with their logical and material translations:

$ab = a$	*Logical translation* Every possible world in which A is true is a possible world in which B is true or equivalently A logically implies B *Material translation* It is not the case that A is true and B is false or equivalently A materially implies B
$a = 1$	*Logical translation* A is a tautology *Material translation* A is true
$xx = x$	*Logical translation* The conjunction of any proposition with itself is logically equivalent to itself *Material translation* The truth-value of the conjunction of any proposition with itself is the truth-value of the proposition
$x(1 - x) = 0$	*Logical translation* The conjunction of any proposition with its negation is a contradiction *Material translation* The truth-value of the conjunction of any proposition with its negation is false

6. CONCLUSION

Though Boole is widely believed to be the founder of mathematical logic, there is no detailed study of his original texts. His work is often taken to be a break from the Aristotelian tradition in logic, but a study of his texts shows that it influenced his work in important ways. Well-known histories of logic have not appreciated this influence, partly because their account of his work is anachronistic. The traditional reduction of hypotheticals to categoricals is adopted by Boole and this leads to conflicts in his first logical work, MAL, which do not seem to have been resolved even later—namely, conflicts about the nature of the hypothetical universe. Though in many parts of his work on hypothetical propositions Boole appears to be dealing with material relations, if the hypothetical universe is taken as containing possible worlds, as it seems to be in other parts of his work, Boole's equations can be translated into English sentences about logical relations between propositions.

Boole is recognized as the founder of modern logic, and the modern tradition of mathematicians working in logic can be traced back to him. The use of equations made works considered to be a part of logic appear more "mathematical" as well. However, it is difficult to specify precise, substantial reasons for attributing the origin of mathematical logic to Boole or to specify precisely the sense in which it was mathematical. It is clear that Boole regarded himself as a follower of Aristotle, and his work as bringing to full completion the project first undertaken by Aristotle two thousand years ago. It is also clear, as indicated in the epigraph to this paper, that Boole believed that he had shown that logic is a part of mathematics.

To understand Boole's project—his presuppositions and his motivations—it is vital to understand the tradition which he saw himself as completing. The philosophy of logic that Whately made explicit in order to defend logic, along with some of his implicit presuppositions, seems to be what oriented Boole's approach to logic. Boole accepted some of Whately's claims and modified others but his overall project was shaped by Whately's views on traditional logic.

It is instructive to compare Boole's logic, which he believed was a development of Aristotle's ancient logic, with modern interpretations of the ancient logic (Corcoran 1974 and 1994). It is surprising to find that Boole's work, in many respects, was a step backwards from Aristotle's. Aristotle's language and his deductions were much more carefully specified. In fact, it is not an exaggeration to clam that Aristotle had a formal language, but it is undoubtedly inaccurate to claim this of Boole, despite his use of equations. Aristotle's study of his system was detailed and systematic. He provided criteria for determining whether any given argument in his system was valid or invalid. Though his language is by comparison simple, his logical methodology is substantially similar to modern logical methodology: he used deductions, both direct and indirect, to show that an argument is valid and counterarguments to show that an argument is invalid. Aristotle's logic is complete and his study was so thorough that some modern interpreters even believe that he had a completeness proof (Smiley 1994). In addition, Aristotle reveals a sophisticated understanding of the logical relations between propositions in the square of opposition. Boole on the other hand seems confused about the distinction between logical and material relations. This confusion reveals itself, for example, in his account of contradiction.

A reason for Aristotle's achievements in logic may be the clarity of his philosophical purpose, namely, the study of proof (especially in mathematics—a subject that yields many examples of proofs). The underlying purpose of Boole's project seems to have been set by later interpreters of Aristotle. I have traced the influence of a traditional conception of the relation between categorical and hypothetical propositions on Boole and the confusions this gave rise to. Nevertheless, there are many original features of Boole's work which are a part of modern logic—for example, the notion of a decision procedure and the use of logical axioms. While Boole did not recognize some of the distinctive features of modern logic such as truth-functions, the quantifier, and the predicate, his work was clearly influential on the modern tradition and he deserved to be acknowledged as the founder of mathematical logic. However, evaluating his influence is a difficult task and recognizing the influence of traditional logic, which Boole saw himself as completing, is a step towards accomplishing it.

ACKNOWLEDGMENTS

A version of this paper was presented to the Buffalo Logic Colloquium and I thank the audience for their criticisms, which led to many improvements, especially those of my teachers John Corcoran, John Kearns and Richard Vesley. John Corcoran first inspired my interest in logic and has continued since then to be the most important influence on my education in logic. I thank Randy Dipert and José Sagüillo for their comments. I would like to thank the organizers and participants of the Boole 1997 Conference, especially James Gasser, for their comments and hospitality. This paper benefited from the hours of patient and careful feedback I got from Ky Herreid and it is much more intelligible than the first draft due to my attempts to meet his high standards of clarity.

BIBLIOGRAPHY

The date immediately following the author's name indicates the year when the book or article was first published. A second date separated from the first by a '/' is the date of a later edition which is being used. The page numbers in a reference always refer to the later date unless there is another date at the end of the entry.

Boole, George. 1847. *The mathematical analysis of logic: being an essay towards a calculus of deductive reasoning*. Cambridge: Macmillan, Barclay and Macmillan.

———. 1854. *An investigation of the laws of thought on which are founded the mathematical theories of logic and probability*. London: Walton and Maberley.

Bradley, Raymond, and Norman Swartz. 1979. *Possible worlds: an introduction to logic and its philosophy*. Indianapolis: Hackett.

Bynum, Terrell W., ed. 1972. *Gottlob Frege: conceptual notation and related articles*. Oxford: Clarendon Press.

Church, Alonzo. 1983. Dictionary entry on George Boole, in Runes (1983).

Corcoran, John. 1974. Aristotle's natural deduction system. In *Ancient logic and its modern interpretations*, edited by John Corcoran. Dordrecht: Reidel.

———. 1987. Review of MacHale (1985). *Mathematical Reviews* 87c:01030.

———. 1994. The founding of logic. *Ancient Philosophy* 14: 9–24.

Corcoran, John, and Michael Scanlan. 1981. Review of Englebretsen (1980). *Mathematical Reviews* 81k:03014.

Corcoran, John, and Susan Wood. 1980. Boole's criteria for validity and invalidity. *Notre Dame Journal of Formal Logic* 21: 609–38. [Reprinted in this volume, pp. 101–28.]

Dawson, John W. 1997. *Logical dilemmas: the life and work of Kurt Gödel*. Wellesley, Mass.: A.K. Peters.

Englebretsen, George. 1980. Noncategorical syllogisms in the Analytics. *Notre Dame Journal of Formal Logic* 21: 602–8.

Frege, Gottlob. 1882. On the scientific justification of a conceptual notation, in Bynum (1972), 83–9.

———. 1883. On the aim of the "Conceptual Notation", in Bynum (1972), 90–100.

Gasser, James. 1989. *Essai sur la nature et les critères de la preuve*. Cousset: Editions DelVal.

Hailperin, Theodore. 1984. Boole's abandoned propositional logic. *History and Philosophy of Logic* 5: 39–48.

Jongsma, Calvin. 1982. Richard Whately and the revival of syllogistic logic in Great Britain in the early nineteenth century. Ph.D. Dissertation, University of Toronto.

Kennedy, Hubert C., ed. 1973. *Selected works of Giuseppe Peano*. Toronto: University of Toronto Press.

Kneale, William. 1948. Boole and the revival of logic. *Mind* 57: 149–75.

Kneale, William, and Martha Kneale. 1962. *The development of logic*. Oxford: Clarendon Press.

Lewis, Clarence I., and Cooper H. Langford. 1932/1959. *Symbolic logic*. 2nd ed. New York: Dover.

MacHale, Desmond. 1987. *George Boole: his life and work*. Dublin: Boole Press.

Peano, Giuseppe. 1889. *Principles of arithmetic, presented by a new method*, in Kennedy (1973), 101–34.

Prior, Arthur. 1948. Categoricals and hypotheticals in George Boole and his successors. *Australasian Journal of Philosophy* 27: 171–96.

Quine, Willard V.O. 1940/1951. *Mathematical logic*. Revised ed. Cambridge, Mass.: Harvard University Press.

———. 1950/1959 *Methods of logic*. Revised ed. New York: Holt, Reinhart and Winston.

Rosenbloom, Paul C. 1950. *Elements of mathematical logic*. New York: Dover.

Runes, D.D., ed. 1983. *Dictionary of philosophy*. Littlefield: Adams Quality Paperback.

Russell, Bertrand. 1903. *Principles of mathematics*. New York: Norton (1964).

Smiley, Timothy. 1994. Aristotle's completeness proof. *Ancient Philosophy* 14: 25–39.

Tarski, Alfred. 1941/1994. *Introduction to logic and to the methodology of the deductive sciences*. 4th ed. New York: Oxford University Press.

Van Evra, James. 1984. Richard Whately and the rise of modern logic. *History and Philosophy of Logic* 5: 1–18.

Whately, Richard. 1827. *Elements of logic*. Delmar, N.Y.: Scholars' Facsimiles & Reprints (1975).

Department of Philosophy, State University of New York at Buffalo,
Buffalo, New York 14260 USA

BÉATRICE GODART-WENDLING

THE CONCEPTUALIZATION OF TIME IN BOOLE'S ALGEBRAIC LOGIC

In the *Laws of Thought*, Boole abandoned the interpretation of secondary propositions based on the notion of 'conjunctures of circumstances' (or in terms of the 'universe of cases') as proposed in *Mathematical Analysis of Logic*,[1] and developed an original semantical theory based upon temporal relations between propositions.[2] This new approach emphasizes the notion of duration in order to give a temporal interpretation of several propositions and thereby to determine the respective periods of their duration.

This temporal interpretation was disapproved of by Frege,[3] who saw no 'organic relation' between Boole's primary and secondary propositions due to the different ways of interpreting them.[4] This criticism, which has to do with an issue that was of no real concern to Boole, fails to recognize that Boole nevertheless presents us with the first completed representation of one of the most important features of language—namely, time (or tense)—for which contemporary semantic theories are still seeking an adequate formalization.

This paper is presented as follows: section 1 emphasizes that the temporal interpretation chosen by Boole is accounted for by the peculiar nature of his semantics, which does not interpret propositions, but only relations between propositions. The formalization used to indicate the temporal relations between propositions is briefly recalled. In section 2 we discuss consequences of the non-symmetrical nature of Boole's logical equations—in particular the impossibility of relating the durations of two secondary propositions if they are not in a transitive relation. In section 3 we show that transitivity between two secondary propositions is necessary in order to investigate their common dependence upon the idea of time or the universe of discourse. In sections 4 and 5 we discuss the specific properties of the qualitative and quantitative logical symbols associated with the notion of time and we show the result of their specific properties on the two major processes of Boole's formal logic—namely, reduction and development of logical functions. Finally, we propose a simplification in the treatment of quantitative logical symbols.

1. THE TEMPORAL INTERPRETATION OF SECONDARY PROPOSITIONS

The use of the temporal interpretation to account for secondary propositions can be explained by the fact that Boole's purpose in the *Laws of Thought* is not to provide a semantics for propositions in themselves. His goal is thus not to determine (as in modern semantic theories) the semantic values for each primary proposition *X, Y, …* that occurs in a secondary proposition. Instead, Boole accepts as given the truth or falsehood of primary propositions for the purpose of studying relations between

J. Gasser (ed.), A Boole Anthology, 241–255.

secondary propositions. He examines examples such as the following (cf. 1854, 163):

(a) 'Either the proposition X is true, or the proposition Y is true'
(b) 'If the proposition X is true, the proposition Y is true'
(c) 'The propositions X and Y are jointly true'

Boole is interested, then, in the semantics of relations between propositions, defined as 'relations of coexistence' of truth or falsehood. The search for such a coexistence justifies the use of the temporal interpretation, implying as it does that the respective periods during which propositions are accepted as true or as false must be known in order to determine the overlapping period during which they coexist as true (or as false). This peculiarly relational semantics is more closely related to Carnap's 'transformational rules'[5] than to contemporary semantics. According to Boole, a 'relation of coexistence' is sufficient for purposes of logical inference but does not necessarily exhaust the whole meaning included in each proposition:

> Now, in considering any such relations as the above [examples (a), (b), (c)], we are not called upon to inquire into the whole extent of their possible meaning (for this might involve us in metaphysical questions of causation, which are beyond the proper limits of science); but it suffices to ascertain some meaning which they undoubtedly possess, and which is adequate for the purposes of logical deduction. Let us take, as an instance for examination, the conditional proposition, 'If the proposition X is true, the proposition Y is true'. An undoubted meaning of this proposition is, that the *time* in which the proposition X is true, is *time* in which the proposition Y is true. This indeed is only a relation of coexistence, and may or may not exhaust the meaning of the proposition, but it is a relation really involved in the statement of the proposition, and further, it suffices for all the purposes of logical inference (1854, 163).

Using the capital letters X, Y, Z to denote elementary propositions and small letters x, y, z to denote that portion of time for which the propositions X, Y, Z are true, Boole establishes a system of notation for the expression of secondary propositions and shows that these symbols are logical variables subject to the fundamental law of thought expressed by the relation $x^2 = x$. For secondary propositions, the values 0 and 1 represent, respectively, '*nothing* in reference to the element of time', and 'the universe, or whole of time, to which the discourse is supposed in any manner to relate' (1854, 166). These two values are used when different assertions of the unconditional truth or falsehood of propositions are given as the premises of a logical solution.

Examination of secondary propositions analysed by Boole reveals that in most cases their truth values correspond to durations of unspecified length. If, for example, the premises contain conditional propositions such as:

'If X is true, Y is true'

then the truth values of such a proposition are associated with an indefinite portion of time denoted by the logical symbol v, so that the above conditional proposition is written as an equation:

$$x = vy$$

where the symbol v represents the whole, or an indefinite part, or no part, of the whole time y. On this interpretation, then, the symbol v is a quantitative logical variable.

The symbol v is also used frequently in the developments carried out by Boole in order to obtain interpretations of secondary propositions. It is particularly noteworthy that this symbol can enter into the development of expressions in which it does not occur explicitly at the outset. Suppose for example that the interpretation of X be sought in the equation

$$y = xz.$$

From this equation, which means that the time during which the proposition Y is true is the same as that during which the propositions X and Z are together true, we obtain the development

$$x = \frac{y}{z} = f(y,z) = yz + \frac{1}{0}y(1-z) + 0(1-y)z + \frac{0}{0}(1-y)(1-z),$$

which can be resolved into the following two equations:

$$x = yz + v(1-y)(1-z)$$

$$y(1-z) = 0$$

The first is interpreted as meaning that the time during which X is true is the same as that of the whole time during which Y and Z are together true as well as an indefinite portion of the time during which Y and Z are together false. The second equation is interpreted to mean that there is no time when proposition Y is true but proposition Z is false.

2. READING LOGICAL EQUATIONS

The presence of the quantitative logical variable v in logical equations leads to a closer and more careful reading and therefore to greater awareness of the lack of symmetry in such relations. Let us consider, for example, the two conditionals:

'If Y is true, X is true' and 'if X is true, Y is true'

which differ only in the transposition of the propositional variables. In symbols, we have:

(1) $$y = vx$$

(2) $$x = v'y$$

These are interpreted as follows:

(1) the time y for which the proposition Y is true corresponds to an indefinite portion of time x for which the proposition X is true.

(2) the time x for which the proposition X is true corresponds to an indefinite portion of time y for which the proposition Y is true.

If, following Boole's 'process of limitation' (1854, 228), these two equations be multiplied by vv' in order to keep them algebraically equivalent, we obtain (taking into account the duality law for v and v'):

(1′) $vv'y = v'vx$

(2′) $v'vx = v\,v'y$

which are interpreted as follows:

(1′) the indefinite portion of time for which Y is true is an indefinite portion of time x for which X is true

(2′) the indefinite portion of time for which X is true is an indefinite portion of time y for which Y is true.

These are not equivalent for secondary propositions. It follows, then, that the algebraical equivalence of (1′) and (2′) does not guarantee that the logical interpretations will be synonymous. In particular, the logical relations obtained from secondary propositions involving quantitative variables are not symmetrical as in classical algebra but must be read from left to right.

This requirement is confirmed in the case of a system of secondary propositions involving a transitive relation. Let us first consider the system of propositions 'if X then Y' and 'if Y then Z', written as:

(3) $x = vy$

(4) $y = v'z$

whence by substitution:[6]

(5) $x = vv'z$

This conclusion expresses the transitivity between the premises (3) and (4). It is obtained by a 'left-to-right' reading for the two premises.

Let us next consider a system of secondary propositions involving between them an algebraic transitive relation which would require, in order to be expressed in Boole's algebra:

- a 'left-to-right' reading for the first equation, and
- a 'right-to-left' reading for the second.

Thus:

(6) $x = vy$

(7) $z = v'y$

This algebraically transitive system is not a logically transitive system. In order to make this clear, let us try, as in the previous example, to eliminate y by substitution. We obtain from (6) $y = \frac{x}{v}$ and from (7) $y = \frac{z}{v'}$. After substitution, the following two relations are obtained:

$$z = v'\frac{x}{v} \quad \text{and}$$

$$x = v\frac{z}{v'}$$

These are interpreted, respectively, as follows: the time z for which Z is true is an indefinite portion of the time x for which X is true, and the time x for which X is true is an indefinite portion of the time z for which Z is true. These two relations are contradictory; this is the first problem that we meet with.

The same result can be obtained in another way by reducing the system (6) and (7) by the elimination of y through the process of addition.[7] This elimination gives the relation:

(8) $vz - v'x = 0$

which means that no time remains when we take away from the indefinite portion of time vz for which Z is true that indefinite portion of time $v'x$ for which X is true.[8] This interpretation shows that the indefinite portions of time $v'x$ and vz are of identical duration.

A similar interpretation would have been given for the relation (5) if we had used the process of addition. We should have obtained:

(5′) $x - vv'z = 0$

which gives, by limitation:

(5″) $vx - vv'z = 0$

This relation is to be compared with that given in footnote 8:

(8′) $v'x - vz = 0$

If now we compare the relation (5′) with (8) or (8′) we meet with a second problem—namely the pairing-up of qualitative and quantitative variables. In (5′), v' is connected with z, as in premise (4). In relations (8) and (8′) the quantitative variables v and v' are connected respectively with z and x, whereas in premises (6) and (7) they are both connected with y. The change of connection in (8) and (8′) breaks the logical rule requiring that 'the symbol v, when used in the sense of "some", applies to that term only with which it is connected in the premises' (1854, 233).[9]

This violation of a logical rule prevents a logical interpretation of the relations (8) and (8′). Moreover we cannot rewrite relation (8′) as

(9) $v'x = vz$

because the elimination of y does not justify inclusion of the time $v'x$ in the time vz. This can be made clearer if we develop (9) with respect to x. We obtain:

$$x = v'vz + \frac{0}{0}(1-v)(1-v')z$$

The interpretation of this logical equation shows that x is not—as in relation (5)—simply included in z.

All the above arguments show the impossibility of finding a transitive relation between the variables x and z when the middle term $v'y$ occurs in the second premise in the position of an extreme term. The non-transitive relation between x and z justifies the requirement for any logical equation that it be read from left to right. It has also been established that a logical relation such as $v'x - vz = 0$ does not follow that $v'x = vz$ unless there be a transitive relation in the premises.

3. THE RELATIVE POSITIONS OF DURATIONS AND TRANSITIVITY

We should now like to show that the possibility of calculating the relative positions (in the time of the discourse) of the durations of two secondary propositions depends on their being linked by a transitive relation. Let us consider first the following system of secondary propositions:

(10) $vx = v'y$

(11) $y = wz$

which is transitive and which yields, by Boole's rule of inference 'equate the extremes' (1854, 234), the relation $vx = wz$. This result is in agreement with the inclusions indicated by the premises, namely $vx < v'y < wz$. The relative positions of the durations of propositions X, Y, and Z can also be represented in a diagram:[10]

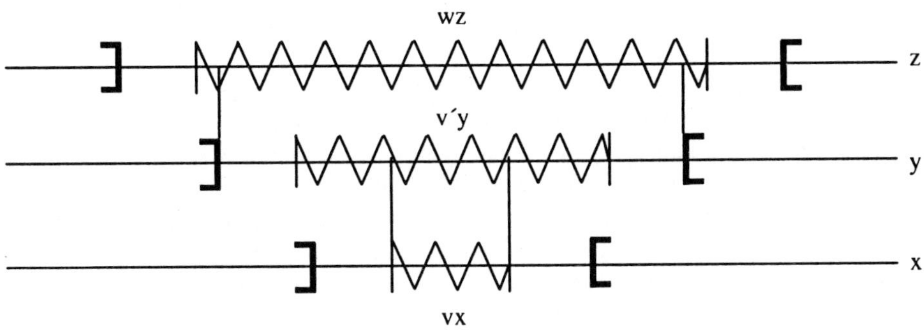

Let us now consider a new system:

(12) $vy = v'z$

(13) $wx = w'z$

the algebraic transitivity of which does not extend to the algebraic logic of Boole (in which no relation between wx and vy is deducible). In this case we are not able to locate the relative durations x and y for the propositions X and Y:

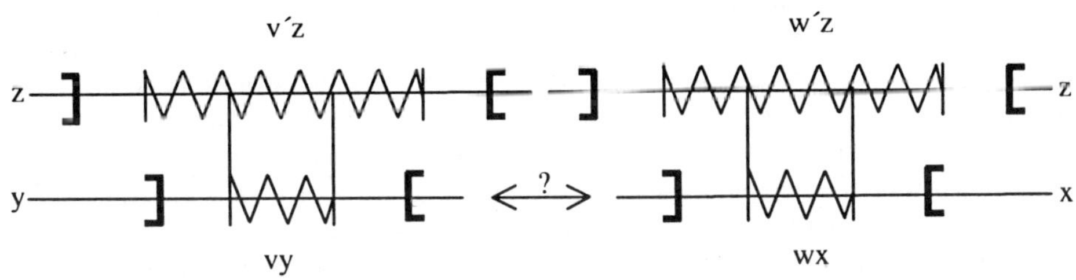

From these two examples, we can conclude that secondary propositions must be in a transitive relation (their durations must be included one within the other) in order for it to be possible to give the relative positions of the duration in the time of discourse.

4. THE PROPERTIES OF LOGICAL VARIABLES

In the chapter of the *Laws of Thought* devoted to Aristotelian logic,[11] Boole (1854, 232–3) distinguishes two kinds of logical variables—qualitative and quantitative—fundamentally different in character.

Qualitative variables such as x, y, ... indicate the time during which truth or falsehood may be attributed to the propositions X, Y, ... These variables can be associated with any possible combination of truth values of given premises, and so can be used to express these interactions with respect to propositions X, Y, ...

The quantitative variables v, v' quantify the duration of x, y, ... Thus the term vx may be understood to represent the whole ($v = 1$) or an indefinite part ($0 < v < 1$) or no part ($v = 0$) of the time x.

When these two kinds of variables are used together, as in the expression vx, every combination of quality and quantity can be represented.

Moreover, Boole maintains that each quantitative variable is specifically linked to a given qualitative variable. He writes: 'the symbol v, when used in the sense of "some", applies to that term only with which it is connected in the premises' (1854, 233). This is an important sentence, because if the same quantitative variable could be assigned to different qualitative values x and y, such as v in $vx = vy$, it would no longer be possible to obtain all the quantitative combinations represented by the relation $vx = v'y$. The quantitative variable v applied to x thus represents an indefinite part of the duration x and is therefore entirely distinct from a variable v' applied to a duration y different from x. It follows that if v is applied to x in the premises, vy will be uninterpretable, and that it will not be possible to apply v to $(1 - x)$ or to any other term but x.

Although Boole never explicitly formulated the consequences of his distinction between qualitative and quantitative variables, two differences can be observed which concern their properties and their analysis.

The first is the non-homogeneity of these variables: variables such as x and y represent durations, and thus have dimension, whereas quantitative variables merely act on durations, and have no dimension themselves. As any proposition is homogeneous with respect to a duration, the variables v and v' are necessarily associated with qualitative variables in such a way as to be homogeneous with respect to a duration. It follows that there can never be an isolated variable v in the development of a logical function ('$f(x) = yz + v + \ldots$' is nonsense)—or, as will be shown in section 5.2 below, that it is impossible to develop v in a relation involving durations.

The analysis of the mental operation of multiplication is also very different in the case of qualitative and quantitative variables. Thus, the product xy is interpreted as the common portion of time for which the propositions X and Y are both true (Boole 1854, 165). The interpretation of an expression like $vv'y$ does not correspond to the notion of the common part, however, but involves taking into account the quantitative variable applied to y—so that the term $vv'y$ will be interpreted in the same way as the term $v'y$.

We propose in the last section of this paper to extend this simplification procedure (i.e., the reduction of $vv'y$ to $v'y$) to the case where the quantitative variables are applied in quotients.

5. THE TREATMENT OF LOGICAL VARIABLES

It is now possible to investigate the effects of the specific properties associated with qualitative or quantitative variables on the two fundamental processes of analysis presented by Boole—namely those of reduction and development.

5.1. Reduction

Let us examine again the following system:

(3) $x = vy$

(4) $y = v'z$

The elimination of y will allow us to discover a relation between x and z. This can be accomplished by various means.

Firstly, we can apply the rule of elimination suggested by Boole in order 'to remove [...] any possible difficulty' (1854, 178). This rule consists in carrying out, in order, the following operations:

> Eliminate separately from each equation in which it is found the indefinite symbol v.
>
> Eliminate the remaining symbols which it is desired to banish from the final solution: always before elimination reducing to a single equation those equations in which the symbol or symbols to be eliminated are found. Collect the resulting equations into a single equation $V = 0$.
>
> Then proceed according to the particular form in which it is desired to express the final solution (1854, 178).

According to this rule, the variables v and v' in the equations (3) and (4) are eliminated by using the logical rule for the elimination of variables—namely $f(v \text{ or } v' = 1)f(v \text{ or } v' = 0) = 0$ applied to the equations:

(3′) $f(v) = x - vy = 0$

(4′) $f(v') = y - v'z = 0$

The following are obtained by elimination of v and v':

(3″) $x(1 - y) = 0$

(4″) $y(1 - z) = 0$

We reduce these equations to a single equivalent equation:[12]

(5) $x(1 - y) + y(1 - z) = 0$

We subsequently eliminate y from this equation by using $f(1)f(0) = 0$ and obtain:

(6) $(1 - z)x = 0$

This relation is interpreted to mean that there is no portion of time for which the proposition Z is false and the proposition X is true.

Secondly, we can change the order of elimination of the variables in order to test whether the behaviour of the qualitative variables x, y, z is similar to that of the quantitative variables v and v' with respect to the process of elimination. Thus we first eliminate y in the system (3) and (4) and obtain from (3′) and (4′) the relations:

$$(3^*) \qquad x(x - v) = 0$$

$$(4^*) \qquad (1 - v'z)(-v'z) = 0$$

This new system is uninterpretable, because the variable v is not associated with any qualitative variable in (3*) and also because multiplication by the negative term $-v'z$ in (4*) confronts us with the unresolved problem of the meaning of an expression such as $-v'$. The procedure fails, then, when we try to eliminate a qualitative variable before the elimination of the quantitative variables.

It is, however, possible to eliminate y first by using a method of reduction by addition.[13] We can obtain:

$$(5_2) \qquad x = vv'z$$

which can be rewritten as:

$$x = v'z$$

since only v' was applied to z in the premises.

Now we can eliminate v' by the rule $f(1)\,f(0) = 0$ with $f(v') = x - v'z = 0$. We finally obtain:

$$(6_3) \qquad x(1 - z) = 0$$

This equation is equivalent to (6); both methods for eliminating variables thus lead to the same result. The reduction rule proposed by Boole is thus a sufficient but not a necessary condition for the elimination of variables (since elimination can also succeed by the method of addition).

Moreover we observe that the reduction of the quantitative variables can be carried out only if the order of operations specified by Boole's rule be respected. This rigidity in the order of the steps for elimination shows that properties specific to quantitative and qualitative variables determine the possible ways of treating them.

5.2. Development

Now let us consider the following equation:

$$vx = wz$$

If the quantitative variable v had the same properties as qualitative variables, we would be able to express v in terms of the other variables:

$$v = \frac{wz}{x} = f(x,w,z)$$

Expanding the second member by the usual method, we would obtain, after rejecting terms whose coefficients are zero:[14]

$$v = xwz + \frac{1}{0}(1-x)wz + 0x(1-w)z + \frac{0}{0}(1-x)(1-w)z$$

In examining this development we face two difficulties. The first is the meaning of the term xwz. Since x and wz are durations, xwz should be interpretable as their common part; but this is impossible because the duration of x cannot be determined from the initial equation $vx = wz$. Consequently the common part between x and wz cannot be found from the expansion given above. The second difficulty is the presence of the term with coefficient $\frac{0}{0}$ which involves $(1 - w)z$. Given the premises which stipulate that z can have no other quantitative variable than w, this is uninterpretable.

These kinds of difficulties are systematically encountered when we attempt to carry out the development of an indefinite variable v. This shows that the expansion of a quantitative variable alone does not lead—unlike what happens in the case of qualitative variables—to an interpretable result.

In conclusion we can say that the difficulties involved in the two processes of analysis (reduction and expansion) when they are applied to quantitative variables reinforce the idea that Boole's algebraic logic is based on two classes of variables, each of which has its own truly distinct properties.

6. ATTEMPT AT A SIMPLIFICATION OF THE TREATMENT OF SECONDARY PROPOSITIONS

The simultaneous occurrence of two classes of logical variables (with specific properties) in the formalization of the notion of duration makes the procedures of analysis as well as the interpretation of secondary propositions particularly complex. Consider for example the following premises:

(14) $$vx = v'y$$

(15) $$w'y = wz$$

If we seek a relation between x and z, we can first eliminate y by the method of addition.[15] We obtain:

(16) $$x = \frac{v'w}{vw'}z = f(v,v',w,w',z)$$

where x is a function depending on five arguments.

The interpretation of x can be obtained (following Boole's rules) by an expansion of the above function (the denominator vw' prevents a direct interpretation). Boole (1854, 233) gives the following development for the relation (16):

$$x=\left[vv'ww'+\frac{0}{0}\{vv'(1-w)(1-w')+ww'(1-v)(1-v')+(1-v)(1-w)\}\right]z+\frac{0}{0}\{vv'(1-w')+(1-v)\}(1-z) \tag{17}$$

Since the interpretation of the relations (16) and (17) must be coherent for the duration x, the coefficients of $(1 - z)$ in (17) must be cancelled. This is achieved by introducing an additional condition on the variables v, v' and w'. In order to do this, Boole develops vx and obtains:

$$vx=\left[vv'ww'+\frac{0}{0}vv'(1-w)(1-w')\right]z+\frac{0}{0}(1-w')(1-z) \tag{18}$$

Since the coexistence of z and $(1 - z)$ is not coherent with (16) multiplied by v, Boole once again eliminates $(1 - z)$ by posing that $w' = 1$. He obtains:

$$vx = vv'wz \tag{19}$$

Boole then enunciates the rule of inference 'equate the extremes' and applies it to premises (14) and (15):

$$vx = wz$$

This conclusion respects the properties of quantitative variables mentioned in section 4 of this paper (only w is connected to z in premise (15)).

The condition $w' = 1$ is necessary and sufficient to eliminate the term $(1 - z)$ in (18) but it is not sufficient to eliminate the same term in (17). In order to do that, Boole adds a new condition $v = 1$.

With these two conditions $w' = 1$ and $v = 1$, we obtain for x the following development:

$$x = v'wz = wz \tag{20}$$

which is the final solution.

Let us remark that the two relations (19) and (20) are consistent, as (19) is deduced from (20) by the process of limitation.

Boole's analysis shows that the development of functions containing many variables can be long and fastidious. It would seem that this could be avoided by using, from the very first step of the analysis, the fact that the quantitative variable w is applied to z in a premise. Is it possible to justify this procedure in the case of a quotient, such as in (16)? Let us rewrite the relation (16), taking into account the condition $w' = 1$, $x = \frac{v'w}{v}z$ in the form:

$$x = \frac{w}{v}z = f(v, w, z)$$

With this form, x becomes a function with only three arguments (instead of five) which can be more easily expanded than those given without this simplification.

We shall thus have:

$$\text{(21)} \qquad \begin{aligned} x = &\left[vw + \frac{1}{0}(1-v)w + 0v(1-w) + \frac{0}{0}(1-v)(1-w) \right] z \\ &+ \left[0vw + \frac{0}{0}\{(1-v)w + (1-v)(1-w)\} \right] (1-z) \end{aligned}$$

whence:

$$\text{(22)} \qquad x = \left[vw + \frac{0}{0}(1-v)(1-w) \right] z + \left[\frac{0}{0}\{(1-v)w + (1-v)(1-w)\} \right] (1-z)$$

Now if we eliminate the term $(1 - z)$ by using, with Boole, $v = 1$, we obtain Boole's relation (17).[16] The simplification procedure that involves respecting the connections of the variables in the premises can thus be extended to the most complex quantitative variables.

7. CONCLUSION

The study of the properties necessary to a proper interpretation of the notion of duration in Boole's algebraical logic has established that the Boolean model of time is based on two kinds of variables which produce their own relations inside a single temporal class. The variables in secondary propositions, being connected with only one class (namely that of durations), are fundamentally different from those used in primary propositions, which are in relation to many distinct classes of objects. Thus the formalization of the primary proposition 'the sun shines' as $x = vy$ involves two logical variables which denote two different classes of objects: x denoting the class with only one term (the sun) and vy the indefinite part of the class of shining objects. The same symbolic equation $x = vy$ can be used to translate secondary propositions such as the conditional 'if the sun shines, the day will be fair'; in this case, the equation involves variables which place durations in relative positions on the same temporal axis (temporal class). These relative positions inside the same class can lead to difficulties in the treatment and interpretation of secondary propositions, which motivate the simplification procedures proposed in this paper.

NOTES

[1] Boole explains the reasons for this new approach as follows: 'I was led to interpret the symbol 1 in secondary propositions as the universe of "cases" or "conjunctures of circumstances"; but this view involves the necessity of a definition of what is meant by a "case", or "conjuncture of circumstances"; and it is certain, that whatever is involved in the term beyond the notion of time is alien to the objects, and restrictive of the processes, of formal Logic' (1854, 176). We recall that this assertion was contested by Hailperin who wrote (1986, 174): 'Indeed we shall see that, precisely contrary to what Boole says, it is the notion of time which is "alien to the objects, and restrictive of the processes, of formal logic" and that, correctly conceived "cases" have nothing to do with time. That Boole had a promising approach in *Mathematical Analysis of Logic* has already been noted in the literature, e.g. by Kneale and Kneale [1962, p. 414], and there is a fairly extensive appraisal in Prior [1948, pp. 176–182]'.

[2] Let us recall that Boole distinguishes between primary and secondary propositions as follows: the former express relations among things (thus the primary proposition 'the sun shines' relates the sun and things that shine); the latter express relations among primary propositions. These are of two kinds: (i) propositions asserting the truth or falsehood of primary propositions, such as 'It is true that the sun shines', (ii) propositions asserting relations between primary propositions, such as conditionals ('if the sun shine the day will be fair', in which the truth of the primary proposition 'the day will be fair' depends on that of 'the sun shines') and disjunctions ('either the sun will shine, or the enterprise will be postponed' in which the truth of one of the two primary propositions excludes that of the other).

This classification has been criticized by Hailperin as a 'semantic confusion': 'Contemporary semantic theory however would relegate example (i) to a meta-language and indeed, with Boole, as an assertion about the truth of a proposition; on the other hand example (ii) would be in the language proper and not an assertion *about* propositions of the language' (Hailperin 1986, 184).

[3] See Frege's article of 1880/81 'Boole's Logical Calculus and the Concept-script' in Frege's *Posthumous Writings* (1979, 14).

[4] The interpretation of primary propositions is given in terms of classes of things, whereas that of secondary propositions involves the idea of time. The formal laws as well as the procedures applicable to primary and secondary propositions remain the same however.

[5] Although the transformational rules given by Carnap in *Logische Syntax der Sprache* have nothing to do with the notion of time, they can be used, like the approach adopted by Boole, to identify the logical relations between propositions.

[6] See Boole (1854, 231).

[7] By this method we can also obtain another relation: (8′) $v'x - vz = 0$.

[8] See Boole (1854, 165) for the interpretation of the sign '–'.

[9] The use of the variables v and v' in (5′′) does however respect this rule.

[10] The symbols ']' and '[' indicate the open-ended durations of the propositions. The zigzags represent the indefinite portions of time vx, $v'y$, wz and symbolize the possibility of variation between 0 and 1.

[11] Most of the examples mentioned in this paper are drawn from this chapter (Chapter XV), which concerns the syllogism of primary propositions (categoricals). The extension we have made to secondary propositions can be justified by passages from Boole, who wrote, for example: 'After the discussion of secondary propositions in a former part of this work, it is evident that the forms of hypothetical syllogism must present, in every respect, an exact counterpart to those of categorical syllogism. *Particular* Propositions, such as, "Sometimes if X is true, Y is true", may be introduced, and the conditions and rules of inference deduced in this chapter for categorical syllogisms may, without abatement, be interpreted to meet the corresponding cases in hypotheticals' (Boole 1854, 241).

[12] This single equation must be a logical equation satisfying the law of duality. For that it is necessary to verify that the system of propositions from which this equation is deduced is independent, that is to say a system in which 'it is not possible to deduce from any portion of the system a conclusion deducible from any other portion of it' (Boole 1854, 154).

[13] It will be recalled that elimination of a variable by addition is always possible in a system of linear equations. It is therefore possible to apply this procedure to Boolean logical equations, which are always first order, since they are subject to the principle of duality. As a matter of fact, the only difference involving the elimination of variables in the algebraic logic of Boole is that there is no need, as in ordinary algebra, for a variable to be repeated in the equations of the system in order for it to be eliminated. It should also be noted that this method cannot be used to eliminate a quantitative variable first.

[14] Cf. Boole (1854, 88) for details of this development.

[15] For the elimination of y by Boole's rule before the elimination of the symbols v, v', w, w', see the previous section.
[16] On the condition that $v = 1$, $x = wz$ can be obtained directly from (21). Boole obtains the same result in two steps: $x = v'wz$ followed by $x = wz$.

REFERENCES

Boole, G. 1847. *The mathematical analysis of logic, being an essay towards a calculus of deductive reasoning*. Cambridge: Macmillan, Barclay and Macmillan. Reprinted in Boole (1952), 45–119.

———. 1854. *An investigation of the laws of thought on which are founded the mathematical theories of logic and probabilities*. London: Walton and Maberley.

———. 1952. *Studies in logic and probability*. Edited by R. Rhees. La Salle: Open Court.

Carnap, R. 1934. *Logische Syntax der Sprache*. Wien: Julius Springer.

Frege, G. 1979. *Posthumous writings*. Oxford: Blackwell.

Hailperin, T. 1986. *Boole's logic and probability*. 2nd ed. Amsterdam: North-Holland.

Kneale, W., and M. Kneale. 1962. *The development of logic*. Oxford: Clarendon Press.

Prior, A.N. 1948. Categoricals and hypotheticals in George Boole and his successors. *The Australasian Journal of Philosophy* 27: 171–96.

*Université de Paris VII, UMR 7597, CNRS, case 7034, 2 place Jussieu,
75 251 Paris cedex 05, France*

GÉRARD BORNET

GEORGE BOOLE AND THE SCIENCE OF LOGIC

1. INTRODUCTION—LOGIC AS A SCIENCE

It is my aim in this paper to take an especially broad look at the logic of Boole. This does not mean that I think details are not important, but the more you go into them the more you get caught at a particular spot in space and time—say Lincoln in the year 1847—and Boole seems correspondingly further away from us. If the details of Boole's logic are left aside, what remains is a science. The question therefore arises of how much Boole's concept of a science determined his logic. This is the vital question of the following exposition.

Let me illustrate this objective with a passage from *The Mathematical Analysis of Logic*[1] where Boole considers the possibility that there are laws of logic other than those he actually takes as a basis for his system. There he says (MAL 6):

> ... it may with confidence be asserted, that if [the laws of logic] were other than they are, the entire mechanism of reasoning, nay the very laws and constitution of the human intellect, would be vitally changed. A Logic might indeed exist, but it would no longer be the Logic we possess.

There could, then, be other logics,[2] and the look at Boole's logic that I intend to take is broad enough to include those other logics alluded to in this prophetic-sounding passage. This broad look is possible because we need know nothing about the other laws that come under consideration; it is sufficient for our purpose to assume that they form the basis of a science. It is only in order to give a concrete example of such a science that details of Boole's logic are mentioned here.

But is it not possible that the quotation given above is really a slip of the pen 'In a hasty and (for this reason) regretted publication',[3] as Boole was to call *The Mathematical Analysis of Logic* four years later? I don't think so. It is true that Boole writes as if the beings he allows for the sake of an argument—beings with a 'vitally changed mechanism of reasoning'—are, like Martians, only figments of the imagination. And it is quite clear that Boole no more *believes* in the existence of other logics than we believe in Martians. A plurality of logics is for him only a theoretical possibility—but it is a possibility.[4] To see this helps to remove one of the great stumbling-blocks in the way of an adequate understanding of Boole's work, because it raises the question of the determinants of his logic.

If what Boole calls, in the year 1847, 'the Logic we possess' is roughly identified with Aristotelian logic, it is easily recognised to be a result of our history and a part of our changeable *culture*. Yet Boole writes as if logic forms part of our unchange-

J. Gasser (ed.), A Boole Anthology, 257–270.

able *nature*. It seems therefore that the subsequent development of logic has disproved Boole, just as Aristotle's geocentric cosmology was refuted by the further development of astronomy.

Against this picture I would like to set another. I would like to show that Boole's conception of science is restricted, in the sense that it allows a view of the world only from *inside* science. *This* is the reason why the validity of a science—in particular the science of logic—is universal and *therefore* that it seems to us that Boole speaks of nature, whereas in fact he refers only to culture. As a member of this culture he was blind to certain peculiarities which however, 150 years later, are conspicuous to us.

The main part of this investigation will be devoted to an attempt to give a criterion of identity for sciences, in order to make the notion of being *inside* or *outside* a science clearer. I will then conclude the discussion of Boole's concept of a science by speaking among other things of the intimate connection of the science with the scientist which depends on the notion of evidence. This is again to interpret Boole's conception of science as restricted, in that it allows us to construct a science only relative to a group of people who share evidence.

Much of what strikes modern logicians as odd when reading Boole will be seen to be simply a consequence of these two restrictions. At the end of this paper I shall therefore try to show that there are situations in which such restrictions make sense in the field of logic although the general tendency today is to neglect them altogether.

2. UNITY, HARMONY AND ORDER—BOOLE'S PHILOSOPHY OF SCIENCE

2.1. The Source Material

Boole did not develop or explicitly present a philosophy of science; there are only isolated remarks on this topic scattered throughout his work. At first sight they give the impression of eclecticism in the bad sense of the word, but, on rereading, the outline of a consistent whole appears. I suppose that Boole might have worked out a philosophy of science of his own had he lived longer, but as a matter of fact he did not. There is only one work by him on the subject, entitled *The Claims of Science—especially as founded in its Relations to Human Nature*.[5] This is a lecture which Boole gave as Dean of the Science Division of the Faculty of Arts in Queen's College, Cork, when he addressed the Faculty at the beginning of the session 1851–1852.[6] As his biographer Desmond MacHale points out, Boole was at this time working on his major work in logic, *The Laws of Thought*, which is also the most important source of his philosophy of science.

It would be an interesting task to specify the various sources of Boole's views in the philosophy of science and to put them into relation to thinkers of his time, notably John Stuart Mill (1806–1873)[7] and William Whewell (1794–1866).[8] But this would lead us too far away from our theme which is only concerned with one general feature of his philosophy of science. Moreover, this feature will be presented from a modern point of view by means of a kind of rational reconstruction. I say 'a kind' because of the tentative character of the undertaking, which has yet to prove its fruitfulness in a comprehensive interpretation of Boole's philosophy of logic.

2.2. Unity

To put Boole's philosophy of science in a nutshell, one can say that according to him the progress of science has the aim of reaching *unity*. Unity is accomplished by bringing the different stages and branches of science into *harmony*. And the key to harmony is the cognition of *order*. In this way the three catchwords 'unity', 'harmony' and 'order' are interdependent—I will begin the discussion with the one that is most important for our consideration: 'unity'.

The proviso made above that we will not exactly reflect the standpoint of Boole, but take a look at him from the outside so to speak, applies above all to the expression 'unity'. Boole uses it chiefly as a synonym for the symbol 1 to represent the universe (MAL 15) and not to refer to the 'unity of science' as I will do. That he believes in the unity of science is *implied* by what he says and is not expressly stated. It is not quite the concept of the unity of science which a century later became the slogan of Rudolf Carnap and other logical positivists. Whereas their unity is synchronical and lacks a historical component entirely, Boole's is essentially diachronical.[9] References to the past are characteristic of Boole's writings, and they are always used as confirmations. What I want to say about this is best shown by an anecdote told by Boole's wife Mary.[10]

> Some one [Robert Leslie Ellis] wrote to my husband to say that in reading an old treatise by Leibniz ['Addenda ad Specimen Calculi Universalis']... he had come upon the same formula which the Cambridge people call 'Boole's Equation' [i.e. the index law]. My husband looked up Leibniz and found his equation there, and was perfectly delighted: he felt as if Leibniz had come and shaken hands with him across the centuries.

'Shaking hands across the centuries': this is precisely what Boole was always seeking in history. If he had found nobody who expressed the same idea as he had discovered, I am convinced that he would have distrusted his discovery. Of course the idea would have been expressed in a form appropriate to its time; he would not have expected Aristotle to have written in a now lost paper his 'Mathematical Analysis of Logic'.

Aristotle's presence is prominent from the very beginning. Right on the frontispiece of Boole's epoch-making book there is the following citation from *Posterior Analytics*:[11]

> All the sciences have common ground in the common axioms; I mean the elements used as premises, not the subjects or attributes of which demonstrations are made.

Here Aristotle expresses nothing other than the doctrine of the unity of science[12] and Boole makes this doctrine his slogan. Why? I think because, as a rule of thumb, synchronical depth and diachronical depth go hand in hand. This is similar to archaeology, where the deeper you dig in the present the more you reach into the history of man. If you define man as a being who uses tools, you can be sure to find tools from the very beginning of his history and in like manner if you define man as a thinking being[13] you can be sure to find the laws of thought at work from the first manifestations of man. This is just analytical. Now, if in reversal of this train of thought you seek not man but a definition of him, it is very helpful to go back into history to look at what was always there.

The following quotation from Boole's unpublished writings, which expresses his idea of unity very clearly, should therefore be given a historical dimension:[14]

> The perfect idea of Logic is ... that ... of a philosophy from which as from a common stem all sciences whose method is deductive are developed and with which they all stand in vital connection.

Accordingly, logical considerations mark the beginning of all sciences. And for Boole everything originated historically from the idea of unity—more exactly, from the idea that the universe is a unity (CoS 207).

2.3. Harmony

From our discussion so far, it appears that the unity of science is the attitude of stressing the common elements between various stages of development rather than the differences. Whereas certain historians of science like to talk of revolutions, there are no breaks in history leading to unbridgeable gulfs for Boole. Boole stressed the continuity of history in one of his addresses as president of a society which was ironically named after the French Baron Cuvier, the founder of catastrophism:[15]

> Each generation as it passes away bequeaths to its successor ... the truths which it has won, and the ideas which it has learned to conceive ... This perpetual transmission of the light of knowledge ... has been compared to those torch races of antiquity in which a lighted brand was transmitted from one runner to another until it reached the final goal.

I suppose this final goal[16] is a kind of 'New Atlantis' as described by Francis Bacon, where people satisfy their needs without conflict. Boole gives a prominent place to Bacon in *The Claims of Science* (CoS 189), although in another connection. It is important to note that for Boole the 'happy life ... to which reason ... point[s]' (CoS 205) is not an end in itself but a means to reach a truly moral life (cf. CoS 205). But the religious aspect, important as it is for Boole, must here be left out because it goes beyond the scope of our discussion. Religion is a possible but not a necessary motive for advocating the sort of philosophy of science which we are after here.

The following description of the role of science for man I find rather revealing of the harmony-loving spirit of Boole (CoS 204):

> ...the more [the multiplied industry of man] is under the control of science, the more does it consist in simply arranging the train of natural circumstances; the inherent and impassive forces of matter ever offering themselves as substitutes for animal toil and animal suffering.

This is not the usual metaphor of the *dominion* of man over nature. It is a question of *arrangement*; accordingly nature is not *forced* to do such and such but offers its services *voluntarily*. It seems that, for Boole, man comes to live in harmony with nature *thanks to* scientific progress. Times have changed: many people today have come to think otherwise and see science as something which alienates us from nature.

'Harmony' is a favourite expression of Boole's. When he talks about what makes a method perfect, he says, for example, that it is the fact that it manifests in all its parts and processes 'a certain unity and harmony' (LT 150). Or he says that the courses of nature are 'settled in a consistent harmony' (CoS 192). And elsewhere (MAL 14) he expresses the conviction that 'with the advance of our knowledge of all true science, an ever-increasing harmony will be found to prevail among its separate branches'. It is easy to see that the concept of harmony goes hand in hand with the concept of unity: unity excludes disharmony between the parts, and to harmonize means to unify.

What has been said so far should elucidate two points. First, Boole had a clear concept of the progress of science. Accordingly, he calls steady growth a characteristic of science (LT 2) and even affirms (CoS 203, emphasis mine), 'that the peculiar order of the development of human thought which *preceded* the rise and growth of modern science, was not an arbitrary thing, but is in its main features susceptible to explanation'. Secondly, science keeps its identity throughout time. What existed before was an earlier stage of the *same* science which in the meantime has matured.

If we look for a possible criterion of identity for a science, the key lies in the notion used by Boole to explain how harmony can be produced, namely that of order.

2.4. Order

First I want to make a terminological remark. The philosophical position which in our century comes nearest to the standpoint of Boole is in my view that of Rudolf Carnap (1891–1970) as expressed in his famous *Der logische Aufbau der Welt.*[17] There he defends the thesis that all scientific statements are statements about structures. The concept of structure which is at work here was given by Bertrand Russell (1872–1970) in his *Introduction to Mathematical Philosophy.*[18] For the logical positivists, a scientific theory consists of an uninterpreted structure given in axiomatic form (i.e. a calculus) plus an empirical interpretation. This is exactly the same idea as Boole's (see LT 1), although he complements the admissible means of interpretation with that of introspection.

When I speak of 'structuralism' in the philosophy of science I mean a general variety of structuralism which is related to structuralism in linguistics, semiotics, ethnology, anthropology or sociology and *not* the specific variety of recent origin connected with the names of Joseph D. Sneed (1938–) and Wolfgang Stegmüller (1923–1991). Their standpoint, which is also called the 'non-statement view' is in a certain sense *opposed* to that of the logical empiricists. Also, the sense in which a modern mathematician speaks of 'structures' is not quite the same as the one that Russell introduced in 1919. There, this notion served as an explication of what is meant by the 'likeness of relations'[19]—a phrase Boole uses literally to explain what he means by 'analogy'.[20] It is possible in this sense to speak for instance of the structure of a railroad-map[21]—an object which a modern mathematician would not be likely to include in his mathematical world.

In short: there is a broad sense of structure or structuralism and many different narrower ones. The reason why I want here to use the broad sense is obvious: I want to make comparisons possible and the many different narrow senses which are also in use are introduced for exactly the opposite reason, namely to exclude all other varieties.

If the concept of structure is conceived in this way, it becomes nearly synonymous with the concept of order and it is easy to see how both notions are connected with the concept of law. If a realm possesses an order, then it has a structure. If a law is valid in a certain realm, then there is order in this realm and it has a structure.

After these preliminaries I turn to the following principle of Boole (LT 39):[22]

> The object of science, properly so called, is the knowledge of laws and relations.

This is for me Boole's principle of structuralism. Among other things it excludes the possibility of a science of men, because, as Boole remarks, 'men cannot be divested of their individuality' (CoS 196).[23] Correspondingly 'the social and economical *sciences* ... regard men not as individuals, but as members of a community' (CoS 196, emphasis mine). Individuality is opposed to generality and the business of science is to investigate laws (LT 24), not to state facts. This is structuralism.

For Boole, laws are what conflicting parties can agree upon. An important condition for this is that 'the laws in question are really deduced from observation' (LT 40). If this point is met, then divergencies between parties can be due only to the *explanation* of these laws.[24] So Boole can say (LT 39):

> They who believe, and they who refuse to believe, that there is more in the relation of cause and effect than an invariable order of succession, agree in their interpretation of the conclusions of physical astronomy. But they only agree because they recognise a common element of scientific truth [i.e. a law of nature], which is independent of their particular views of the nature of causation.

Boole similarly asserts that science removes 'the contradictions of sense' and establishes 'the consistent uniformity of natural law' (CoS 201). Thus a contradiction is abolished by a law.

Why is it that the recognition of order unites men? As already mentioned, the religious views of Boole[25] are outside the scope of this paper, which shall be confined to epistemic issues. In the epistemology of Boole, all science begins with experience based on observation (cf. CoS 188f.). This is achieved either by our senses, which put us in contact with the external world, or by introspection, which involves the world of our imagination. But this is only the beginning (CoS 189f., slightly paraphrased):

> All the knowledge which the senses have ever communicated to man, has been a mere collection of facts; [...] With such a state of information, however, the mind does not rest satisfied. It ... reduce[s] the scattered details of its knowledge into form and order. It begins to compare and to classify, and to arrange. [...] Thus, from the contemplation of facts, the mind rises to the perception of their relations. While in the former state it is little more than a passive recipient of the impressions of the external world [or of the imagination], in the latter it exercises an unborrowed activity. The faculties of judgment, of abstraction, of comparison, of reason, are an agency of strength and power from within, which it brings to bear upon the lifeless elements before it, shaping them into order, and extracting from them their hidden meaning and significance.

This lengthy extract makes plain that there is a division of labour between the (inner and outer) senses and the mind, and that it is the task of the latter to detect the structure of the world. I suggest Boole believed that it is because there is only one Mind (or because the Mind is one) that men recognize laws as the elements of structures. Irrespective of Boole's special reasons, the role of science as a neutral arbiter between conflicting parties, especially political ones, remains undisputed today. One can reverse the described train of thought and say that because science is a neutral arbiter and is about structure, the recognition of structure and therefore of laws must be what conflicting parties can agree upon.

3. BOOLE RECONSTRUCTED—A CRITERION OF IDENTITY

3.1. The Criterion Stated

Boole has given himself a place in the development of the science of logic. He calls the Aristotelian theory (SMLP 136):[26]

> ... rather a Natural History of the forms in which human thought has been actually developed than an inquiry into the possible—the universal—forms in which thought admits of being developed in virtue of the constitution of the human mind.

'Scholastic logic' is for him (LT 241) 'not a science, but a collection of scientific truths, too incomplete to form a system of themselves'. The founder of the science of logic is—of course—Boole himself. If one compares the last two quotations, it is striking how Boole parallels the development of the history of science with the history of acquisition of knowledge in a single mind: for example both start with a 'collection of facts'.

Now Boole does not simply rank himself among other logicians; he also argues that his system compares favourably with others. Thus, he suggests in a discussion of Aristotelian logic that his own laws are more fundamental because of 'the nature and extent of the structure which they are capable of supporting' (LT 240), after having already defined as 'fundamental those laws and principles from which all other general truths of science may be deduced, and into which they may all be again resolved' (LT 5). Without going into details, one can say that Boole's system is more general than Aristotle's, which it *contains*.[27] For example, Aristotle needs a middle term to draw a conclusion whereas Boole does not (MAL 41f.).

Speaking in terms of structure and of science in general we can say that for Boole the development of a science brings about a more general structure in which the old one is *embedded*. I will develop this idea into a criterion of identity.

We have seen in the foregoing discussion that Boole seems to believe that embedding is a general tendency for any development of science. At the end of this process there must be a 'superstructure', so to speak, which embeds all other sciences. And indeed, behind the science of logic as developed in the *Mathematical Analysis of Logic* and *The Laws of Thought*, which can be considered as a development of the notion of class (SMLP 129), there lurks in Boole's unpublished writings a 'higher Logic' (SMLP 143; cf. CLW 226), 'considered as the Science of all Thought that admits of expression' and which 'can only be apprehended through the particular manifestations of them which are founded in the Logic of Class, the Logic of Number and the other sciences into which it may be developed'.

It is difficult to imagine what this 'Science of all Thought' could be. From the passage quoted it seems that it cannot be understood for itself—which can mean either of two things. It can mean that—like all other sciences—it cannot be developed a priori (cf. SMLP 129) or that it cannot be represented in symbols. I do not dispute that the latter assumption makes sense relative to the epistemological and religious background of Boole.[28] But we need not share the assumption that there really *is* a 'superstructure' described by the 'Science of all Thought'; there is a way to make a part of this assumption explicit which allows us to handle the case hypothetically.

To see what I am trying to say, consider the question of what happens if a further stage of development does *not* contain or embed the earlier one. I think it most natural in such a case to speak of the later stage as of a *new* science and to describe the step from one stage to the other as a *revolution*. If we use embedding as a criterion of identity, then revolutions are excluded—*by definition—within* the development of a science. Furthermore, it is *analytic* that the evolution of *a* science occurs by embedding.

This leaves us with the following result: for reasons which we do not need to share, Boole believed that there are no revolutions in science at all and that the natural aim of evolution is a 'unity of science'. We can abstain from this general presupposition and nevertheless share in his results with the proviso that we regard the development of a single science *from within*. I think it is more than a coincidence (although less than a proof of the view I am developing here) that Boole uses the same image in the preface to *The Mathematical Analysis of Logic* when he says that his plan is to regard logic 'from *within*' (MAL 1).

Now for most of us—though not for Boole—it is, to say the least, quite imaginable that no further 'internal' development of the logic which dominated around the year 1847 is possible and that the need which was satisfied by the old science must be satisfied by an entirely new one. But *if* a development is possible, *then* we can predict or expect certain features of the new stage of the science. To some of these features I would like to turn now. Their presence in the logical work of Boole underlines the suitability of the proposed criterion of identity.

3.2. What Was New with Boole

— When we look at the contribution of Boole to the science of logic and assume it to be an evolution rather than a revolution, we do not expect to find a refutation of his predecessors but an extension of their results. Indeed, Boole asserts that he discusses the system of Aristotle 'in no narrow or harshly critical spirit' (LT 226). And analyzing some works of Samuel Clarke (1675–1729) and Benedict Spinoza (1632–1677) he promises the reader the most important inferences which the 'actual' premises of these two philosophers involve 'in addition to the particular inferences actually drawn by the authors' (LT 185).

— When we learn that the science of logic made a step forward we expect according to our criterion of identity of a science a deeper analysis of the central notions of the science. The central notion analyzed by Boole was conception, upon which he built up a theory of judgment and reasoning.[29]

— We would further expect to find a reconstruction of the old stage of the science in terms of the new one. None of Boole's printed work on logic lacks a treatment of syllogism. Whereas in *The Mathematical Analysis of Logic* the aim was simply to show that syllogisms are expressible in the new system, in *The Calculus of Logic*,[30] published one year later, he demonstrated the power of his new approach by greatly simplifying the expression of the laws of syllogism.

— We would judge the progress made as significant if it then became possible to unify hitherto unconnected realms; we could even equate progress with such unifications. Boole shows that he can give to the logics of class, of propositions and of probability the same basic structure. I think the achievement of Boole in presenting a calculus with *three* different interpretations is the most underestimated part of his contribution to logic. For instance, nobody has to my knowledge *defended* Boole's preference[31] for the exclusive 'or' on the grounds that it is a prerequisite of this threefold interpretation. Instead, it is usual to point out the advantages of the inclusive 'or' for predicate or propositional logic without mentioning the logic of probabilities at all—even though this would mean that it could no longer be conceived as a direct interpretation of the same calculus.

— And last but not least, if a science made a step forward in the sense described above, one would expect to find in the new stage steps which are *uninterpretable* in terms of the old stage. I say 'last but not least' because it is this point which prompted me to carry out the present exposition. It is characteristic of Boole's logic that only the conclusions of his symbolic method need be interpretable and not the intermediate steps as well. An adequate exegesis of Boole's writings must in my view be able to explain why he did *not* see this as a defect of his method but as the price to pay for a 'general method in Logic'—and not only in logic but in 'every developed form of human reasoning which is based upon the employment of a symbolic language' (LT 67).

In *The Laws of Thought* (LT 69), Boole gives 'The employment of the uninterpretable symbol $\sqrt{-1}$, in the intermediate processes of trigonometry' as an illustration, but in his unpublished writings there are examples which are much simpler.[32] Theodore Hailperin comments that to a 'modern mathematician' the correctness of the procedure cited by Boole 'is justified by an appeal to the *embeddability* of the reals in the complex numbers'[33] and in general by the embeddability of one structure in another. I think he has hit upon Boole's idea of scientific progress.

4. SCIENCE QUO VADIS?

4.1. Boole's Philosophy of Science Completed

In the preceding part of this investigation I have tried to make plausible the idea that Boole's philosophy of science allows him only to account for scientific progress which never breaks with the past. Progress is on this view a refinement of old ideas, a penetration of the historically given with one's intellect.[34] In conclusion I want to show that this idea of continuity also appears if one looks not into the *history* of science but into the *life* of a scientist. To use a dichotomy introduced above in connection with the concept of unity, the problem I face now is no longer diachronical but synchronical. Because of limited space, I shall have to confine myself to just a few allusions.

In Boole's philosophy of science, the scientist must be taken into account whenever the evidence becomes crucial, i.e. whenever the laws are no longer empirical and probable but necessary because they are 'laws of the mind'. The truth of such a law 'is made manifest in all its generality by reflection upon a single instance of its application' (LT 4). There is thus a gulf between that which comes from experiment and observation and that which is necessarily true: to come nearer to certainty from the side of the senses one has to increase the number of cases, but with this the distance from the one single case which is needed to get absolute certainty also increases.[35]

Evidence also plays a role when it comes to the refutation of scepticism, for instance in the question of the external reality of space and time. Boole does not bring forward any argument to prove this reality but says (LT 418):

> Whenever, in the view of the intellect, different hypotheses are equally consistent with an observed fact, the instinctive testimony of consciousness as to their relative value must be allowed to possess *authority*.

The well-known objection to the admission of the concept of evidence into a philosophy of science is that it makes the result subjective. But there is an easy defence against this objection. The validity of the result is restricted to that group of people who share this evidence. Together with the other restriction connected with the proposed criterion of identity, a concept emerges which in analytical philosophy has slowly been gaining ground in recent years because it has been made popular by John R. Searle, namely that of social reality.[36] What Searle presents is only a variety of this concept—a much more common noun is 'culture'. We can roughly say that Boole's scientific methods are valid if one stays *within* a culture—or, in other words, the results of these methods are valid for people who share the same present and the same history as the scientist who uses these methods.

There is, apart from evidence, another feature in Boole's philosophy of science which stands in opposition to current views: the admission of ordinary language. As with evidence, this admission is probably restricted to cases where the laws of mind are in question, so I no longer speak of science in general (although one can argue with good reason that the more a science progresses the more the laws of mind play a role). Be that as it may, in the realm of logic it is the declared intention of Boole (LT 41):

> ...to avail myself as far as possible of the language of common discourse, without regard to any theory of the nature and powers of the mind which it may be thought to embody.

Behind this strategy there are two reasons. First, there is the insight that 'We can in no other way describe mental operations than by terms material in their origin' (SMLP 76) and, secondly, that the position of structuralism leads from itself to a theory of metaphor which allows us the use of ordinary language in science. For instance, instead of the symbols 0 and 1 of the dual algebra we could also use the symbols *u* and *n* or even 'men', 'golden mountains' or 'Utopia', says Boole, provided they remain subject to the same formal laws as 0 and 1 (cf. SMLP 94f.). In other words, it does not matter *how* a structure is expressed—as long as only the structure is taken into consideration, one can even use ordinary language.

4.2. The Progress of Science

To sum up: I have left aside the details of Boole's logic and investigated the concept of science which backed it up. A totally different concept of science from what is usual today has appeared and with it another idea of what the progress of science consists in. Although Boole used a mathematical tool which only few were able to understand, he pursued a concept of science in which the scientist is connected with his tradition, with the grounds on which his beliefs are founded and with the world he lives in as an ordinary being. The usual concept which we form of science does not take these things into account. This has paid off, in a certain sense: it has taken man to the moon and made it possible to drive around on Mars. But it seems to me that to restrict oneself to this kind of progress is to deprive oneself of an understanding of science as a very valuable means of cultivation. We are literally going further and further into the outer world, but we are not thereby coming nearer to a better understanding of humanity. As a consequence of the common disregard of history, which is also part of the usual concept of science, the beginner is flooded with countless systems of logic, each newer than the other and each different. This is more a hindrance to the spreading of logic than a help. The boast of the authors of modern logic books not to be concerned with evidence because this could spoil the purity of their systems has its price, namely that their readers are not touched at all and logic is in danger of degenerating into a mere manipulation of symbols. And to take no account of ordinary language, as is the usual attitude of today's scientists, brings about the effect that the application of logic to everyday problems is severely hampered or is not even attempted.

If 150 years after *The Mathematical Analysis of Logic* we ask ourselves how far the science of logic has advanced since, then according to one's viewpoint there are *two* answers. We can say that the progress is overwhelming if we speak about the power of expression of our artificial languages and the multitude of available calculi. But we should be modest if with progress is meant a better understanding of ourselves as rational beings—and this is what Boole was concerned with.[37]

NOTES

[1] Macmillan, Barclay and Macmillan, Cambridge 1847 (referred to hereafter as MAL). Reprinted in *Collected Logical Works by George Boole*. Vol. I. Studies in Logic and Probability, Rush Rhees (ed.), The Open Court, La Salle Ill. 1952 (referred to hereafter as CLW), pp. 45–124. MAL was recently reprinted in *From Kant to Hilbert: A Source Book in the Foundations of Mathematics*, W.B. Ewald (ed.), Clarendon Press, Oxford 1996, vol. I, pp. 451–509, and also as a monograph (with an introduction by J. Slater) by Thoemmes Press, Bristol 1998. References are to the original 1847 pagination.
[2] The idea still haunted Boole when he wrote *An Investigation of the Laws of Thought, on which are founded the Mathematical Theories of Logic and Probabilities*, Walton and Maberley, London 1854 (referred to hereafter as LT). There, in a footnote (LT 406), he recognizes the possibility of sciences erected on fundamental ideas which 'might be wholly different from those with which we are at present acquainted'.
[3] 'On the Theory of Probabilities, and in particular on Mitchell's Problem of the Distribution of Fixed Stars', *The Philosophical Magazine*, Series 4, vol. i, 1851, pp. 521–30. Reprinted in CLW, pp. 247–59. The quotation is on p. 252 of the reprint.
[4] At LT 50 Boole takes up the theme of the possibility of other logics and stresses the hypothetical character of this train of thought. It does not imply, he points out, that our logic 'is the product either of chance or of arbitrary will'. This remark is also true, by the way, if logic is considered as part of our culture.
[5] Taylor and Walton, London 1851 (referred to hereafter as CoS). Reprinted in CLW, pp. 187–210; references are to the pagination of the reprint.
[6] From this circumstance one might expect an uplifting speech lacking depth, but Boole 'took considerable pains over the matter of the lecture' because he intended to publish it and was not willing 'to labour for a merely temporary object' as he wrote in a letter to Augustus De Morgan. See Desmond MacHale, *George Boole—His Life and Work*, Boole Press Dublin 1985 (referred to hereafter as MacHale), pp. 97f. The letter from which the foregoing quotations are taken is also reprinted on pp. 54f. of G.C. Smith, *The Boole-De Morgan Correspondence 1842–1864*, Clarendon Press, Oxford 1982.
[7] His *System of Logic, Ratiocinative and Inductive* appeared in 1843 and is referred to by Boole more than once, beginning with the preface to MAL.
[8] His *History of the Inductive Sciences* and *The Philosophy of the Inductive Sciences* appeared in 1837 and 1840 respectively. Boole refers to the latter work at LT 406.
[9] I borrow these terms from Ferdinand de Saussure, who uses them to distinguish two types of linguistics (i.e. 'linguistique synchronique' and 'linguistique diachronique').
[10] The following quotation from her writings is taken from Luis María Laita: 'Boolean Algebra and Its Extra-logical Sources: The Testimony of Mary Everest Boole', *History and Philosophy of Logic* 1, 1980, pp. 37–60. The quotation is on p. 58; additions are mine. The anecdote is treated exhaustively by Volker Peckhaus in his *Logik, Mathesis universalis und allgemeine Wissenschaft—Leibniz und die Wiederentdeckung der formalen Logik im 19. Jahrhundert*, Akademie Verlag, 1997, pp. 222ff. In the Nachlass of Boole there is a fragment in which he deals with 'a recently published tract of Leibnitz', drawing parallels to his own laws of logic. This fragment is printed in George Boole, *Selected Manuscripts on Logic and Its Philosophy*, Ivor Grattan-Guinness and Gérard Bornet (eds), Birkhäuser Verlag, Basel 1997 (referred to hereafter as SMLP), pp. 185ff.
[11] Book 1, Chapter xi. Boole cites Aristotle in Greek; the above translation is by A.E. Wardman.
[12] It is not my intention here to give an interpretation of Aristotle but to elucidate Boole's concept of 'unity'. It is true that Aristotle criticized the Platonist conception of a science 'which embraces everything' on the grounds that 'being is said in many ways' (*Metaphysics* Book 1, Ch. 9). But whether this is enough to make him unsuitable as an exponent of the doctrine of the 'unity of science', as Prof. Elisabeth Schwartz argued in the discussion of this paper, depends exactly on the concept of unity which one applies. Boole was probably thinking of the principle of contradiction as an example of an 'axiom' which is common to all sciences. He discusses it in light of his reading of Aristotle (*Metaphysics* Book 4, Ch. 3) at LT 49f. and also at SMLP 164ff. (N.B. *Metaphysics* Book 4 was formerly referred to as Book 3; this explains why Boole refers to Book 3, Ch. 3 in his footnote at LT 49.)
[13] It depends of course on how (correct) thinking is defined. Boole did nothing other than what we would do today: he consulted the leading scientists of his time. This is, in my opinion, the proper role to assign for instance to Richard Whately (1787–1863) and even to Immanuel Kant (1724–1804). But this has

become obscured because the role has long since been passed on to others. The result of this confusion is that Boole is blamed for having formalized Aristotelian logic. But his greatness lies not in *what* he formalized but in *how* he did so.

[14] SMLP 126. The quotation is also found in an excerpt at CLW 14.

[15] *The Social Aspects of Intellectual Culture*, George Purcell, Cork 1855, pp. 8f. MacHale, pp. 122–5, contains an extensive extract from this small brochure; the passage quoted is on p. 123.

[16] Portrayed at some length at CoS 204f.

[17] Weltkreis, Berlin 1928 (referred to hereafter as Aufbau).

[18] George Allen & Unwin, London 1919 (referred to hereafter as Introduction).

[19] Introduction 61.

[20] SMLP 120. Boole speaks there (pp. 119–20) of the analogy 'between the operations of thought in Logic and its operation within a particular sphere of the science of number'.

[21] Carnap uses this example to introduce the concept of a structural description in his Aufbau, p. 17.

[22] LT 320 contains a different but cognate determination of the 'final object of science'.

[23] Compare also LT 404: '...if the process of reasoning be carefully analyzed, it will appear that abstraction is made of all peculiarities of the individual to which the conclusion refers...'.

[24] For instance in the last chapter of LT at 415f. Boole says in a survey of philosophical systems of all ages: 'The attempts of speculative minds [...] have differed less in the forms of theory which they have produced, than through the nature of the interpretations which have been assigned to those forms'.

[25] I suppose that Boole would have accepted a proof of the existence of God which would conclude from the existence of an orderly creation to the existence of a being which caused this order. Thus he says (CoS 191): 'Science ... may ... be compared to some stately temple, whose materials have been brought together from many distant regions ... in whose fair proportions and goodly order, we read the traces of the designing mind'. Boole is careful not to make the intellect of men the cause of order in nature according to the strategy of Kantian transcendental idealism. 'The native powers of the mind, cast abroad amid a world of mere chance and disorder, could never have realised the conception of law' (CoS 190).

[26] This passage is also quoted by Mary Hesse, 'Boole's Philosophy of Logic', *Annals of Science* 8, 61–81 on p. 66. The phrase 'Natural History of Thought' is also used at SMLP 53.

[27] 'We may by restricting the canon of interpretation confine our expressed results within the limits of the scholastic logic; but this would only be to restrict ourselves to the use of a part of the conclusions to which our analysis entitles us' (MAL 34).

[28] Epistemologically the clue for the incomprehensibility may lie for Boole in the capability of the *human* mind to grasp general conceptions. This capability is restricted because humans *need* signs to have something perceptible—a 'sensuous element, without which the conception could not have been thought' (CLW 226; see also SMLP 68)—and their mental power is *finite*. Compare also what Boole says about the intelligibility of God at SMLP 200–202. The reason why the existence of a single all-uniting 'superstructure'—the subject of this 'higher logic'—is plausible for Boole could lie in the fact that for him nature and mind were designed by the same supreme being. One should be careful to note that for Boole this does not mean that the laws of the mind could be *reduced* to the laws of nature. They are still *normative* in the sense that they are subject to an authority which does not reign by blind necessity as in the realm of nature but 'with reference to an ideal standard and a final purpose' (LT 421). A last remark: it is very tempting to draw the border between class logic and the 'higher logic' with the help of the difference between an object-language and a metalanguage. But this idea is alien to Boole and his time. Boole knew, however, the difference in logic between operations which depend upon 'laws of combination' and those which depend upon 'interpretation', as he makes clear in a letter to Arthur Cayley (SMLP 192). This allows for the possibility of modestly identifying the 'higher logic' as that logic which comprises 'laws of combination' only.

[29] See SMLP 125, where Boole claims that no one before him has given 'a complete statement of the elementary operations involved in Conception ...[and]... an analysis of the laws of that faculty'.

[30] *The Cambridge & Dublin Mathematical Journal* 3, 1848, pp. 183–98. Reprinted in CLW, pp. 125–40. The passage specifically referred to is on p. 136 of the reprint.

[31] In the discussion of this paper Theodore Hailperin pointed out that in spite of Boole's well-known adherence to the exclusive 'or' in logic, his calculations nevertheless contain steps which are not allowed on an exclusive understanding of 'or'.

[32] In the world of things, Boole says (SMLP 146), the operation of subtraction 'does presuppose the condition that when a number x is to [be] diminished by another number y, x should be greater than y'. But, Boole goes on, in algebra this restriction does not hold. At SMLP 43 he proposes a thought experiment in which it is imagined that we could conceive of particular numbers only, such as 10, 20, 30 etc. and of no others. If we were to investigate the laws of these numbers, we could make calculations as usual, as long as the results were interpretable, although it would mean 'the intermediate steps being uninterpretable *for these numbers*' (Boole's emphasis).
[33] *Boole's Logic and Probability*, North-Holland, Amsterdam 1986 (2nd ed., revised and enlarged), pp. 89f. (emphasis mine, slightly paraphrased). I want to point out that Hailperin does not find the talk of embeddability quite appropriate for Boole and that I have chosen a broader and somewhat different concept of structure here than he does.
[34] It is significant that Boole places the symbolical methods at the end of his *Treatise on Differential Equations* (Macmillan, Cambridge 1859); see there pp. vi–viii. Likewise he places his *Mathematical Analysis of Logic* at the *end* of the historical development of logic and emphasises that without the 'aids of symbolical language' we would have 'to abandon the hope of further conquest' in the field of logic (MAL 10).
[35] The collection of cases is not useless though; it can help to find the one single instance which is necessary to find absolute truth.
[36] Cf. his *The Construction of Social Reality*, The Free Press, New York 1995.
[37] This is the text of the talk as distributed at the conference entitled 'Boole 1997: One Hundred Fifty Years of Mathematical Analysis of Logic', held at Lausanne in September 1997, with some additions in the footnotes which reflect the discussion. I thank Elaine Lerf for proofreading, James Gasser for careful editing and Andreas Bächli for advice in connection with the objection of Prof. Elisabeth Schwartz dealt with in footnote 12. This paper has benefited considerably from the results of research carried out by the author with support from the Swiss National Science Foundation during the period 1991–1995.

Philosophisches Seminar, Université Miséricorde, 1700 Fribourg, Switzerland

VOLKER PECKHAUS

WAS GEORGE BOOLE REALLY THE 'FATHER' OF MODERN LOGIC?*

1. INTRODUCTION

Was George Boole really the 'father' of modern logic? I confess that I regard this question as rather academic, because it is not possible to answer it convincingly. A typical answer to questions of this kind is: 'it depends'. In our case it depends on what is meant by the 'father of a scientific discipline', it depends on the understanding of modern logic, and it even depends on one's own attitude towards the notion of scientific development.

The question was posed in a 'virtual' discussion between Desmond MacHale, author of the 'most distinguished' (Grattan-Guinness 1997, lxii, note 1) book-length biography on Boole (MacHale 1985), and Willard Van Orman Quine, undoubtedly one of the major living representatives of modern logic, who reviewed MacHale's book for *The Times Literary Supplement* in the same year (Quine 1985, 1995). It might therefore be worthwhile taking up this question and using it as a starting point to reconsider Boole's place in the history of logic.

The first section will give an account of this virtual debate, discussing Quine's and MacHale's arguments, but keeping the debate open for others to step in. The father of something is often identified as the founder of the same. The participants in this debate seem to be of the opinion that it is always possible to isolate a single point of departure of a development. They argue from the perspective of priority being primarily concerned with first formulations and the initial occurrence of seminal ideas. This perspective of priority will be called the 'originator interpretation' of fatherhood. I would like to confront it with the 'initiator perspective' in which a father is essentially responsible for instigating a new development.

In the second section an alternative approach taking a more genetic perspective will be suggested.[1] Taking into account the fact that authors of logic have adopted ideas which were propounded at earlier periods, I intend to set Boole's algebra of logic into its relationship with Leibnizian logic, with the German algebraists of logic (Graßmann, Schröder), with Frege and with the logical research of the Hilbert school. This will lead me to a new interpretation of fatherhood in logic.

2. THE VIRTUAL DISCUSSION

The question of fatherhood is one of the main points of criticism in Quine's review of MacHale's biography. His criticism is that MacHale 'falls in with the tradition [...] of representing Boole as the father of modern logic. It is a tradition that had its be-

J. Gasser (ed.), A Boole Anthology, 271–285.

ginnings almost in Boole's day, but we are able nowadays to view matters in a truer perspective' (Quine 1995, 252). Quine characterizes Boole's logic as an algebra of classes modeled on arithmetic (ibid.). He mentions Boole's precursors (ibid.):

> Leibniz had already been tinkering with symbols to somewhat the same effect before 1700, and in the intervening six generations the idea had been taken up by Bernoulli, Castillon, Holland, Lambert, Ploucquet, Segner, and Tönnies in varying degrees.

As a source for this observation Quine refers to C.I. Lewis's *Survey of Symbolic Logic* (Lewis 1918). The rationalistic precursors of Boole in the tradition of Leibnizian and Wolffian logic had been known, however, in the English speaking world at least from the time of John Venn's *Symbolic Logic* (Venn 1881), which was a milestone in the early logicians' reflection on the historical roots of 'symbolic logic' as Venn preferred to call the new logic. One can therefore justly say that the 'truer perspective' also had its beginnings about the time of Boole.

But who is Quine's alternative candidate as the father of mathematical logic? It is Gottlob Frege, the inventor of the modern quantification theory. Quine writes (1995, 254):

> The avenue from Boole through Peirce to the present is one of continuous development, and this, if anything, is the justification for dating modern logic from Boole; for there had been no comparable influence on Boole from his more primitive antecedents. But logic became a substantial branch of mathematics only with the emergence of general quantification theory at the hands of Frege and Peirce. I date modern logic from there.
>
> Frege got there first. His initial and succeeding contributions were richer and more refined, moreover, than any other logical work until the present century, when the current from Boole through Peirce and the current from Frege converged. Thus I have long hailed Frege as the founder of modern logic, and viewed Boole, De Morgan, and Jevons as forerunners. John Venn and Lewis Carroll also belong back with them, though coming on the scene only after the great event.

Quine combines his understanding of the essentials of modern logic with the originator interpretation which causes him to select Frege. He admits to Boole's strengthening position in the initiator interpretation, but says this interpretation is secondary.

Before discussing Quine's alternative suggestion, let us see what MacHale actually wrote (1985, 71–2). MacHale stressed that in his *The Mathematical Analysis of Logic* Boole

> unwittingly [...] highlighted a topic that has come to influence virtually every aspect of present-day life—the storage and processing of information, which in turn has led to the development of computer science. Not alone is Boole's algebra the 'correct' and most economical tool for handling information, but the electronic machines which now do the work actually operate according to principles determined by that self-same algebra. Boole has been called the 'Father of Symbolic Logic' and the 'Founder of Pure Mathematics', but he is just as deserving of the title, 'Father of Computer Science'.

Taking this quote word for word it must be observed that MacHale obviously wanted to stress that Boole was the father of computer science, an opinion which can be justified by indicating his impact on Claude E. Shannon's correlation of electric circuits with Boolean truth functions (cf. Shannon 1938). He is more careful as far as the fatherhood of symbolic logic is concerned, referring to anonymous authors who had called Boole the 'father of symbolic logic'. He had good reason for being careful because, as indicated above, in 1985 it had been clear for more than a century that Boole did not invent symbolic logic.

It is therefore easier to find authors who dispute the assertion that Boole was the father of modern logic[2] than authors who explicitly refer to him in this way. One of the latter is P.L. Heath, who writes in the respective section of the article 'Logic, History of' in Edwards' *The Encyclopedia of Philosophy:* 'In working out this idea [that the laws of algebra can be formally stated without regard for any particular interpretation], which he did most fully in *An Investigation of the Laws of Thought* (London 1854), he became the first to devise a practicable symbolic algebra and is thus reckoned (next to Leibniz) as the father of modern logic' (Prior [ed.] 1967, 542). The association with Leibniz irritates. Obviously in Heath's eyes Boole's fatherhood is only a partial one.

It is more common to call Boole an originator. The Belgian logician Robert Feys can be mentioned as an example (Feys 1957), and we should not forget Bertrand Russell who claimed in his paper 'Mathematics and the Metaphysicians' (1951, 74) that

> pure mathematics was discovered by Boole, in a work which he called *The Laws of Thought* (1854). This work abounds in asseverations that it is not mathematical, the fact being that Boole was too modest to suppose his book the first ever written on mathematics. He was also mistaken in supposing that he was dealing with the laws of thought [...] His book was in fact concerned with formal logic, and this is the same thing as mathematics.

This view was commented on by William Kneale as follows (1948, 174–5): 'Like many other epigrams, this is a gross exaggeration; but anyone who remembers Mr Russell's account of mathematics can understand what prompted him to make this remark'. Although Kneale is rather critical of this remark it is evident that both Russell and Kneale advocated an originator interpretation in regard to Boole.

I would like to mention now supporters of the initiator interpretation. Karel Berka and Lothar Kreiser maintain in their commented selection of logic texts that the period of modern logic began with Boole's *The Mathematical Analysis of Logic* (Berka and Kreiser 1983, 23). J.M. Bocheński goes even further in his selection *Formale Logik* by saying that Boole was the founder of a school and that he represented, at the same time, the beginning of the continuous development of mathematical logic (Bocheński 1956, 312). I'm very sceptical whether one should accept Bocheński's opinion that Boole was the founder of a school. Although terms like 'the Boolean period in Logic' or 'the Booleans' are very common, Boole was never the founder of a scientific school in the sociological sense. He shared this fate with most other contemporary and subsequent logicians. Only Giuseppe Peano in Italy and Charles S. Peirce at the Johns Hopkins University were able to gather schools. To some extent David Hilbert's group of mathematicians studying the foundations of mathematics could also be called a school. Be that as it may, the authors of both selections mentioned above correctly identify a period of dynamic development of logic in the second half of the 19th century. This development concerned formal logic, and therefore only a part of the old philosophical omnibus discipline which had been almost completely ignored by philosophers of the time. This period obviously started with George Boole's *The Mathematical Analysis of Logic* (1847).

Some commentators confuse originator and initiator interpretations. An example can be found in C.I. Lewis and Harold Cooper Langford's *Symbolic Logic* (1932). They credit Aristotle as being the founder of symbolic logic, but: 'George Boole was really the second founder of symbolic logic. His algebra, first presented in 1847, is the basis of the whole development since' (Lewis and Langford 1932, 9).[3]

Wolfgang Lenzen, the expert on Leibniz's logic, makes Lewis and Langford's assessment the target of a polemic argument against accepted views in the history of logic. He proposes striking Boole off the list of possible fathers of modern logic and replacing him by Leibniz (cf. Lenzen 1984). He argues that Leibniz had already presented a complete axiomatization of a Boole-Schröderian algebra of sets. He proves this claim by showing that Leibniz's logical system, especially that of the *Generales Inquisitiones* of 1686 (cf. Leibniz 1982) is isomorphic to an axiom system of an algebra of sets as presented, e.g., in Lewis and Langford's *Symbolic Logic*. In rejecting Louis Couturat's opinion (cf. Couturat 1901, 387), the first editor of the *Generales Inquisitiones,* Lenzen claims that Boole was not the first to develop an algorithmical, i.e. exact and rigorous logic, but Leibniz. To speak of Leibniz's final failure should be regarded as one of the worst jokes in the history of logic (1984, 203).

Lenzen's opinion of setting Leibniz up as the father of modern logic leads us to possible alternatives with Boole, and thus back to Quine's decision to claim that modern logic begins with Frege. Of course this opinion has been disputed as well. Ivor Grattan-Guinness' well known distinction between the two traditions in the history of logic should be brought to mind: the algebraic tradition starting with Boole and the tradition of mathematical logic starting with Peano (see e.g. Grattan-Guinness 1988). Two points in his argument are noteworthy: (i) Boole supported the algebraic side and therefore had nothing to do with mathematical logic, and (ii) Frege is absent from the list of principal figures in the history of logic because of his lack of effectiveness. Although critical of some of Ivor Grattan-Guinness' interpretations—I doubt, e.g., that the gap between the algebra of logic and mathematical logic was as impenetrable as he claims, and think that one could well dispute the question of whether other logicians besides Frege were really effective at their time—I have sympathy with the method he uses. He combines the initiator interpretation with investigations of the effects caused by the initiatives. The view that an initiator was present at the beginning of a development is associated with the question as to what extent the initiator really influenced this development. Answers to this question are far from evident, and the method sometimes leads to unexpected consequences. As to Quine's criterion, that Frege's most important innovation of logic was his invention of first-order logic, Geraldine Brady, e.g., refers to the independent (but later) invention of general quantification theory in Peirce's school of logicians, and claims:[4]

> The restriction to first-order logic [i.e. Peirce's restriction to 'first-intentional logic of relatives' in his paper of 1885] has been enormously fruitful in modern mathematical logic, and although modern foundation literature ascribes first-order logic primarily to Frege's seminal paper [which is, of course, a book] (1879), we support the case that it was Peirce's work, as systematized and extended by Schröder (1895), that was a primary influence on Löwenheim (1915) and Skolem (1920, 1923), as reflected by their notations and methods.

This example shows how difficult it can be to identify a proper starting-point for a certain development.[5] It is even more difficult to decide fatherhood in logic, once and for all. Speaking with metaphors is a difficult thing. The more you think about the meaning and the consequences of such metaphors, the more they lose their illustrative power.

3. BOOLE'S PLACE IN THE HISTORY OF LOGIC

I would now like to reconsider Boole's place in the history of logic. This will be done by discussing four pairs of relationships between Boole and other major figures in the history of logic, namely Gottfried Wilhelm Leibniz, Ernst Schröder, Gottlob Frege, and David Hilbert.

3.1. Boole and Leibniz

Let me start with considerations on the relationship between Boole and Leibniz. As early as 1833, during his stay at Doncaster, where he worked as an usher, i.e., as an assistant teacher in Mr. Heigham's School, Boole first contemplated the idea of expressing logical relations in symbolic or algebraic form. According to MacHale's report (1985, 19) 'the thought flashed upon him suddenly one afternoon as he was walking across a field'. But it was 14 years later in (1847) that this idea was published in *The Mathematical Analysis of Logic.* Boole himself hinted at the controversy between the philosopher William Hamilton and the London mathematician Augustus De Morgan on the quantification of the predicate as the initial motive for the publication. The mathematical contexts, in which his thoughts emerged, however, were more important. These contexts were determined in particular by the symbolical algebra of the Cambridge mathematicians around George Peacock, and above all by the calculus of operations created by Duncan F. Gregory who died in 1844. Symbolical algebra, for Peacock, dealt with interpretations of those symbols which stand for numbers in arithmetic. Gregory, on the other hand, proposed a more general view of algebra. He defined symbolical algebra as 'the science which treats of the combination of operations defined not by their nature, that is, by what they are or what they do, but by the laws of combination to which they are subject' (Gregory 1840). With this definition he stressed the possibility of interpreting the symbols of symbolical algebra as operations which can be applied to other objects (cf. Schlote 1987, 49).

In his praised paper 'On a General Method of Analysis' (Boole 1844) for which he obtained the Royal Society's Medal, Boole took up Gregory's calculus of operations as the main methodological tool for analysis. He referred to Gregory when presenting the principles of his method:

> There are a number of theorems in ordinary algebra, which, though apparently proved to be true only for symbols representing numbers, admit of a much more extended application. Such theorems depend only on the laws of combination to which the symbols are subject, and are therefore true for all symbols, whatever their nature may be, which are subject to the same laws of combination.[6]

Among these 'much more extended applications' of the calculus of operations Boole included logic. I would like to give prominence to the first words of Boole's introduction to *The Mathematical Analysis of Logic.* 'They who are acquainted with the present state of the theory of Symbolical Algebra', he starts (1847, 3–4),

> are aware, that the validity of the processes of analysis does not depend upon the interpretation of the symbols which are employed, but solely upon the laws of their combination. [...]
>
> But the full recognition of the consequences of this important doctrine has been, in some measure, retarded by accidental circumstances. It has happened in every known form of analysis, that the elements to be determined have been conceived as measurable by comparison with some fixed standard. The predominant idea has been that of magnitude, or more strictly, of numerical ratio. [...]
>
> That to the existing forms of Analysis a quantitative interpretation is assigned, is the result of the circumstances by which those forms were determined, and is not to be construed into a universal condition of Analysis. It is upon the foundation of this general principle, that I purpose to establish the Calculus of Logic, and that I claim for it a place among the acknowledged forms of Mathematical Analysis, regardless that in its object and in its instruments it must at present stand alone.

According to this opinion, logic is applied algebra. Boole's book primarily served a mathematical purpose: Boole wanted to provide further evidence for the calculus of operations. I would even like to go so far as to maintain that *The Mathematical Analysis of Logic* is not a book on the reform of logic, but a book supporting the break-through of a new algebraic method. This focus on mathematical aims, however, changed in Boole's later writings on logic which became increasingly philosophical.

Leibniz's name had not been mentioned at that stage. Boole was not aware of the German polymath's ingenious anticipations of his own logical system. Nevertheless, it would have been possible to have known them at that time, although most of Leibniz's logical tracts were hidden amongst his papers in Hanover until the French algebraist of logic Louis Couturat published the *Opuscules et fragments inédits de Leibniz* (Leibniz 1903). In Johann Eduard Erdmann's edition of Leibniz's *Opera philosophica* published in two volumes in 1839 and 1840 (Leibniz 1839/40), his programme of a general science consisting of a *characteristica universalis,* a *logica inventiva,* a *calculus ratiocinator* and an *ars iudicandi,* together with his algebraic-combinatorial methods and above all one of the most important fragments on algebraic logic entitled 'Non inelegans specimen demonstrandi in abstractis' became available. The latter fragment contains the so-called 'calculus K XIX' or 'plus-minus calculus'.

Boole came across Leibniz only after the publication of his *Laws of Thought* of (1854). This event is reported by his wife Mary Everest Boole (1905, 1142):

> Some one wrote to my husband to say that, in reading an old treatise by Leibnitz (who lived at the same time as Newton) he had come upon the same formula which the Cambridge people call 'Boole's Equation'. My husband looked up Leibnitz and found his equation there, and was perfectly delighted; he felt as if Leibnitz had come and shaken hands with him across the centuries.[7]

'Boole's equation' is his 'Law of Duality' $AA = A$. It was Robert Leslie Ellis, Fellow of Trinity College, Cambridge, who acquainted Boole with Leibniz's anticipations. It could have been that Ellis himself had recognized the connection whilst working on the edition of Francis Bacon's *Novum Organum* for James Spedding and Douglas Denon Heath's collection of Bacon's works. There is a remark by Bacon on an analogy between mathematics and logic concerning the principle of substitution, which motivated Ellis to write the following footnote (Bacon 1858, 281, note 1):

> The importance of the parallel here suggested was never understood until the present time, because the language of mathematics and logic has hitherto not been such as to permit the relation between them to be recognised. Mr. Boole's *Laws of Thought* contain the first development of ideas of which the germ is to be found in Bacon and Leibnitz; to the latter of whom the fundamental principle that in logic $a^2 = a$ was known (v. Leibnitz, *Philos. Works*, by Erdmann, 1840, p. 130). It is not too much to say that Mr. Boole's treatment of the subject is worthy of these great names.

Thus Ellis was responsible for the introduction of Leibnizian logic in Great Britain. Boole began to study Erdmann's edition of Leibniz's *Opera Philosophica*, a fact for which Ivor Grattan-Guinness and Gérard Bornet give evidence in their recent edition of Boole's *Selected Manuscripts on Logic and Its Philosophy.*[8] After Boole's death Robert Harley, his friend and first biographer, published an article on Leibniz's anticipations of Boole's calculus in which he quoted Ellis' remark in toto and reported on his search for Leibniz's version of Boole's equation which could not be found where specified by Ellis (Harley 1867).

William Stanley Jevons, who modified the Boolean calculus to a Boolean algebra in (1864), added some words in the second edition of his seminal *Principles of Science* (1877) giving the reasons for his own ignorance of Leibniz's anticipations of the substitution of the similars which Jevons had made the basis for his philosophy of science. He gave credit to Robert Adamson, his follower in the chair of logic at Owens College, Manchester, who had given him the decisive hint about Leibniz.

In the 1880s John Venn's *Symbolic Logic* (1881) proved that British logicians had become sensitive to the pre-history of the Boolean algebra of logic. Venn argued that after the suggestions by Ellis and Harley, Leibniz had occasionally been called a precursor of Boolean logic, but that he wanted to show in his book that there had been many more perfect attempts to establish symbolic logic after Leibniz. He particularly pointed out the logic of Johann Heinrich Lambert. Venn's logic can be seen as one of the first indications of historical interest in the history of symbolic logic.

3.2. Boole and Schröder

The relationship between Boole and the German algebraist and logician Ernst Schröder will now be discussed. Schröder's work is regarded as a culmination point in the algebra of logic. He ranks as the person who brought the Boolean period in logic to its completion (Bocheński 1956, 314). His close association with Boole is evident upon opening his first book on logic, the *Operationskreis des Logikkalkuls* (1877). In this booklet he interpreted and modified the Boolean calculus, stressing the duality of logical addition and logical multiplication by means of a two-column setting. But this book was not the first of his publications to contain considerations in logic. The first was his *Lehrbuch der Arithmetik und Algebra* (1873) in which he not only gave an introduction to elementary arithmetic for mathematics teachers, but also examined the formal laws of algebraic operations with 'general numbers', independent of any interpretation. Schröder designed an 'absolute algebra' in the spirit of the British symbolic algebraists, but without any knowledge of these precursors. His sources were the combinatorial analysis of Carl Friedrich Hindenburg, the head of the Combinatorial School in Leipzig, Martin Ohm's algebraic analysis (see Ohm 1853), and Hermann Günther Graßmann's arithmetic (1861) and his *Lineale Ausdehnungslehre* (1844). The latter contained a 'general doctrine of forms' which dealt with a general theory of connective operations. This theory was elaborated by Hermann Hankel in his *Vorlesungen über complexe Zahlen* (1867), another important source of Schröder's algebra. Incidentally, in this work Hankel gave a short report on Peacock's symbolic algebra (1867, 15) which didn't have any proven effect on Schröder. He revised Peacock's 'principle of the permanence of equivalent forms' formulating the heuristic principle of the permanence of formal laws (1867, 11).

Schröder was driven to considerations on logic whilst thinking about the symbolic expression of the relationship between an ambiguous expression like $\sqrt{a}$ and its values, e.g. the relationship between $\sqrt{9}$ and its bases +3 and –3 (Schröder 1873, 27). He characterized this relationship as similar to those between a general value and its particular values, or between a general term and the terms subsumed to it. In order to express the relationships between various extensions he introduced five logical relations, among them the subsumption relation, which was of central importance for his later writings on logic. While writing a later section of his book, he took a further step towards symbolic logic by discovering that the operations used to associate concepts, i.e. collective combination and cut, obeyed the laws expressed in formal algebra, and that 'logical' addition and 'logical' multiplication could therefore be expressed by using the customary algebraic signs (Schröder 1873, 145–7). This was not obtained from Boole, but from Hermann Günther Graßmann's brother Robert who had published his *Formenlehre oder Mathematik* in (1872) in which he presented an elaborate symbolic logic without having any prior knowledge of Boole or his successors.

Like Boole when writing his *The Mathematical Analysis of Logic* Schröder was not interested in a reform of logic at this stage of development. His considerations were aimed at mathematical purposes. Ernst Schröder discovered the duality of logical addition and logical multiplication in Robert Graßmann's *Formenlehre oder Mathematik.* This discovery allowed him to associate the algebraic structures treated in his absolute algebra with the structure of logic. From the school programme pamphlet *Über die formalen Elemente der absoluten Algebra* published in (1874), we learn that he had only shortly before become acquainted with Boole's classic *An Investigation of the Laws of Thought.* An analysis of Schröder's later writings shows that all these works followed the programme expressed in his *Lehrbuch.* Logic was treated as a model for absolute algebra. This is also true of his masterpiece, the 'algebra of relatives' published in (1895), which Schröder regarded as an initial step towards a scientific universal language responsible for the semantical association of the structures of absolute algebra with their possible representatives.

In the 'general doctrines of forms' by Hermann Günther Graßmann and Hermann Hankel, with the logical calculus by Robert Graßmann and Ernst Schröder's 'absolute algebra', I have named four important precursors of lattice theory who had no association with Boole. There is, however, even an association between Hermann Günther Graßmann and Garrett Birkhoff by way of Alfred North Whitehead's *Universal Algebra* (1898),[9] and Whitehead knew Graßmann's *Lineale Ausdehnungslehre* very well.[10]

3.3. Boole and Frege

Let me now add a few words about the relationship between Boole and Frege. It is hard to believe that Frege had no knowledge of the Boolean algebra of logic when he published his *Begriffsschrift* in (1879). There is not, however, a single reference to the fact that he had ever seen Boole's writings before this publication. Even if he had seen them, Frege may have regarded the Boolean calculus as irrelevant for his own programme of the philosophy of mathematics. He was not interested in logic as a model of an algebra but in a symbolic logic which could serve as a suitable device for founding mathematics on logic. The Boolean notation was in danger of confusing the logical and mathematical side of this programme.

Nevertheless, Schröder was right in criticizing Frege for his ignorance of contemporary symbolic logic. 'First of all', he writes in his review of Frege's *Begriffsschrift* (1880, 83, transl. Frege 1972, 220),

> I consider it a shortcoming that the book is presented in too isolated a manner and not only seeks no serious connection with achievements that have been made in essentially similar directions (namely those of Boole), but even disregards them entirely.

Schröder used the opportunity to eloquently advertise the new logic by presenting the Boolean version of symbolic logic. He praised its notational advantages, but simply did not grasp—at least at that time—the superiority of Frege's quantificational theory.

Frege was also right arguing in his response to Schröder published in the paper 'Ueber den Zweck der Begriffsschrift' (1883, transl. Frege 1972, 90)

> that the Boolean formula language, in the more than twenty years which have passed since its invention, has in no way had such a smashing success that leaving out the foundation it established would, as a matter of course, necessarily appear foolish, and that only a further development could be considered.

Boole had to die before his logic found a broader audience. It was William Stanley Jevons in his *Pure Logic* (1864) who made a considerable step forward in Boolean logic. One of the marks of Jevons' logic was that he left close analogy to mathematical notation and used symbolic logic for ends in the philosophy of science. His *Principles of Science* (1874) became the driving force behind a broader reception of Boolean logic outside Great Britain. It motivated the French philosopher Louis Liard to write two papers on the logical systems of Boole and Jevons in (1877a, b) and a booklet entitled *Les logiciens anglais contemporains* published in (1878). The book ran to five editions by 1907 and was translated into German as early as (1880). In Germany wider acceptance began in 1877, not through Ernst Schröder's *Operationskreis des Logikkalkuls* but through a paper 'Die englische Logik der Gegenwart', in which the Austrian philosopher Alois Riehl primarily described Jevons' logic (Riehl 1877). It was published in the first issue of the newly founded journal *Vierteljahrsschrift für wissenschaftliche Philosophie* and was thus presented as a paradigm for scientific logic.

3.4. Boole and Hilbert

It is one thing to look for the genesis of a new field of knowledge by gathering those pieces together which constitute this field as a whole, and another to search for the procedures which lead to the constitution of this field of knowledge as a scientific discipline of its own.

It is amazing to see that most 19th century logicians were not professional logicians in the sense that they had chairs or professorships of logic. Most were mathematicians, members of an international community, but too weak to gain wide academic recognition which could have been measured by the number of paid academic positions for logicians. The process of the institutionalization of modern logic only began in the 20th century. This process will be described under the heading 'Boole and Hilbert', because David Hilbert was its initiator at least in Germany.

Hilbert and Boole were not contemporaries. Hilbert was born two years before Boole died. But Hilbert entered the scene when the new logic transcended the inner circle of logicians, whether algebraists or mathematical logicians, during that great period of foundational studies in Göttingen. David Hilbert began as a mathematician who didn't concern himself with logic. In (1899) he presented his axiomatic foundation of Euclidean geometry which seemed to solve the former epistemological problems of an intuitional foundation. But new questions concerning the nature of mathematical objects and the essence of mathematical truth were raised. Hilbert tried to answer these questions with the help of his concept of consistency. He was able to prove the consistency of the axioms of geometry presupposing the consistency of arithmetic. The presupposition turned out to be far from trivial. This was reason enough to include the consistency proof for the axioms of arithmetic in his famous 'Mathematical Problems' presented at the Second International Congress of Mathematicians in Paris in 1900 (Hilbert 1900, transl. Hilbert 1902). In the beginning Hilbert believed that the tools of traditional logic would be adequate for his aims, but he suffered a severe setback when the paradoxes were published in 1903. Hilbert's conception was not really affected by the fact that Cantorian set theory was shown to be inconsistent. He regarded set theory as competing with his axiomatic programme to found arithmetic. Furthermore, he and his collaborator Ernst Zermelo had earlier discovered similar paradoxes in Cantor's theory. It was alarming, however, that with Frege's system of the *Grundgesetze der Arithmetik* (Frege 1893/1903), the most elaborate system of logic in existence at the time, logic was affected as well. Hilbert now saw the need to include logic in his axiomatic programme and to base his mathematical axiom system on axiomatized logic and set theory. With this revision of his axiomatic programme Hilbert formulated the mathematicians' claim for logic. In Germany Hilbert was responsible for the emergence of symbolic logic as a mathematical subdiscipline.[11]

It is astonishing to see how little knowledge of logic the Göttingen mathematicians had when they started their own research on logic in 1905. I'm sure that Hilbert knew Frege's logic, but that he did not take it seriously because of notational reasons. Paradigm logic for Hilbert and his collaborators was the algebra of logic of the Boolean, Ernst Schröder. Ernst Zermelo, e.g., used a mixture of Schröder's and Peano's notation for his famous axiomatization of set-theory in 1908.[12] Hilbert's attempts were based on the algebraic systems of logic, but he wanted to revise these conceptions by binding the theory of logic closer to mathematics. Obviously misconceiving the aims which Schröder had had in mind when he introduced his version of the algebra of relatives as a universal language for all sciences, Hilbert criticized these logical systems as being 'too formal', only concerned 'with the formal construction of a calculus for the presentation of the old logical inferences', without 'far-reaching aims or tasks' (Hilbert 1905, 216), i.e. an interest in making use of logic in the foundation of mathematics (cf. Peckhaus 1994).

It took a long time for Hilbert and his circle to obtain the necessary competence in logic to master their programme. It is instructive to browse through the reports on lectures held at the meetings of the Göttingen Mathematical Society, the forum for discussion on new trends in mathematics.[13] On May 12, 1903, Ernst Zermelo gave a report on Frege's *Grundgesetze der Arithmetik* of which the second volume had just been published. According to the report, the paradox, published in the epilogue of Frege's book, was not a topic. On January 19, 1904, Hermann Fleischer introduced Peano's *Arithmetices principia nova methodo exposita* (Peano 1889), 16 years after it had been published. On February 7, 1905 William Young gave an exposition of Russell's paradox as presented in Russell's *The Principles of Mathematics* (1903). These notes show that the Göttingen mathematicians were far from up-to-date as far as logic was concerned. In the beginning they stumbled, creating their own alternative concepts, but then, after 1910 they took up their logical research again on the comparatively firm basis of Whitehead's and Russell's *Principia Mathematica* (1910–1913).

Despite these peculiarities, the institutionalization of symbolic logic in Germany began in Göttingen but proceeded very slowly. The first official lectureship for 'Mathematical Logic and Related Fields' was granted to Ernst Zermelo in 1908, and it was not before 1943 that the first German chair for Mathematical Logic and Foundations was created in Münster on the initiative of Heinrich Scholz. Only then was the German process of the emergence of symbolic logic as a scientific discipline of its own completed. The modern distinction between certain currents or traditions in the history of logic played no significant role in the processes just described. In 1908, e.g., David Hilbert, in his application for granting the lectureship of mathematical logic to Ernst Zermelo (cf. Peckhaus 1990a, 1990b, 1992), used in one breath the terms 'logical calculus', 'algebra of logic', and 'mathematical logic' synonymously, and mentioned as representatives Russell, Whitehead, Frege, Schröder, Peirce and Venn.

4. FATHERHOOD REVISITED

I hope to have shown that the development of logic cannot be seen as a tree growing from a single seed, or as two trees joining after a while to form one single trunk. The story of the development of logic can only be written as one of multiple creations, independent of one another, and in most cases independent from the development of traditional logic. The period of innovation started in the 17th century and began to form an interrelated current only in the last quarter of the 19th century. There is no doubt that a period of continuous development in symbolic logic started with Boole. Unlike Lambert or Gergonne he had his followers. But let us consider whether

Boole's part in this development was indispensable. Of course this is an obscure question. Answers can only be deeply speculative. Nevertheless I have the feeling that Boole was first only by historical accident. Given the foundational problems of the various fields of mathematics such as geometry, number theory and analysis, and the traditional role of logic as responsible for the normative rules of correct reasoning, the association between logic and mathematics was topical, an association which could not be established by philosophers who preferred to work on the philosophical foundation of logic and on the establishment of a philosophy of science, an outcome of traditional applied logic.

Multiple creation would mean multiple fatherhood. Terms like 'multiple fatherhood' or 'collective fatherhood' indicate that the metaphor of fatherhood gives a distorted picture when applied to the history of logic. We should therefore stop searching for the father of modern logic!

NOTES

* Lecture delivered on 26 September 1997 at the conference 'Boole 1997: One Hundred Fifty Years of Mathematical Analysis of Logic' (26–27 September 1997) at Lausanne. I would like to thank Christian Thiel and James Gasser for their comments, and Mairi Barkei for her efforts to improve my English.

[1] For a comprehensive discussion of the emergence of symbolic logic from Leibniz to Russell see my book *Logik, Mathesis universalis und allgemeine Wissenschaft* (Peckhaus 1997a).

[2] E.g. Michael Dummett (1959, 205): 'Boole cannot correctly be called "the father of modern logic"'; another example is Arthur N. Prior (1949, 171).

[3] A similar position was already held by Louis Liard, who made the British algebra of logic popular outside Britain: 'Pourtant, si la réforme poursuivie par les logiciens anglais du XIXe siècle doit prévaloir, c'est à Boole que sera due la nouvelle constitution de la science' (1877b, 285).

[4] In her paper 'From the Algebra of Relations to the Logic of Quantifiers' (Brady 1997, 173).

[5] The identification of starting-points is especially important if the history of logic is reconstructed in terms of a revolutionary process in the Kuhnian sense. See my criticism of this approach in Peckhaus (1997b).

[6] Boole (1844, 225), quote from Gregory (1841), 2nd ed. 1846, 237.

[7] Quote according to Laita (1976, 243).

[8] Grattan-Guinness and Bornet, eds. (1997), fragment 4 (185–7) and fragment 6 (188–9).

[9] I follow the discussion of Fearnly-Sander (1982).

[10] See Whitehead's 'Note on Grassmann' (1898, 573–5).

[11] For a detailed presentation of these developments see Peckhaus (1990b, 1995).

[12] Zermelo (1908). For the early logical calculi of Hilbert and Zermelo see Peckhaus (1994).

[13] These reports were printed in the second section of the *Jahresbericht der Deutschen Mathematiker-Vereinigung*.

REFERENCES

Bacon, Francis. [1858] 1963. Novum Organum sive indicia vera de interpretatione naturae. In *The works of Francis Bacon*, edited by James Spedding, Robert Leslie Ellis, and Douglas Denon Heath, 14 vols 1858–1874. London: Longman. Vol. 1 (1858), 149–365. Reprint, Stuttgart-Bad Cannstatt: Friedrich Frommann Verlag Günther Holzboog.

Berka, Karel, and Lothar Kreiser. 1983. *Logik-Texte. Kommentierte Auswahl zur Geschichte der modernen Logik*. 4th ed., enlarged. Berlin: Akademie-Verlag.

Bocheński, Joseph Maria. 1956. *Formale Logik*. Orbis Academicus, III, 2. Freiburg/München: Alber. (4th ed. 1978.)

Boole, George. 1844. On a general method in analysis. *Philosophical Transactions of the Royal Society of London for the Year MDCCCXLIV*, pt. 1, 225–82.

———. [1847] 1951. *The mathematical analysis of logic. Being an essay towards a calculus of deductive reasoning*. Cambridge: Macmillan, Barclay and Macmillan; London: George Bell. Reprint, Oxford: Basil Blackwell.

———. [1854] [1958]. *An investigation of the laws of thought, on which are founded the mathematical theories of logic and probabilities*. London: Walton and Maberley. Reprint, New York: Dover.

Boole, Mary Everest. [1905] 1931. Letters to a reformer's children. In: Boole, M.E., *Collected works*, 4 vols, edited by E.M. Cobham. London: C.W. Daniel. Vol. 3, 1138–63.

Brady, Geraldine. 1997. From the algebra of relations to the logic of quantifiers. In *Studies in the logic of Charles S. Peirce*, edited by Nathan Houser, Don D. Roberts, and James Van Evra. Bloomington: Indiana University Press.

Couturat, Louis. [1901] 1961, 1969. *La logique de Leibniz d'après des documents inédits*. Paris: Alcan. Reprint, Hildesheim: Olms.

Dummett, Michael. 1959. Review of Boole (1952), *Journal of Symbolic Logic* 24: 203–9. [Reprinted in this volume, pp. 79–85.]

Fearnly-Sander, Desmond. 1982. Hermann Grassmann and the prehistory of universal algebra. *American Mathematical Monthly* 89: 161–6.

Feys, Robert. 1957. Boole as a logician. *Proceedings of the Royal Irish Academy* 57: sect. A, no 6, 97–106.

Frege, Gottlob. 1879. *Begriffsschrift, eine der arithmetischen nachgebildete Formelsprache des reinen Denkens*. Halle: Louis Nebert. Reprinted in Frege (1977).

———. 1883. Ueber den Zweck der Begriffsschrift. *Jenaische Zeitschrift für Naturwissenschaft* 15: supplement: *Sitzungsberichte der Jenaischen Gesellschaft für Medicin und Naturwissenschaft für das Jahr 1882*, 1–10. Reprinted Frege (1977), 97–106.

———. [1893] 1962. *Grundgesetze der Arithmetik. Begriffsschriftlich abgeleitet*, vol. 1. Jena: Hermann Pohle. Reprint, Hildesheim: Olms.

———. [1903] 1962. *Grundgesetze der Arithmetik. Begriffsschriftlich abgeleitet*, vol. 2. Jena: Hermann Pohle. Reprint, Hildesheim: Olms.

———. 1972. *Conceptual notation and related articles*. Translated and edited with a biography and introduction by Terrell Ward Bynum. Oxford: Clarendon Press.

———. 1977. *Begriffsschrift und andere Aufsätze*, 3rd ed. Edited by Ignacio Angelelli. Darmstadt: Wissenschaftliche Buchgesellschaft.

Graßmann, Hermann Günther. [1844] 1878. *Die lineale Ausdehnungslehre ein neuer Zweig der Mathematik dargestellt und durch Anwendungen auf die übrigen Zweige der Mathematik, wie auch auf die Statik, Mechanik, die Lehre vom Magnetismus und die Krystallonomie erläutert*. 2nd ed. Leipzig: Otto Wigand.

———. 1861. *Lehrbuch der Arithmetik für höhere Lehranstalten*. Berlin: Th. Chr. Fr. Enslin. (= Graßmann, *Lehrbuch der Mathematik für höhere Lehranstalten*, pt. 1).

Graßmann, Robert. 1872. *Die Formenlehre oder Mathematik*. Stettin: R. Grassmann. Reprinted with an introduction by J.E. Hofmann. Hildesheim: Georg Olms.

Grattan-Guinness, Ivor. 1988. Living together and living apart. On the interactions between mathematics and logics from the French Revolution to the First World War. *South African Journal of Philosophy* 7: 73–82.

———. 1997. Editor's introduction. Part 1: Boole's quest for the foundations of his logic. In Grattan-Guinness and Bornet (1997), xiii–xlvii, notes lxii–lxiv.

Grattan-Guinness, Ivor, and Gérard Bornet, eds. 1997. *George Boole: Selected manuscripts on logic and its philosophy*. Science networks—historical studies, no 20. Basel: Birkhäuser.

Gregory, Duncan Farquharson. 1840. On the real nature of symbolical algebra. *Transactions of the Royal Society of Edinburgh* 14: 208–16.

———. 1841. *Examples of the process of the differential and integral calculus*. Cambridge: Deighton.

Hankel, Hermann. 1867. *Theorie der complexen Zahlensysteme insbesondere der gemeinen imaginären Zahlen und der Hamilton'schen Quaternionen nebst ihrer geometrischen Darstellung*. Leipzig: Leopold Voss. (= Hankel, *Vorlesungen über die complexen Zahlen und ihre Functionen*, pt. 1).

Harley, Robert. 1867. Remarks on *Boole's* mathematical analysis of logic. *Report of the thirty-sixth meeting of the British Association for the Advancement of Science; held at Nottingham in August 1866*. London: John Murray.

Hilbert, David. 1899. Grundlagen der Geometrie. In *Festschrift zur Feier der Enthüllung des Gauss-Weber-Denkmals in Göttingen*. Edited by the Fest-Comitee. Leipzig, 1–92.

———. 1900. Mathematische Probleme. Vortrag, gehalten auf dem internationalen Mathematiker-Kongreß zu Paris 1900. *Nachrichten von der königl. Gesellschaft der Wissenschaften zu Göttingen. Mathematisch-physikalische Klasse aus dem Jahre 1900*, 253–97.

———. 1902. Mathematical problems. Lecture delivered before the International Congress of Mathematicians at Paris in 1900. Translated by Mary Winston Newson. *Bulletin of the American Mathematical Society* 8: 437–79.

———. 1905. *Logische Principien des mathematischen Denkens*. Lecture course, summer term 1905, notes by Ernst Hellinger. (Library of the Mathematical Seminar of the University of Göttingen.)

Jevons, William Stanley. 1864. *Pure logic or the logic of quality apart from quantity with remarks on Boole's system and the relation of logic and mathematics*. London: Stanford. (Reprinted in Jevons, *Pure logic and other minor works*, edited by Robert Adamson and Harriet A. Jevons. London: Macmillan 1890 (re-issued by Thoemmes Press, Bristol 1991), 3–77.)

———. 1874. *The principles of science. A treatise on logic and scientific method*. 2 vols. London: Macmillan [New York 1875].

———. 1877. *The principles of science. A treatise on logic and scientific method*. 2nd ed. London and New York: Macmillan. (3rd ed. 1879; 'stereotyped edition' 1883.)

Kneale, William. 1948. Boole and the revival of logic. *Mind* 57: 149–75.

Laita, Luis Maria. 1976. *A study of the genesis of Boolean logic*. Ph.D. thesis, University of Notre Dame.

Leibniz, Gottfried Wilhelm. 1839/40. *God. Guil. Leibnitii opera philosophica quae exstant Latina Gallica Germanica omnia*, edited by Johann Eduard Erdmann. 2 parts. Berlin: Eichler. (Pt. 1, 1840; pt. 2, 1839.)

———. 1903. *Opuscules et fragments inédits de Leibniz. Extraits des manuscrits de la Bibliothèque royale de Hanovre*, edited by L[ouis] Couturat. Paris: Alcan.

———. 1982. *Generales Inquisitiones de Analysi Notionum et Veritatum. Allgemeine Untersuchungen über die Analyse der Begriffe und Wahrheiten*, edited by Franz Schupp. Philosophische Bibliothek, no 338. Hamburg: Felix Meiner. (2nd ed. 1993.)

Lenzen, Wolfgang. 1984. Leibniz und die Boolesche Algebra. *Studia leibnitiana* 16: 187–203.

Lewis, Clarence Irving. [1918] 1960. *A survey of symbolic logic*. Semicentennial Publications of the University of California. Berkeley: University of California Press. New edition (shortened) New York: Dover.

Lewis, Clarence Irving, and Cooper Harold Langford. 1932. *Symbolic logic*. The Century Philosophy Series. New York: The Century Co.

Liard, Louis. 1877a. Un nouveau système de logique formelle. M. Stanley Jevons. *Revue philosophique de la France et de l'étranger* 3: 277–93.

———. 1877b. La logique algébrique de Boole. *Revue philosophique de la France et de l'étranger* 2: 285–317.

———. 1878. *Les logiciens anglais contemporains*. Paris: Germer Baillière. (5th ed. 1907.)

Löwenheim, Leopold. 1915. Über Möglichkeiten im Relativkalkül. *Mathematische Annalen* 76: 447–70.

MacHale, Desmond. 1985. *George Boole: His life and work*. Profiles of Genius Series, no 2. Dublin: Boole Press.

Ohm, Martin. 1853. *Versuch eines vollkommen consequenten Systems der Mathematik*. Pt. 1: *Arithmetik und Algebra enthaltend*. 3rd ed. Nürnberg: Korn'sche Buchhandlung.

Peano, Giuseppe. 1889. *Arithmetices principia, nova methodo exposita*. Torino.

Peckhaus, Volker. 1990a. 'Ich habe mich wohl gehütet, alle Patronen auf einmal zu verschießen'. Ernst Zermelo in Göttingen. *History and Philosophy of Logic* 11: 19–58.

———. 1990b. *Hilbertprogramm und Kritische Philosophie. Das Göttinger Modell interdisziplinärer Zusammenarbeit zwischen Mathematik und Philosophie*. Studien zur Wissenschafts-, Sozial- und Bildungsgeschichte der Mathematik, no 7. Göttingen: Vandenhoeck & Ruprecht.

———. 1992. Hilbert, Zermelo und die Institutionalisierung der mathematischen Logik in Deutschland. *Berichte zur Wissenschaftsgeschichte* 15: 27–38.

———. 1994. Logic in transition: The logical calculi of Hilbert (1905) and Zermelo (1908). In *Logic and Philosophy of Science in Uppsala. Papers from the 9th International Congress of Logic, Methodology and Philosophy of Science*, edited by Dag Prawitz and Dag Westerståhl. Synthese Library, no 236. Dordrecht: Kluwer.

———. 1995. Hilberts Logik. Von der Axiomatik zur Beweistheorie. *NTM. Internationale Zeitschrift für Geschichte und Ethik der Naturwissenschaften, Technik und Medizin* n.s. 3: 65–86.

———. 1997a. *Logik, Mathesis universalis und allgemeine Wissenschaft. Leibniz und die Wiederentdeckung der formalen Logik im 19. Jahrhundert*. Berlin: Akademie-Verlag.

———. 1997b. The way of logic into mathematics. *Theoria* 2nd ser., 12: 39–64.

Peirce, Charles S. 1885. On the algebra of logic. A contribution to the philosophy of notation. *American Journal of Mathematics* 7: 180–202.

Prior, Arthur N. 1949. Categoricals and hypotheticals in George Boole and his successors. *Australasian Journal of Philosophy* 27: 171–96.

Prior, Arthur N., ed. 1967. Logic, History of. In *The Encyclopedia of Philosophy*, edited by Paul Edwards. New York: Macmillan and The Free Press; London: Collier-Macmillan. 4: 513–71.

Quine, Willard Van Orman. 1985. In a logical vestibule. Review of MacHale (1985). *Times Literary Supplement* of 12 July 1985, p. 767. Reprinted as Quine (1995).

———. 1995. MacHale on Boole. In Quine, *Selected logic papers*. Enlarged ed. Cambridge Mass.: Harvard University Press.

Riehl, Alois. 1877. Die englische Logik der Gegenwart. *Vierteljahrsschrift für wissenschaftliche Philosophie* 1: 51–80.

Russell, Bertrand. [1903] 1992. *The principles of mathematics*. Cambridge: The University Press [as 'Vol. I']. 2nd ed. London: Allen & Unwin 1937. Reprint, London: Routledge.

———. 1951. Mathematics and the metaphysicians. In Russell, *Mysticism and logic and other essays*. 10th ed. London: George Allen & Unwin. [The edition was first published in 1910, the paper in 1901.]

Schlote, Karl-Heinz. 1987. *Die Entwicklung der Algebrentheorie bis zu ihrer Formulierung als abstrakte algebraische Theorie*. Dissertation (B), University of Leipzig.

Schröder, Ernst. 1873. *Lehrbuch der Arithmetik und Algebra für Lehrer und Studirende*. Vol. 1, *Die sieben algebraischen Operationen*. Leipzig: Teubner.

———. 1874. *Über die formalen Elemente der absoluten Algebra*. Stuttgart: Schweizerbart'sche Buchdruckerei. [Programme pamphlet of the Pro- and Real-Gymnasium at Baden Baden for 1873/74.]

———. [1877] 1966. *Der Operationskreis des Logikkalkuls*. Leipzig: Teubner. Reprint, Darmstadt: Wissenschaftliche Buchgesellschaft.

———. 1880. Review of Frege 1879. *Zeitschrift für Mathematik und Physik*, historisch-literarische Abt. 25: 81–94.

———. 1895. *Vorlesungen über die Algebra der Logik (exakte Logik)*. Vol. 3, pt. 1, *Algebra und Logik der Relative*, Leipzig: Teubner. [Reprinted in the 'second edition', vol. 3. Bronx N.Y.: Chelsea 1966.]

Shannon, Claude Elwood. 1938. A symbolic analysis of relay and switching circuits. *Transactions American Institute of Electrical Engineers* 57. 713–23.

Skolem, Thoralf. 1920. *Logisch-kombinatorische Untersuchungen über die Erfüllbarkeit oder Beweisbarkeit mathematischer Sätze nebst einem Theoreme über dichte Mengen*. Skrifter utgit av Videnskapsselskapet i Kristiana. I. Matematisk-naturvidenskabelig klasse no 4.

———. 1923. Einige Bemerkungen zur axiomatischen Begründung der Mengenlehre. *Wissenschaftliche Vorträge gehalten auf dem fünften Kongress der Skandinavischen Mathematiker in Helsingfors vom 4. bis 7. Juli 1922*. Helsingfors, 217–32.

Venn, John. 1881. *Symbolic logic*. London: Macmillan.

Whitehead, Alfred North. [1898] 1960. *A treatise on universal algebra with applications*. Vol. 1. Cambridge: Cambridge University Press. Reprint, New York: Hafner.

Whitehead, Alfred North, and Bertrand Russell. 1910–1913. *Principia mathematica*. 3 vols. Cambridge: Cambridge University Press.

Zermelo, Ernst. 1908. Untersuchungen über die Grundlagen der Mengenlehre. I. *Mathematische Annalen* 65: 261–81.

Institut für Philosophie der Universität Erlangen-Nürnberg, Bismarckstr. 1, 91054 Erlangen, Germany

SHAHID RAHMAN

HUGH MacCOLL AND GEORGE BOOLE ON HYPOTHETICALS

1. INTRODUCTION

Hugh MacColl (1837–1909), the father of formal non-classical logic, reported that he had developed his early formal system—published as papers in the journals *Educational Times*, *Proceedings of the London Mathematical Society* and *Mind* between 1877 and 1893—without having even read the work of George Boole.[1] Later on, however, MacColl felt it necessary to publish further papers in order to clarify the basic lines of his concept of logic as opposed to that of Boole. MacColl's main criticism of Boole may be summarised as follows:

- Pure logic was devised as an aid for practical uses in argumentation. For MacColl this practical character is due to its abstract nature, which in turn makes generalisations possible. Since pure logic is abstract, its categorical symbols do not stand for numbers or for classes of objects or temporal states, but more generally for *statements* or *propositions*. Boole's logic, in which propositional operations are translated into class operations, is basically not general enough and should be replaced by a *Calculus of Equivalent Statements* in which propositional connectives occur instead of Boole's equations.[2]

- It could well be that for some sciences the existing argumentation demands a type of logic which is not applicable in others. When constructing a symbolic system for a particular type of logic, the corresponding expressions in use should therefore be taken into careful consideration. In MacColl's opinion, for example, Boole's logic coincides neither with the propositions in the theory of probability nor with those in use in natural language, since both the theory of probability and natural language require a strongly connected if-then.[3]

In this paper some of the consequences of this criticism of Boole will be studied, using the concept of hypotheticals. It must be pointed out at this stage that MacColl's criticism cannot be placed within the discussion between formalism and antiformalism. MacColl described himself as a peacemaker who wanted to unite the philosophical tradition of logic with the new instruments of formal logic.[4] This is particularly the case in the formulation of the if-then, which was the principal connective of the hypothetical in MacColl's system. In answer to the question whether it was even possible to give a formal transcription of the intuitive understanding of the if-then in natural language, he said that it was and that such a transcription would agree with

J. Gasser (ed.), A Boole Anthology, 287–310.

the philosophical tradition but not with material implication. MacColl believed that his formulation of the if-then could forge a bridge between supporters and opponents of formalisation because it is closer to usage or application than that of Boole. This brings us to the main topic of the next section, where I would like to describe MacColl's construction of the most important pillar of such a bridge.

2. THE TRADITIONAL CONCEPT OF 'HYPOTHETICALS'

MacColl's first formulations of the if-then make use of the expression $A = AB$. The most well-known and probably also the most enigmatic lines are the following:

> The equation $A = AB$ asserts either that the statement A implies the statement B, or else that B is true necessarily (MacColl 1877a, 91).
>
> Rule 2.—Let A be any statement whatever, and let B be any statement which is implied in A (and which must therefore be true when A is true, and false when A is false), or else let B be any statement which is admitted to be true independently of A; then (in either case) we have the equation $A = AB$ (MacColl 1877b, 10).

The equation $A = AB$, which the Booleans understood as a class equation and used to express the form A of the Boethian square, is transcribed in MacColl's system of logic, which he presented in the series of articles 'The Calculus of Equivalent Statements (and Integration Limits)', as the (propositional) biconditional $A \leftrightarrow (A \wedge B)$.[5] The texts consider two possible readings of this equivalence:

1. The case where B is true but A can be both true and false. This means that the truth of B is independent of the truth value of A. Thus, B is either a formally true proposition (tautology) or any empirically (materially) true proposition. In other words, if B^1 (i.e. if B is true), then either $A^1 \leftrightarrow (A^1 \wedge B^1)$ or $A^0 \leftrightarrow (A^0 \wedge B^1)$.
2. The second case describes a connective relation between A and B through which the truth of A necessarily leads to B being true. Similarly, the value false of A necessarily leads to B being false. MacColl describes this propositional connection with the expression 'is implied in'. Thus, A and B have the same truth value, where the truth value of B (necessarily) depends on the truth value of A—we will come back to this concept of dependence soon. I call this implicative connective 'connexive implication' (from the Stoic συναρτησις). In symbols, $A \Rightarrow B$ means either $A^1 \Rightarrow B^1$ or $A^0 \Rightarrow B^0$. Thus, either $A^1 \Leftrightarrow (A^1 \wedge B^1)$ or $A^0 \Leftrightarrow (A^0 \wedge B^0)$ is valid. Firstly it is clear that the connexive implication is not the material implication, because of the non-validity of $A^0 \Rightarrow B^1$. Secondly, a biconditional formulation of the connexive implication is prevented by the dependence of B on A.

In his second paper 'Calculus of Equivalent Statements' MacColl strengthens the relation between the propositional reading of the equation $A = AB$ and the connective 'if-then': the biconditional $A \leftrightarrow (A \wedge B)$ and the connective 'if-then' should be logically equivalent formulas:

> Def. 12.—The symbol $A : B$ (which may be called an implication) asserts that the statement A implies B; or that whenever A is true B is also true.
>
> Note.—It is evident that the implication $A : B$ and the equation $A = AB$ are equivalent statements (See Rule 2). (MacColl 1878, 177.)

The connective 'if-then' should be understood here as the material implication, that is the case 1 includes now $A^0 \leftrightarrow (A^0 \wedge B^0)$. Actually it is not clear until the Appendix to this second paper that MacColl is thinking of the material implication. In this Appendix MacColl explicitly states the truth conditions of the classical conditional:

> Since $\alpha : \beta$ asserts that β is a factor of α [i.e. conjunctively connected with α, because of $\alpha \rightarrow \beta \leftrightarrow (\alpha \wedge \beta \leftrightarrow \alpha)$], the implication $\alpha : 0$ asserts that α is false [i.e. $\alpha : 0$ or $\alpha \rightarrow \beta^0 \leftrightarrow (\alpha \wedge \beta^0 \leftrightarrow \alpha^0)$], and the implication $1 : \alpha$ asserts that α is true [i.e. $1{:}\alpha$ or $\beta^1 \rightarrow \alpha \leftrightarrow (\beta^1 \wedge \alpha^1 \leftrightarrow \beta^1)$]; but the implications $0 : \alpha$ and $\alpha : 1$ make no assertion as to whether α is true or false [i.e. $(\beta^0 \rightarrow \alpha) \leftrightarrow ((\beta^0 \wedge \alpha^?) \leftrightarrow \beta^0)$ and $(\alpha \rightarrow \beta^1) \leftrightarrow ((\alpha^? \wedge \beta^1) \leftrightarrow \alpha^?)$] (MacColl 1878, 183).

These quotations contain MacColl's understanding of 'hypotheticals'. MacColl, who, as already noted, defined himself as a peacemaker between the Booleans and the traditionalists, searched for a formal translation of the natural semantics of traditional hypotheticals. MacColl's biconditional was intended to express the general form of conditionals, including the classical and several non-classical forms of conditional. The stricter forms of conditionals were then supposed to give a new formal basis for traditional hypotheticals.

If we wish to obtain a fair evaluation of MacColl's contribution to the concept of hypotheticals, it seems sensible to consider the traditionalistic and the Boolean views of this concept first.

2.1. Boethius and the Traditionalistic View of 'Hypotheticals'

2.1.1. Boethius

The traditionalists of the 19th century called 'hypothetical' a proposition of which the truth is stated under certain conditions. This view of the hypothetical contrasts with categorical propositions, the truth of which is stated unconditionally. Or to put it in other more traditional words, in the pre-Boolean tradition hypotheticals were propositions in which a certain predication is conditioned by another predication.

A.M.S. Boethius' work on hypotheticals seems to represent the most important historical source for the traditionalistic logicians of the 19th century. Boethius included in the expression *hypothetica propositio* not only if-then propositions, which he called conditional or connected, but also disjunctive propositions.[6] He further distinguished two types of connected propositions:

- *secundum accidens*: e.g. 'If the fire is warm, the sky is spheroidal', which expresses only a temporal casual connection, and
- *secundum naturam*: such propositions express a strong, i.e. a necessary, intensional and causal, connection.

Boethius also distinguished two types of connected propositions *secundum naturam*,

- connected propositions *secundum naturam per terminorum positionem*, i.e. propositions whose connection agrees with the positions their terms are in, and
- propositions connected without any such condition

The traditionalistic logicians of the 19th century did not explicitly use this distinction between two types of connected propositions. I therefore feel I can disregard such a distinction in this paper. Instead, we should consider more precisely the nature of strongly connected propositions. The following example of Boethius' is very interesting in this respect:

> *Si est* a, *cum sit* b, *est* c [...]
> *atqui cum sit* b *non est* c
> *non est igitur* a[7]

I.e.:

If a, then if b then c	$a \Rightarrow (b \Rightarrow c)$
(and) if b then not-c	$b \Rightarrow \neg c$
Therefore, not-a	$\neg a$

If we keep in mind that the conditionals $b \Rightarrow \neg c$ and $b \Rightarrow c$ are mutually incompatible,[8] it is possible to reconstruct Boethius' reasoning as follows: since the conditionals $b \Rightarrow \neg c$ and $b \Rightarrow c$ are mutually incompatible, one can be considered as the negation of the other.[9] Thus, because of the reciprocal incompatibility of these two conditionals, Boethius can apply modus tollens in the conjunction $((a \Rightarrow (b \Rightarrow c)) \wedge (b \Rightarrow \neg c))$, obtaining the conclusion $\neg a$. The argumentation for the incompatibility of $b \Rightarrow c$ and $b \Rightarrow \neg c$ can be seen as based in the connexive validity of $\neg(a \Rightarrow \neg a)$: if we apply contraposition and double negation in $b \Rightarrow \neg c$, we obtain $c \Rightarrow \neg b$. From $b \Rightarrow c$ and $c \Rightarrow \neg b$ follows $\neg b \Rightarrow b$. But this is impossible, because of the validity of $\neg(b \Rightarrow \neg b)$. This reading of Boethius links his strongly connected propositions with the Stoic concepts of connection (συναρτησις) and incompatibility (μαχητης).[10] It is the Stoic interpretation which gives a sensible argument for the thesis $\neg(a \Rightarrow \neg a)$: the Stoic reading of $a \Rightarrow \neg a$ is 'it is impossible that a and $\neg\neg a$'. But since this conjunction, at least in the logical reading of it, cannot be impossible, we obtain $\neg(a \Rightarrow \neg a)$. Actually Aristotle had already used a similar argumentation in the second book of the *Analytica priora* (57a36–b18) by showing that the incompatibility of $a \Rightarrow b$ and $\neg a \Rightarrow b$ follows from the thesis $\neg(\neg a \Rightarrow a)$.[11] Like Storrs McCall (1966, 415), I call the formulae $\neg(A \Rightarrow \neg A)$ and $(A \Rightarrow B) \Rightarrow \neg(A \Rightarrow \neg B)$ the first and the second Boethian theses of connexivity, and the formulae $\neg(\neg A \Rightarrow A)$ and $(A \Rightarrow B) \Rightarrow \neg(\neg A \Rightarrow B)$ the first and the second Aristotelian theses of connexivity.

William and Martha Kneale found no evidence of Stoic logic in Boethius' work. This led them to think that Boethius was wrong when he used modus tollens in the conjunction $((a \Rightarrow (b \Rightarrow c)) \wedge (b \Rightarrow \neg c))$ of the example just mentioned (Kneale and Kneale 1962, 191). Jan Łukasiewicz (1951, 50) and Günther Patzig (1969, 200–7) reasoned similarly against Aristotle's argument for the incompatibility of $a \Rightarrow b$ and $\neg a \Rightarrow b$.

Let us return to Boethius and this time examine his disjunctive propositions. Boethius' disjunctive propositions are sometimes inclusive and sometimes exclusive, but more important than this distinction is his now well-known formulation of disjunctive propositions with the aid of conditional propositions, the last of which might be regarded as *secundum accidens*:

> Then, saying: 'Either not-*A*, or not-*B*' we express: 'If *A*, then not-*B*' and 'If *B*, then not-*A*'.[12]

It is important to observe that Boethius considered hypotheticals as a genus of complex propositions. The elements of a hypothetical are, in Boethius' view, no longer hypothetical but categorical propositions. Categorical propositions have a predicative grammatical form where the verb 'to be', but no other particle, has a copulative function. It is this copulative predication which distinguishes categoricals from hypotheticals. The principal predication in hypotheticals makes use of other propositional particles than the copula 'to be'. The proposition 'Humans are animals', for example, is categorical. The proposition 'If it is human, (then) it is (an) animal', on the contrary, is hypothetical. The necessity of distinguishing between copulative categorical and connective hypothetical predication arises, in Kant's words, from the difference in their modality: hypothetical predication is a conditioned predication, whereas categorical predication is not. Indeed, categorical predication is an unconditionally stated predication.[13] A categorical proposition can thus be obtained as a result of partitioning a hypothetical proposition, but it is not possible to obtain a hypothetical proposition from a single categorical proposition.

Boethius' terminology fluctuates between conceptual and propositional language. This is especially the case when he speaks of the conditioned predication of conditional hypotheticals, where sometimes it seems that it is the concept rather than the proposition of the if-part which conditions the then-part. We shall meet this variation in the terminology of hypotheticals again in the work of the traditionalist logicians of the 19th century, particularly in the work of Sir William Hamilton (1788–1856), who was one of the most influential logicians of the last century and whose anti-mathematical attitude was explicitly criticized by Boole.[14]

2.1.2. *Hamilton*

Hamilton, who quotes Boethius, underlines the importance of hypothetical propositions *secundum naturam*. He uses the word 'conditional' instead of 'hypothetical' as a general term.[15] Hamilton's conditional propositions therefore include exclusive disjunctive, hypothetical and dilemmatic propositions. In Hamilton's view, hypotheticals are necessarily connected conditionals, while dilemmatic propositions are such hypothetical propositions in which at least one disjunctive proposition occurs. For the sake of uniformity I am not going to follow Hamilton's terminology. I will instead use the word 'hypothetical' as a general term.

Hamilton's reception of Boethius had a remarkable influence on the logic literature of the 19th century. The most important points can be summed up as follows:

1. Like Kant, and in contrast to Boethius, Hamilton rejects any kind of reduction between hypotheticals and categoricals.[16]
2. Hamilton considered that the connexivity feature of conditional, disjunctive and dilemmatic propositions was the result of the interaction of the mind. The mind produces an inseparable conceptual connection between the different parts of those propositions, and this is reflected on the language level by their conditioned predication.
3. Hamilton's classification of hypothetical propositions follows from the subject-predicate structure of propositions: if the condition lies more proximately in the subject, we have conditional propositions. If the condition lies more proximately in the predicate, we have disjunctive propositions. Finally, if there is a twofold condition—the one belonging to the subject, the other to the predicate, we have dilemmatic propositions.[17]

Point 1 follows from point 2. Because of the mind's interaction producing a strong and inseparable connexivity, which determines Hamilton's hypotheticals, their parts are inseparable from their corresponding whole. Since the parts of Hamilton's hypothetical propositions are, to put it in modern words, 'dependent parts', hypothetical propositions cannot be separated in order to obtain categorical ones. This relation of dependence is stated for the conditional propositions as a causal necessary relation of dependence.[18] Hamilton's reasoning for the irreducibility of dilemmatic and disjunctive propositions is a little more complicated and is virtually another version of Kant's argument, posed when he discussed this issue in his *Logik* of 1800.[19] Hamilton's argument can be divided into two steps and comprises the examination of an example which was proposed by Kant and borrowed by Boole:

a) *The disjunction in the grammatical predicate is bound up by the subject.* In disjunctive propositions, as already mentioned, the condition lies more proximately in the predicate. Hamilton's example, which he borrowed from Kant, is the following: 'Men are either black or white'. We speak here of men under the condition that they are either black or white. Such a conditioned predication of the subject expresses that the domain of the subject's predicator 'Men' (for the concept 'Men' Hamilton does not use the word 'domain' but the direct translation of the Kantian term: 'sphere') binds up the exclusive disjunction of the predicate into one single unity.[20] In other words, men are either black or white, and within the context of this proposition there is no other predication of men to be considered. Hamilton's use of the expression 'binds up' relates to the interaction of the mind producing the unity in the grammatical predicate already mentioned.
b) *If the disjunctive proposition is regarded as a complex proposition, we have to give up the binding force of the domain of the subject's predicator.* The disjunction thus separates into two predications, each including the whole of the domain of the subject's predicator. This allows the complex proposition to be divided into two single categorical ones. The proposition 'Either men are black or they are white', for example, is therefore a complex categorical proposition which can be divided into the propositions 'Men are black', 'Men are white'. In Hamilton's view, which contradicts Boethius' theory of propositions here, two single categoricals build a complex categorical, not a complex hypothetical.

It should be remembered that the contemporary idea of the domain or extension is very different from Hamilton's. He seems to identify what we nowadays call the domain of quantification of the proposition with the (extensional) interpretation of the predicator in this domain. We would argue that the domain of quantification of the proposition 'Either (all) men are black or they are (all) white', such as the set {European}, binds up the predicators 'men', 'black' and 'white' in a similar way as the extension of the predicator 'men' binds up the predicators 'black' and 'white' in the proposition 'Men are either black or white'. The problem with this anti-Hamiltonian argument is, as once remarked by Narahari Rao, that it produces an inaccurate picture of the binding and condition connection in Hamilton's conditional propositions: for the Kantian Hamilton, the connection conditioning either the subject or the predicate is a conceptual and not an extensional one. Extensions are, to put it in Rao's words, 'possible individual instances of a concept'.[21] Hamilton would possibly answer to the objection, based on the distinction between the domain of the quantification and interpretation of the predicators in a domain, as follows: the set {European} can only be understood as the set of possible instances of the concept European, which should be explicitly included in the proposition. The correct formulation of the proposition in Hamilton's view would be: 'Either (all) European men are black or they are (all) white'. At this point Hamilton can repeat his argument for the complexity of this proposition. It is this conceptual connection which forms the grounds of Hamilton's strongly connected conditional and which he misses in the formal treatment of logic.

For Booleans, on the other hand, the price paid for such a conceptual understanding of hypotheticals is too high. Hamilton abandons Boethius' principle of compositionality, where complex hypothetical or conditional propositions were built up of simple categorical ones. To give up this compositionality principle means giving up any possible relation between hypothetical propositions and categorical ones in favour of an uncertain interaction of the mind, and in the Boolean view this is definitely too much to pay for having concepts instead of extensional domains.

2.2. Boole and the Booleans on 'Hypotheticals'

2.2.1. Boole

In Boole's *The Mathematical Analysis of Logic* we find a definition of hypotheticals which is nearer to Boethius' and Abelard's view than to Hamilton's. In the first lines of his chapter on 'hypotheticals', where he attempts to formalise propositional logic, Boole picks up on Boethius' compositionality principle: complex hypotheticals are built up of simple categoricals (cf. Boole 1847, 48). In this chapter Boole uses the same symbols introduced in his chapter on syllogisms, but obviously with different interpretations: in the context of this chapter capital letters like X, Y, Z stand for propositions, whereas lower-case letters like x stand for elective symbols which represent an elective operation,

> [that] shall select those cases in which the Proposition X is true,

from the hypothetical universe 1,

> [that] shall comprehend all conceivable cases and conjunctures of circumstances (Boole 1847, 49).

Thus, the hypothetical universe 1 is thought to include all those possible combinations of truth values which Ludwig Wittgenstein later expressed with the help of truth tables.[22]

This allows Boole to introduce the elements of a truth-functional definition of connectives.

In this outline it is already possible to recognise Boole's understanding of hypotheticals, which differs radically from the traditionalist one:

1. Boole transports the problematic modality of traditional hypotheticals to a hypothetical universe containing all the possible truth values. Suppose a given formally true proposition, i.e. a proposition whose truth value is 'true' for any possibility contained in the hypothetical universe. In this case the formal truth of this proposition is independent of the actual truth of any of its elementary propositions. Each of the truth values of these elementary propositions can thus be seen as problematic, because it is always possible to change this value for another one.
2. Instead of the conceptual, causal, and intensional connections of traditional hypotheticals, Boole refers to truth-functional connectives. Actually Boole explicitly rejects the causal interpretation of the conditional in his postscript to *The Mathematical Analysis of Logic*, where he argues against Hamilton and Aristotle.[23]
3. While distinguishing categorical from hypothetical propositions, Boole replaces the traditional opposition of conditioned/unconditioned with the opposition between an ontological and a propositional interpretation of symbolic language: categoricals follow from an ontological interpretation, in which the symbols are taken to represent classes of objects and operations between these classes. Hypotheticals, on the other hand, follow from a propositional interpretation, in which the symbols are taken to represent propositions and operations between truth values. Boole illustrates this distinction with Kant's and Hamilton's example of the non-distributivity of the universal quantification over the disjunction: the disjunction in the complex proposition 'Either the inhabitants of the island are all Europeans or they are all Asiatics' *separates* two propositions. According to Boole, and in contrast to Kant and Hamilton, this proposition is a hypothetical one. On the other hand, the disjunction in the proposition 'All inhabitants of the island are either European or Asiatic' separates two classes of individuals, namely the class of Europeans from the class of Asiatics, yielding a categorical proposition (cf. Boole 1847, 58–9).

This position of Boole's involves replacing the strongly connected if-then in the traditional conditional with the truth-functional connective of material implication. He correspondingly reformulates the Stoic definition of the conditional, replacing the modal expressions with truth-functional ones. Boole rewrites the expression 'It is impossible that *A* and not-*B*' as the equation $x(1 - y) = 0$, which should be read as 'It is false that *A* and not-*B*' (cf. Boole 1847, 54). Now, the symbols '1' and '0' are taken to represent the hypothetical universe and its complement. The problem in this case, as once remarked by Arthur Prior, is that Boole's original notation does not allow the expression either of contingent conditionals or, more generally, of any contingent proposition (cf. Prior 1949, 173). Boole's formalisation of modus ponens, for example, should accordingly read as a derivation rule for tautological premises:

$x = 1$	(X *is* [formally] *true*)
$x(1 - y) = 0$	(*If* X *is* [formally] *true*, Y *is* [formally] *true*, [or the conjunction of *X* and not-*Y* yields a contradiction])

$y = 1$

(Cf. Boole 1847, 56)

Boole later felt it necessary to extend his symbolic language and to reassume in part the traditional position on contingent propositions. To this effect he introduced in *An Investigation of the Laws of Thought* his temporal interpretation of hypotheticals which has been criticized so often:

> The distinction must be familiar to every reader of Plato and Aristotle, by the latter of whom, especially, it is employed to denote the contrast between the abstract verities of science, such as the propositions of geometry which are always true, and those contingent or phaenomenal relations of things which are sometimes true and sometimes false. But the forms of language in which both kinds of propositions are expressed manifest a common dependence upon the idea of time; in the one case as limited to some finite duration, in the other as stretched out to eternity (Boole 1854, 163–4).

In this way, the elective function x was thought to select from the temporal universe 1 those temporal states where the proposition X is true (Boole 1854, 166–7). The elective function x represents the operation of selecting those temporal states where the proposition X is contingently true if and only if x cannot be equated either with the temporal universe 1 or with its complement.

In his 'Postscript' to *The Mathematical Analysis of Logic*, Boole writes to the effect that the set of contingent propositions represents the point of intersection between categoricals and hypotheticals (cf. Boole 1847, 82). This comment can be understood in the context of the definition of contingent propositions as follows: in *An Investigation of the Laws of Thought* Boole says that the expression x can be the elective function both of a categorical and a hypothetical proposition. A given assertion which states a class relation between things, such as 'Snow belongs to the class of white things', amounts to a categorical. The proposition 'The assertion "Snow belongs to the class of white things" is true', however, expresses a hypothetical. From this it is evident that in the differentiation between categorical and hypothetical propositions it is not the opposition of conditioned and unconditioned which is of prime importance but the differentiation between the ontological level (i.e. a level of *things*) of categoricals and the linguistic level (i.e. a level of *propositions*) of hypotheticals. Thus the same symbolic expression, e.g. $x = y$, can stand both for the identity of two classes and for the assertion 'The proposition with which the identity of the classes X and Y is stated is true', and this is why Boole prefers to speak of 'primary' and 'secondary propositions' (cf. Boole 1854, 52–4). It should be clear that the precondition for this conversion rule in the case of simple propositions is that the proposition be a contingent.

This conversion method between hypothetical and categorical propositions, by means of which the secondary proposition '*A* is true' is formed from the primary proposition *A*, led, because of the redundancy of the expression 'is true', to the discussion as to whether categorical and hypothetical propositions are equivalent and whether a differentiation between them is therefore superfluous—a discussion MacColl took up later. But first let us consider the groundwork achieved by the Booleans.

2.2.2. The Booleans

The works of W.S. Jevons, E. Schröder, C.S. Peirce and J. Venn, inter alia, simplified Boole's work and gave him the classical profile that is attributed to his work today. The temporal interpretation of secondary propositions, for example, was generally rejected. Booleans explicitly rejected the traditional modal, causal and problematic characterisation of hypotheticals, arguing that it belongs to the realm of the psychology of language.[24] John Venn writes:

> Briefly put, my own view as to what may be called the fundamental significance of the hypothetical form is best expressed by saying that it (1) implies a connection, of the kind called a uniformity between two or more phenomena; and (2) implies, along with this, some doubt on our part as to the actual occurrence, in a given instance, of the pair or more of events which compose this uniformity.[...]. As regards the limits of this characteristic doubt, it should be observed that it refers only to the actual occurrence, under some given circumstances, of the elements which compose this connection or uniformity (Venn 1881, 210–1).

> My contention is that the really characteristic elements of the Hypothetical,—those which give it the peculiar significance which it possesses in common life and in Inductive Logic,—are of a comparatively non-formal nature. They are to a considerable extent either material or psychological, and therefore the necessity of forcing them into our system will crush much of the life out of them. Such a loss as this is however inevitable (ibid., 243–4).

Furthermore, the distinction between hypotheticals and categoricals should, according to the Booleans, be considered as taking place not within the semantics of formal but of natural language.

One important point in the dissolution of the distinction between hypotheticals and categoricals was the discussion about adequate formalisation of the natural if-then. Venn, for example, writes explicitly that the elementary form of hypothetical propositions should be formulated as the conditional proposition 'If *P*, then *Q*':

> The conclusion then that we have so far come to is that a very large proportion, in fact an overwhelming majority, of ordinary hypotheticals, though they may demand for their convenient expression more than two terms, are nevertheless to all intents and purposes composed of a pair only of elements. Their actual complexity is mainly a matter of linguistic propriety or convenience. They admit therefore of adequate verbal expression in the form already alluded to, viz. If *P* then *Q* (ibid., 221).

I will now describe how Venn came to abandon the hypothetical/categorical distinction through this view of the if-then, which he shared with most of the Booleans. It should be recalled that the equation $A = AB$, once used by Leibniz for the formal translation of universal affirmative propositions, was introduced by Stanley Jevons in the context of Boole's symbolic language for the same purpose.[25] On the other hand, this equation was taken to be the formal expression of hypotheticals, especially of hypothetical conditional propositions, namely $(A \rightarrow B) = 1 \Leftarrow_{df}\Rightarrow ((A = 1) \wedge (B = 0)) = 0 \Leftarrow_{df}\Rightarrow A = AB)$. Because of the fact that not only traditional hypotheticals but also categoricals have the structure of an elementary if-then proposition, the Booleans thought that the distinction between categoricals and hypotheticals corresponds to what is referred to today as the surface grammar of natural language:

> Many hypotheticals,—in fact the vast majority [...], really consist of only two elements in thought: an antecedent and a consequent [...]. In all such cases the hypothetical is essentially of the form, 'If *P* then *Q*'. This we regard as being equivalent to, or rather as being covered by '*P* is *Q*'. In other words, the really simple hypothetical does not (on our rendering) differ in any way from the categorical, nor therefore does it demand any peculiar form of expression (ibid., 237).

Venn bases his rejection of the logical distinction between categoricals and hypotheticals on the truth-functional semantics of the if-then particle in both categoricals and hypotheticals, considering this semantics to be the same in both cases:

> Briefly put, my meaning is this: I fully admit the broad practical distinction between the cases which fall naturally into the [hypothetical] 'If *P* then *Q*', and those which fall into the [categorical] '*P* is *Q*' form; and also the fact that although common speech can sometimes bridge over the distinction by the convenient introduction of a predicate, it far more often cannot do this. But none the less they both display the same essential characteristics of what may be called 'predication' in a generalized sense of the term. They both alike yield the four alternatives which we want: viz. *P* asserted, *Q* asserted; *Q* denied, *P* denied; *P* denied; *Q* (so far) doubtful; *Q* asserted, *P* (so far) doubtful. And this is the essential point for purposes of judgement and inference (ibid., 223–4).

In rejecting the problematic modality in the course of the dissolution of the distinction between categoricals and hypotheticals, the Booleans were then faced with the difficulty of deciding on the appropriate interpretation of the equation $x = 0$: if, according to the Booleans, the symbol x stands for the value true of the proposition X, then what is meant by the equation $x = 0$? In this case Venn seems to reintroduce the problematic modality of propositions. He suggests that in employing the symbol x in such an equation, we propose—one is tempted to add *hypothetically*—the truth of the corresponding proposition, but we do not commit ourselves to the factual truth of it:

> As we have said, x stands for the truth of a proposition: to put $x = 0$ is a way of saying that there is no truth in it, that the truth of it is then and there excluded[1] [in this footnote Venn criticises Husserl, who understands '0' as a symbol for 'impossibility']. If the reader feels any difficulty in thus employing a symbol to stand for the truth of a proposition, without at the same time committing ourselves to the fact that the proposition is true, he must remember that this is simply another version of precisely the same general distinction that we had to face at the outset. [...]. What we have said hitherto was, let x stand for the class denoted by a given term: but this in no way committed us

> to the conclusion that any such class only comes up for decision when we introduce a proposition. If we put $x = 0$, we deny its existence (ibid., 438–9).

It was precisely this point which led MacColl to reassume the traditionalist definition of hypotheticals, placing it, however, within the language of Boole's symbolic system. We should now stop to review MacColl's reformulation of hypotheticals.

3. MACCOLL'S REFORMULATION OF HYPOTHETICALS

3.1. The Problematic Modality of Hypotheticals

MacColl's symbolic language is not an equational system but a 'Calculus of Equivalent Statements' in which the equality symbol should generally be interpreted as standing either for a biconditional or a metalogical equivalence.[26] MacColl's comments on the equations '$x = 1$' and '$x = 0$' are rather short:

> Def. 1.—[...]. The equation $A = 1$ *asserts* [my emphasis] that the statement A is true; the equation $A = 0$ asserts that the statement A is false (MacColl 1877a, 91; cf. also MacColl 1877b, 9).

According to MacColl, these equations should be read as propositions in which the metalogical predicators *false* and *true* occur. In other words, the equation '$x = 1$' expresses the proposition 'x is true', while the equation '$x = 0$' expresses the proposition 'x is false'.

When MacColl calculates the truth value of a given complex proposition, he uses the identity symbol not as an equivalence or a biconditional but as a copula. In his late logical papers, written between 1896 and 1911, MacColl starts to use the expressions x^τ and x^ι (instead of x^1 and x^0) for the propositions 'x is true' and 'x is false'[27] and remarks explicitly on this ambiguity in his use of the identity symbol.[28] What is interesting for our purposes is the alternative interpretation of this symbol. The corresponding reading of the equations in question is the following:

$x = \tau \Leftarrow_{\text{df}} \Rightarrow x \Leftrightarrow x^\tau$, i.e. '$x$ and "x is true" are equivalent propositions'

$x = \iota \Leftarrow_{\text{df}} \Rightarrow x \Leftrightarrow x^\iota$, i.e. '$x$ and "x is false" are equivalent propositions'

One possible consequence of the equivalence $x \Leftrightarrow x^\tau$ would be the following: if these two expressions are equivalent, the metapredicator 'true' is redundant. If the metapredicator 'true' is redundant, we have no means to express the problematic modality, therefore we do better to disregard this modality within the formal language of logic. Actually this is how many Booleans argued, linking Boole's method, mentioned above, for the conversion between primary and secondary propositions with the redundancy of the expression 'true'.

MacColl did not argue in this way at all. He insisted on the difference between stating a proposition hypothetically and asserting or confirming its truth. According to MacColl, a complex formula expresses a hypothetical proposition if the subformulae occurring in it are stated hypothetically. The subformulae are said to be stated hypothetically if the truth value of the complex formula is independent of the actual truth of these subformulae:

> The premises $A : B$ and $B : C$ [i.e. $A \Rightarrow B$ and $B \Rightarrow C$] of the latter [implication: $\{(A \Rightarrow B) \wedge (B \Rightarrow C)\} \Rightarrow (A \Rightarrow C)$] are hypothetical concepts of the mind—concepts which may be true or false (as may also the conclusion), without in the least invalidating the formula (MacColl 1902, 368).

Actually, there is some ambiguity in MacColl's use of the word 'hypothetical'. Instead of describing hypotheticals as having subformulae with a problematic modality he speaks, as already mentioned, of subformulae *stated hypothetically*. In general we can say that when MacColl wishes to stress the problematic modality of propositions he calls them 'statements':

> Def. 6.—Statements represented by letters or any other arbitrary symbols, to which we attach only a *temporary* meaning, are usually statements whose truth or falsehood may be considered an open question, like the statements of witnesses in a court of justice. It is convenient therefore to have an invariable symbol which shall be applicable to any statement whose truth is admitted and unquestioned, and to such a statement only. The conventional symbol used for this purpose is the symbol *1*. For a like reason it is convenient to have an invariable symbol to represent any statement whose *falsehood* is admitted and unquestioned. The symbol used for this purpose is *0* (MacColl 1880, 47).

MacColl distinguishes in this context between *statements*, which are *any arrangement of sounds, signs or symbols, employed to give* [not yet confirmed] *information*, and *propositions*, which are confirmed or asserted statements.[29] The problem is still not solved, however: if, again, 'x' and 'x is true' are equivalent, how should the distinction between statements and asserted statements (i.e. propositions) be expressed within a formal language? MacColl tried different answers in several papers: one type of answer was based on the formal use of (what we nowadays call) propositional variables, the other type stressed the pragmatic uses of the predicator 'true'. MacColl's use of the word 'statement' in these types of answers is unfortunately not always consistent, but as I have argued elsewhere the concept of truth-determining propositional variables can be fruitfully applied for understanding the first type (see Rahman 1999). This concept follows the lines of MacColl's main purposes and provides a good basis for reconstructing his reflections on hypotheticals. Thus I will say that a set of (occurrences of) propositional variables is *truth-determining* for a proposition A iff the truth value of A may be determined as true or false on all assignments of true or false to the set. I will say further that a propositional variable occurring in A is *redundant* iff there is a truth-determining set for A that does not contain this propositional variable. Thus, clearly, the set $\{p\}$ is not truth-determining for $p \rightarrow q$, but the set $\{p_1, p_2\}$ is truth-determining for $p \rightarrow p$. In this way we can reconstruct MacColl's use of elementary statements (that is, elementary propositions with a problematic mood) as truth-determining propositional variables, regarded as truth-determining independently of their actual truth value—only their possible truth values should be considered. Now all this allows the ideas behind MacColl's general framework to be expressed in the following way: The elementary expressions of the logical language are truth-determining propositional variables. Propositions in which the actual truth values of their propositional variables are known are called categoricals. Hypotheticals in which the actual truth value of their propositional variables can not modify the truth of the propositions in which they occur are formally valid. To reason symbolically is to reason under the following condition: propositional variables are used independently of their actual truth. This condition, which reflects the problematic modality of traditional hypotheticals, commits itself to a formal use of propositional variables.

With regard to the second type of answers, it is possible to delineate two complementary strategies by means of which MacColl argued for the irredundancy of truth, namely a pragmatical and a semantical one.

The pragmatical argument: This argument is based on the different functions which the metapredicators true and false can assume in the use of an assertion. On the one hand, MacColl, anticipating the remarks of P.F. Strawson and Grover (et alia) on the pragmatics of truth,[30] regards the metapredicators true and false as a result of context-dependent appropriate acts of speech, namely endorsement, or confirmation, and reversal:

> It may be remarked that any statement A and its denial A' are always of the same degree, whereas A^{τ} and A^{ι} their respective equivalents but not synonyms [...] are of one higher degree. The statement A^{τ} is a revision and confirmation of the judgement A [in his last writings MacColl uses the expression A^{τ} instead of A^{1} and A^{ι} instead of A^{0}], while A^{ι} is a revision and reversal of the judgement A (MacColl 1906, 18).
>
> If A = 'The event a did happen', then A^{τ} asserts that 'it is true that a did happen'.[...] Thus A^{τ} and A^{ι} are not exactly synonymous with A and A' (MacColl 1902, 361).

On the other hand, MacColl, who regards the problematic modality of propositions as prefiguring the formality of logic, seems to think that precisely those uses of the metapredicators 'true' and 'false' which today are called 'laziness' make it possible to disregard the content of an asserted proposition. Thus, in the proposition 'All that Holmes said is true', for example, it is possible for the sake of a formal argument to disregard what Holmes actually did say. In fact, it is necessary to neglect the content of this proposition if it occurs in a formal proof. That it is why the propositions 'x' and 'x is true' are logically, i.e. for the sake of formal argumentation, equivalent, in spite of their different pragmatical functions. This allows MacColl to regard propositional elements occurring in valid propositions as problematic in the same sense we mentioned when we gave Boole's reformulation of hypotheticals, that is, because of the independence of the actual truth value of these elements for the validity of the corresponding complex proposition. By these means MacColl explicitly worked out the problematic modality of propositions stated in Boole's system.

It is important to remark that within the scope of the discussion about the irredundancy of truth MacColl distinguishes between *synonymous* and *equivalent expression* (MacColl 1902, 361). Equivalent expressions are extensionally equivalent ones. Synonymous expressions, on the other hand, are those which can be substituted *salva veritate* in all contexts. In the pragmatical arguments MacColl searches for contexts where the substitution *salva veritate* does not agree with the intended use of the substituted expressions. MacColl also searched for semantical contexts with which he could argue for the irredundancy of truth. This search forms the basis of his semantical argument.

The semantical argument: The search for semantical contexts which could refute the redundancy of truth led MacColl to develop the first formal version of modal logic. Here MacColl reasons that in spite of the extensional equivalence of $A \Leftrightarrow A^{\tau}$, there are intensional contexts, more precisely epistemic contexts, where the two expressions cannot be substituted *salva veritate*: Despite its awkward formulation, we can recognise in MacColl's semantic argument for the irredundancy of truth some ideas which prefigure W. Quine's *opaque contexts*.[31]

Traditional hypotheticals were described not only as having a problematic modality but also as having strongly connected connectives. We should now consider MacColl's reformulation of this second feature of traditional hypotheticals.

3.2. MacColl's Dependent and Independent Hypothetical Connectives

MacColl's hypothetical inherited from Boole's secondary propositions the truth-functional formulation of the traditional dependence connections occurring in hypotheticals. Instead of conceptual dependence connections, Boole speaks of the truth-functional dependence of complex propositions on the truth values of its propositional elements, especially in disjunctions and conditionals:

> Secondary propositions also include all judgments by which we express a relation or dependence among propositions. To this class or division we may refer conditional propositions, as, 'If the sun shine the day will be fair'. Also most disjunctive propositions, as, 'Either the sun will shine, or the enterprise will be postponed'. In the former example we express the dependence of the truth of the Proposition, 'The day will be fair', upon the truth of the Proposition, 'The sun will shine'. In the latter we express a relation between the two Propositions, 'The sun will shine', 'The enterprise will be postponed', implying that the truth of the one excludes the truth of the other (Boole 1854, 160).

In the lines quoted on the first page of this paper, MacColl states precisely those two types of connected hypotheticals, namely disjunctive and conditional propositions, which Boole uses to illustrate dependent connectives, and which bring to mind Boethius' definition of hypotheticals.

Here we shall repeat the two cases I distinguished while relating MacColl's description of the biconditional reading of the equation $A = AB$ in the first chapter:

> The first case where B is true but A can be both true and false. This means that the truth of B is independent of the truth value of A.

I suggest that we read this case as describing a weakly connected conditional, or to put it in Boethius' words, a conditional *secundum accidens*, which can be stated truth-functionally either as a disjunction or as a material implication.

> The second case portrays a connective relation between A and B through which the truth of A necessarily leads to B being true. Similarly, the value false of A necessarily leads to B being false.

This last case describes a strongly connected conditional, which MacColl introduces with the expression 'is implied in', and which I called connexive implication. Thus, B (necessarily) depends in this strongly connected conditional on the truth value of A. MacColl, who is trying to produce a truth-functional formulation of hypotheticals here too, uses in his early work the expressions *possible* and *necessary*, which he explains by means of truth-functional and sometimes metalogical transcriptions. A possible truth means not-inconsistent. In other words, *possible* means in this context that no logical contradiction follows from its truth:

> DEF. 13.—The symbol $A \div B$ asserts that A does *not* imply B; it is thus equivalent to the less convenient symbol $(A : B)'$ [i.e. $\neg(A \Rightarrow B)$].
> Note.—The symbol $A \div B$ thus asserts that the truth of B is not a *necessary* consequence of the truth of A; in other words, it asserts that the statement A is consistent with B', but it makes no assertion as to whether A is consistent with B or not (MacColl 1878, 180).
>
> Note.—The implication $\alpha : \beta'$ asserts that α and β are inconsistent with each other; the non-implication $\alpha \div \beta'$ asserts that α and β are consistent with each other (ibid., 184).
>
> α is a consistent statement—i.e., one which may be true (ibid.).

MacColl employed this concept of consistency while stating what I called the second Boethian thesis of connexivity, that is: $(A \Rightarrow B) \Rightarrow \neg(A \Rightarrow \neg B)$:

> RULE 18—If A (assuming it to be a consistent statement) implies B, then A does not imply B'; in other words, from the implication $A : B$ we deduce the non-implication $A \div B'$, which is equivalent to $B \div A'$ (MacColl 1878, 180).

As I understand it, MacColl's reformulation of the strongly connected conditional of traditional hypotheticals comprises the following conditions:

Consistency of the antecedent part of the conditional or irredundancy of its consequent part: The antecedent part should be compatible or not inconsistent. This prevents the conditional being true because of the falsity of its antecedent part. In other words, the truth value of the antecedent part of the conditional should not yield the truth-functional redundancy of its consequent part. *Ex falso sequitur quodlibet*, for example, does not satisfy this condition as it is not connexive. What is connexive are the formulae $\neg(\neg A \Rightarrow A)$ and $(A \Rightarrow B) \Rightarrow \neg(\neg A \Rightarrow B)$, if we can prove that the antecedent is not inconsistent. The second, Aristotelian connexive thesis—so called because it had already been remarked upon by Aristotle—can be derived from the first one: If $A \Rightarrow B$, then, by contraposition, $\neg B \Rightarrow \neg A$. From $\neg B \Rightarrow \neg A$ and $\neg A \Rightarrow B$ it follows by transitivity that $\neg B \Rightarrow B$. This formula is not possible because of $\neg(\neg A \Rightarrow A)$. Therefore $\neg B \Rightarrow \neg A$ and $\neg A \Rightarrow B$ are incompatible or inconsistent, i.e.: $\neg((A \Rightarrow B) \wedge (\neg A \Rightarrow B))$. Summing up, $(A \Rightarrow B) \Rightarrow \neg(\neg A \Rightarrow B)$ follows from $\neg(\neg B \Rightarrow B)$ because of the inconsistency of $(A \Rightarrow B)$ and $(\neg A \Rightarrow B)$.[32]

Dependence of the consequent on the antecedent part of the conditional or irredundancy of its antecedent part: The consequent part of the conditional should be dependent on its antecedent part. This prevents the conditional being true because of the truth of its consequent part. The formula $a \Rightarrow (a \Rightarrow a)$ for example, does not satisfy this condition.

This theory of connexive conditional (cf. Rahman and Rückert 1998), which I cannot go into here, allows MacColl to recover formally some of the conceptual features of traditional conditional hypotheticals *secundum naturam* by preventing many of what today's relevance logicians call 'meaning disconnections' of material implication.[33]

MacColl even employed this reformulation of conditionals in the formalisation of traditional syllogistic. Unlike both the traditionalist and the Boolean logicians, MacColl thought that the hypothetical features of traditional hypotheticals should characterise not only formal propositional but also syllogistic reasoning. There he defends a hypothetical syllogistic free of ontological commitments.[34] His formulation of the universal proposition 'Logicians are drinkers' would be as follows:

(1) For a representative individual of the class of humans holds: if x is a logician, x is a drinker.

On the other hand, MacColl would formalise the particular proposition 'Some of the logicians are drinkers' as

(2) It is not the case that, for a representative individual of the class of humans holds: if x is a logician, x is not a drinker.

Formally:

(1) $(a \Rightarrow b)$

(2) $\neg(a \Rightarrow \neg b)$,

whereby the lower-case letters indicate that the formulae predicate something about a given arbitrary or representative object. MacColl's subalternation accordingly takes the following form:

$$(a \Rightarrow b) \Rightarrow \neg(a \Rightarrow \neg b).$$

MacColl's subalternation thus has the form of the second Boethian thesis of connexivity, whose validity does not depend on ontological commitments but on the logical consistency of a, which was one of the conditions for reformulating the logically strongly connected conditional of traditional hypotheticals.

In his search for different ways of formalising strongly connected conditionals, MacColl formulated from 1896 onwards those formal systems of modal logic which comprised the definition of a strict implication and which were further developed by C.I. Lewis, although the latter was rather reticent in mentioning his debt to MacColl's work.[35]

MacColl examines modal logic extensively in his book *Symbolic Logic and Its Applications* of 1906, where he introduces a modal interpretation of the expressions true, false, certain, impossible, variable (MacColl 1906, 6–7):

true (τ): 'true in a particular case or instance'.
false (ι): 'false in a particular case or instance'.
certain (ε): 'always true (or true in every case) within the limits of our data and definitions, that its probability is 1'.
impossible (η): 'contradicts some datum or definition, [...] its probability is 0'.
variable (θ): 'neither IMPOSSIBLE nor CERTAIN [...] POSSIBLE but UNCERTAIN', variable or contingent.

The definition of strict implication is accordingly:

$$A \Rightarrow B \underset{\text{df}}{\Leftarrow\Rightarrow} (A \wedge \neg B)^{\eta} \underset{\text{df}}{\Leftarrow\Rightarrow} (\neg(A \wedge \neg B))^{\varepsilon} \quad \text{(cf. MacColl 1906, 8)}$$

or to put it in MacColl's original notation:[36]

$$A : B \underset{\text{df}}{\Leftarrow\Rightarrow} (AB')^{\eta} \underset{\text{df}}{\Leftarrow\Rightarrow} ((A \wedge B')')^{\varepsilon}$$

As already mentioned, Hugh MacColl tried to build a system for quantified propositions which should inherit the general framework of hypotheticals. MacColl thought that this implied not only creating a system of first-order logic with a strongly connected conditional, but also the formulation of a system where the use of propositions stating facts about the elements of the corresponding universe of discourse commits only to a symbolic existence of the objects introduced by these propositions. That is, MacColl tried to formulate a logic where even the use of constants assumes a problematic modality:

> Let e_1, e_2, e_3, etc. (up to any number of individuals mentioned in our argument or investigation) denote our universe of real existences. Let 0_1, 0_2, 0_3, etc., denote our universe of non-existences, that is to say, of unrealities, such as centaurs, nectar, ambrosia, fairies, with self-contradictions, such as round squares, square circles, flat spheres, etc., including, I fear, the non-Euclidean geometry of four dimensions and other hyperspatial geometries. Finally, let S_1, S_2, S_3, etc., denote our Symbolic Universe, or 'Universe of Discourse', composed of all things real or unreal that are named or expressed by words or other symbols in our argument or investigation [...].
>
> When a class A belongs *wholly* to the class e, or *wholly* to the class 0; we may call it a *pure* class [...].
>
> We may sum up briefly as follows: Firstly, when any symbol A denotes an *individual*; then any intelligible statement $\phi(A)$, containing the symbol A, implies that the individual represented by A has a *symbolic* existence; but whether the statement $\phi(A)$ implies that the individual represented by A has *real* existence depends upon the context. Secondly, when any symbol A denotes a *class*; then any intelligible statement $\phi(A)$, containing the symbol A implies that the whole class A has a *symbolic* existence; but whether the statement $\phi(A)$ implies that the class A is *wholly real*, or *wholly unreal*, or *partly real and partly unreal*, depends upon the context (MacColl 1905, 74–7).

But how can i) a symbolic use of constants, ii) classifications in such a symbolic universe of discourse and iii) propositions about flat spheres and round squares be introduced in formal logic?

Well, all these questions can be answered in the context of the dialogical approach to free logic developed recently (see Rahman, Rückert, and Fishmann 2000) in a way which is congenial with MacColl's proposal of formulating a first-order logic which reflects the problematic modality of propositional logic. In a nutshell: in an argumentation, it sometimes makes sense to restrict the use and introduction of singular terms in the context of quantification to a formal (or *symbolic*) use of those terms. That is, the Proponent is allowed to use a constant iff this constant has been explicitly conceded by the Opponent. The symbolic use of constants amounts to allowing the use of these constants under the sole restriction that they name an individual: their ontological characterisation besides individuality does not play any role in logics. This reconstruction has been developed by Rahman in a paper on MacColl's concept of symbolic existence (Rahman 1999). This paper also offers a second way of reconstructing MacColl's ideas on contradictory objects by means of combining the concept of formal use of constants in free logics and that of the formal use of elementary negations in paraconsistent logics.

4. FINAL REMARKS

MacColl's propositional reading of the Boolean equation $A = AB$, as I intended to argue through this paper, throws light on his view of the logical form of both traditional and Boolean conditionals. It comprises a weak and a strong form, which in the early work was formulated as connexive implication and in the late work as strict implication. In the course of MacColl's developing his system, the strongly connected conditional slowly grew into the principal connective, and he essayed different formulations of it, although unfortunately he always used the same symbol for it—the colon. MacColl used his strongly connected conditional connective, as already mentioned, not only for propositional but also for syllogistic reasoning. The traditional distinction between categoricals and hypotheticals in pure or formal logic, where the conditional is a strongly connected connective, can thus only be maintained at the level of the modality of propositions, with hypotheticals having a problematic modality and categoricals an assertoric modality. It cannot be maintained at the level of the different kinds of connectives occurring in propositions. This is the closing point of the circle in which MacColl sought to join Booleans and non-Booleans at the moment of birth of formal logic in the hope of collaboration between them. The fruits of this collaboration have only started to be reaped in the past few decades of this century.

NOTES

[1] The only book on logic that I possessed [in 1877] was Prof. Bain's work; and to this I turned. The resemblance which my method bore to Boole's, as therein described, of course struck me at once (MacColl 1882, 243).

[2] Cf. e.g. The Calculus of Equivalent Statements (and Integration Limits), MacColl (1877b, 9–20).

[3] I admit that in [applied] mathematics (the theory of probability excepted) the disjunctive may be considered equivalent to the conditional or implication [...]. But in the wider field of general logic the assumption is unsafe (MacColl 1908, 453).

[4] The writer of this paper would like to contribute his humble share as a peacemaker between the two sciences, both of which he profoundly respects and admires. He would deprecate all idea of aggression or conquest [...]. Do not Englishmen and Scotchmen alike now both 'glory' as George III said he did, 'in the name of Briton'? Why should not logicians and mathematicians unite in like manner under some common appellation? (MacColl 1880, 46–7).

[5] Cf. Rahman (1997b, Ch. I.2.1).

[6] Boethius (1969, bk I, chap. iii, paragr. 4–7, 216–20).

[7] Boethius (1969, bk II, chap. vi, paragr. 1, lines 1–3, 286).

[8] Boethius (1969, bk II, chap. i, paragr. 7, lines 66–72, 258–60).

[9] Ibid.

[10] Sextus Empiricus put forward such a definition of connection, although it is often attributed to Chrysippus:

> [...] But the following statement seems to be true: 'If atomic elements of things do not exist, then atomic elements do exist' [...]. And those who introduce the notion of connexion [συναρτησις], say that a conditional is sound when the contradictory of its consequent is incompatible with its antecedent. According to them the conditionals mentioned above are unsound, but the following is true: 'If it is day, it is day'. (Translated by Kneale and Kneale (1962, 129). Cf. Sextus Empiricus (1976, B-II-111, 223)).

Here it is asserted that the strongly connected conditional $A \Rightarrow B$ expresses the following: it is not the case that A and not-B, i.e. A and not-B should be *incompatible* (μαχηται), because they form an inconsistency.

[11] It is clear that if the conclusion is false, the premises of the argument must be false, either all or some of them; but when the conclusion is true, it is not necessary that the premises should be true, either one or all, yet it is possible, though no part of the syllogism is true, that the conclusion may none the less be true; but it is not necessitated. The reason is that when two things are so related to one another, that if the one is, the other necessarily is, then if the latter is not, the former will not be either, but if the latter is, it is not necessary that the former should be. But it is impossible that the same thing should be necessitated by the being and by the not-being of the same thing. I mean, for example, that it is impossible that B should necessarily be great since A is white and that B should necessarily be great since A is not white. [...]. If then B is not great A cannot be white. But if, when A is not white, it is necessary that B should be great, it necessarily results that if B is not great, B itself is great (but this is impossible). For if B is not great, A will necessarily not be white. If then when this is not white B must be great, it results that if B is not great, it is great, just as if it were proved through three terms. (Translated by W.D. Ross, in: Aristotle (1963, II.4, 57a36–b18).)

At this point Aristotle is correct in asserting that a consequence which follows from false premises does not necessarily have to be true. Aristotle's explanation of this assertion is based on rejection of the proposition $\neg B \Rightarrow B$. Two implications of the forms $A \Rightarrow B$ and $\neg A \Rightarrow B$ cannot be true simultaneously: from $A \Rightarrow B$ we obtain $\neg B \Rightarrow \neg A$ by contraposition, and from $\neg B \Rightarrow \neg A$ and $\neg A \Rightarrow B$ we then obtain $\neg B \Rightarrow B$ by transitivity, and this is what Aristotle says is impossible (αδυνατον).

[12] Translated by Bocheński (1956, 24.13, 159). The original text is:

> nam cum ita proponitur: 'aut non est *a*, anut non est *b*', ita dicitur: 'si sit *a*, non erit *b*'; et sit *b*, non erit *a*. (Boethius 1969, bk III, chap. xi, paragr. 1, lines 1–4. 384.)

[13] Cf. Boethius (1969, bk I, chap. 5–7, 206–208).

[14] In his introduction to *The Mathematical Analysis of Logic* Boole writes:

> Sir W. Hamilton has contended, not simply, that the superiority rests with the study of Logic, but that the study of Mathematics is at once dangerous and useless.* (**Edinburgh Review*, vol. LXII, p. 409, and *Letter to A. De Morgan, Esq.*) (Boole 1847, quoted from reprint (1965, 11)).

[15] By many of the schoolmen, however, the term *hypothetical* (*hypotheticus*) was used to denote the genus, and the term conditional, to denote the species, and from them this nomenclature has passed into many of the more modern compends of logic [...]. This latter usage is wrong. If either term is to be used in subordination to the other, *conditional*, as the more extensive term, ought to be applied to designate the genus [...] (Hamilton 1866, 236f).

[16] Cf. Kant (1800, A 164; 1983, 536).

[17] I proceed, therefore, to the genus of propositions as opposed to categorical,—viz. the Conditional,—Conditioned. This genus, as stated in the paragraph, comprises two species, according as the condition lies more proximately in the subject, or in the predicate; to which is to be added, either as a third species, or as a compound of these two, those propositions in which there is a twofold condition,—the one belonging to the subject, the other to the predicate. The first of these, as stated, forms the class Hypothetical, the second that of Disjunctive, the third that of Dilemmatic, propositions (Hamilton 1866, 236).

[18] Although, therefore, an hypothetical judgment appear double, and may be cut into two different judgments, it is nevertheless not a composite judgment. [...]. The relatives *if* and *then*, in which the logical synthesis lies, constitute thus an act one and indivisible (Hamilton 1866, 238f).

[19] Cf. Kant (1800, A 167–8; 1983, 538f); cf. also Boole (1847, 58–9).

[20] Now, although in consequence of the multiplicity of its predicates, a disjunctive proposition may be resolved into a plurality of judgements, still it is not on that account a complex or composite judgement. For it is realised by one simple energy of thought, in which the two relatives,—the *either* and the *or*,—are thought together as inseparable, and as binding up the the opposing predicates into a single sphere (Hamilton 1866, 240f).

The interesting point is that Hamilton's concept of an exclusive disjunction appears to be intensional, i.e. it can be rendered by means of a strongly connected if-then:

> The attributes thus disjunctively predicable of the subject, constitute together a certain sphere or whole of extension; and as the attributes mutually exclude each other, they may be regarded as reciprocally reason and consequent (Hamilton 1866, 240).

A systematic reconstruction of this disjunction concept would no longer be compatible with the rule of addition. Cf. Rahman (1997b, II(A).5).

[21] This was proposed by Narahari Rao in a joint seminar on the if-then at the University of Saarland.

[22] Boole's definition of this hypothetical universe can indeed be described as a truth table:

> Thus if we associate the propositions *X* and *Y*, the total number of conceivable cases will be found as exhibited in the following scheme.

Cases	*Elective expressions*
1st *X* true, *Y* true	xy
2nd *X* true, *Y* false.	$x(1-y)$
3rd *X* false, *Y* true	$(1-x)y$
4th *X* false, *Y* false	$(1-x)(1-y)$ (Boole 1847, 50)

[23] This rejection of Boole's is to be found in his Postscript to *The Mathematical Analysis of Logic:*

> It must be added that Aristotle's view is consistent with the sense (albeit an erroneous one) which in various parts of his writings he virtually assigns to the word Cause, viz. an antecedent in Logic, a sense according to which the premises might be said to be the cause of the conclusion (Boole 1847, 82).

[24] This mode of treatment [of secondary propositions], combined with his rather fetched and unnecessary doctrine (as it seems to me) that propositions of this secondary kind were peculiarly interpretable in terms of *time*, has probably contributed to make some writers forget how completely Boole had grasped the possibility of this interpretation of his symbolic methods (Venn 1881, 435).

It is remarkable that Venn makes this criticism despite the fact that as far as propositional logic is concerned he sees the need to also be able to express propositions which are not formally true (cf. Venn 1881, 440–52). A similar (more modern) rejection of the introduction of temporal states is also to be found in Kneale and Kneale's book (1962, 414). What I feel is more to the point than the criticism of 'superfluousness', however, is MacColl's remark that Boole's interpretation of logic does not leave the level of things or object level (cf. introduction).

[25] Cf. Jevons (1864, 52). Venn reports the following concerning Jevons' use of the equation $A = AB$ to formulate universally quantified propositions:

> There is a third form [for the symbolic expression of 'All *A* is *B*'], which was probably first employed by Leibnitz[2] ([2]*Difficultates quedam logicae*. 'Omne *A* est *B*, i.e., aequivalent *AB* et *A*'.). It is discussed by Ploucquet[3] ([3]*Sammlung, &c.* p. 262 [cf. Ploucquet 1776, 262]), and is occasionally made use of by Lambert[4] ([4]*Sammlung*, p. 212; Log. Ab. I. 23 [cf. Lambert 1781, chap. I.23]). [...]. This form is best known at present from its systematic employment by Jevons, who did not seem to be aware that it had been so much employed before him (Venn 1881, 176–7).

[26] Def. 1.—[...] the equation $A = B$ asserts that *A* and *B* are equivalent statements (MacColl 1877a, 91).

[27] The symbol *A* is short for A^{τ}, and may be read '*A* is true' (MacColl 1900, 75).

[28] Other paradoxes arise from the ambiguity of the sign of equivalence (=) (MacColl 1906, 13).

[29] I define a statement as any sound, sign, or symbol (or any arrangement of sounds, signs, or symbols) employed to give information; and I define a proposition as a statement which, in regard to form, may be divided into two parts respectively called subject and predicate. [...] A nod, a shake of the head, the sound of a signal gun, the national flag of a passing ship, and the warning 'Caw' of a sentinel rook, are, by this definition, statements, but not propositions. The nod may mean 'I see him'; the shake of the head, 'I do not see him'; the warning 'Caw' of the rook, 'A man is coming with a gun', or 'Danger approaches'; and so on. *These propositions express more specially and precisely what the simpler statements express more vaguely and generally* [our emphasis]. [...] 'The big stag' and 'The killed stag'. These are not complete propositions; they are merely qualified subjects waiting for their predicates (MacColl 1906, 2–4). This is similar to what may be found in MacColl (1902, 352–3).

[30] Cf. Strawson (1950, 44–50), Grover et al. (1975, 96), Lorenz (1996, 598) and Astroh (1996).

[31] Suppose now that *A* denotes θ_τ, a variable that turns out true, or happens to be true in the case considered, though it is not true in all cases. We get $\varphi(A) = A^\varepsilon = \theta_\tau^\varepsilon = (\theta_\tau)^\varepsilon = \eta$; for a variable is never a certainty, though it may turn out true in a particular case. Again, we get $\phi(A^\iota) = (A^\iota)^\varepsilon = \theta_\tau^\varepsilon = (\theta_\tau^\iota)^\varepsilon = \varepsilon$; for θ_τ^ι means $(\theta_\tau)^\iota$, which is a formal certainty (MacColl 1906, 18).

The problems arising from MacColl's system of notation in his presentation of this argument have been elaborated by Stelzner (1993b).

[32] Cf. footnote 12. Obviously, the Boethian thesis $(A \rightarrow B) \rightarrow \neg(A \rightarrow \neg B)$ can be derived equally well from $\neg(A \rightarrow \neg A)$. Cf. Rahman (1997b, II(A).4.2.2).

[33] MacColl's redundancy concept has been discussed in detail in Rahman (1997b), in the last section of Chapter I.2.2.2 and in Chapter II(A).3.

[34] Cf. MacColl (1878, 191; 1906, 44). Cf. McCall (1963, 1964, 1966, 1967a,b). In this context his paper 'Connexive Implication and the Syllogism' (1967a) is of particular importance. Cf. also Astroh (1993, 1995, 1996).

[35] In the bibliography of Rahman (1997a), Lewis' knowledge of MacColl's work is commented on.

[36] Just as Boole already had (1847, 58), so, too, MacColl formulated a *therefore*-conclusion for causal conclusions with the aid of the modus ponens (MacColl 1897, 55; 1906, 44):

> In my system of notation:
>
> '*A* therefore *B*' $\Leftarrow_{\text{df}}\Rightarrow A \downarrow B \Leftarrow_{\text{df}}\Rightarrow A \wedge (A \Rightarrow B)$
>
> Here it becomes apparent that while $A \downarrow B$ requires *A* and $(A \Rightarrow B)$ to be connected conjunctively, $A \Rightarrow B$ merely requires the consistency of *A*. In MacColl's later work this is expressed modally:
>
> '*A* therefore [necessarily] *B*' $\Leftarrow_{\text{df}}\Rightarrow A \uparrow B \Leftarrow_{\text{df}}\Rightarrow A \wedge (A \succ B) \Leftarrow_{\text{df}}\Rightarrow A \wedge (A \wedge \neg B)^\eta$
>
> [...] asserts that the statement (or collection of statements) *A* is true, and secondly the affirmation of *A* coupled with the denial of *B* constitutes an impossibility (MacColl 1906, 80).

REFERENCES

Aristotle. 1963. *The works of Aristotle translated into English*. Vol. I. Oxford: Oxford University Press.

Astroh, M. 1993. Der Begriff der Implikation in einigen frühen Schriften von Hugh McColl. In Stelzner (1993a), 128–44.

———. 1995. Subjekt, Prädikat und Modalität. Ein Beitrag zur logischen Rekonstruktion einer kategorischen Syllogistik. Typescript.

———. 1996. Präsupposition und Implikatur. In Dascal et al. (1992–96), Vol. II, 1391–1407.

Bain, A. 1870. *Logic*. 2 vols. London: Longmans, Green and Reader.

Bocheński, I.M. 1956. *Formale Logik*. Freiburg/München: Alber.

Boethius, A.M.S. 1969. *De hypotheticis syllogismis*. Edited, commented and translated by Luca Libertello. Brescia: Paideia.

Boole, G. [1847] 1965. *The mathematical analysis of logic, being an essay towards a calculus of deductive reasoning*. Reprint, Oxford: Basil Blackwell.

———. [1854] 1958. *An investigation of the laws of thought on which are founded the mathematical theories of logic and probabilities*. Reprint, New York: Dover.

Dascal, M., D. Gerhardus, K. Lorenz, and G. Meggle, eds. 1992–96. *Sprachphilosophie, Philosophy of Language, La philosophie du langage*, Vol. I (1992), Vol. II (1996), Berlin: Walter de Gruyter.

Grover, D., J.L. Camp, and N.D. Belnap. 1975. A prosentential theory of truth. *Philosophical Studies* 27: 73–125.

Hamilton, W. [1861] 1966. Lectures on logic. In *Lectures on metaphysics and logic*, edited by H.L. Mansel and J. Veitch. Edinburgh: W. Blackwood. Reprint, Stuttgart: Frommann-Holzboog.

Jevons, W.S. 1864. *Pure logic or the logic of quality apart from quantity*. In Jevons (1890), 1–79.

———. [1890] 1971. *Pure logic and other minor works*. Reprint, New York: Burt Franklin.

Kant, I. 1800. *Immanuel Kants Logik, ein Handbuch zu Vorlesungen*. In Kant (1983, vol. 5, 421–582).

———. 1983. *Kant. Werke*. 10 vols. Darmstadt: Wissenschaftliche Buchgesellschaft.

Kneale, W., and M. Kneale. 1962. *The development of logic*. Oxford: Clarendon Press.

Lambert, J. [1781] 1965. *Logische und Philosophische Abhandlungen*, edited by J. Bernoulli. Leipzig: Wendler. Reprint, Hildesheim: Olms.

Lorenz, K. 1996. Artikulation und Prädikation. In Dascal, Gerhardus, Lorenz, and Meggle (1992–96), vol. 2, 1098–1122.

Łukasiewicz, J. 1951. *Aristotle's syllogistic from the standpoint of modern formal logic*. Oxford: Clarendon Press.

MacColl, H. 1877a. Symbolical or abbreviated language, with an application to mathematical probability. *The Educational Times and Journal of the College of Preceptors* 29: 91–2.

———. 1877b. The calculus of equivalent statements and integration limits. *Proceedings of the London Mathematical Society* 9: 9–20.

———. 1878. The calculus of equivalent statements (II). *Proceedings of the London Mathematical Society* 9: 177–86.

———. 1880. Symbolical reasoning (I). *Mind* 5: 45–60.

———. 1882. On the growth and use of a symbolical language. *Memoirs of the Manchester Literary and Philosophical Society* 7: 225–48.

———. 1897. Symbolic reasoning (II). *Mind* 6: 493–510.

———. 1900. Symbolic reasoning (III). *Mind* 9: 75–84.

———. 1902. Symbolic reasoning (IV). *Mind* 11: 352–68.

———. 1905. Symbolic reasoning (VI). *Mind* 14: 74–81.

———. 1906. *Symbolic logic and its applications*. London: Longmans, Green.

———. 1908. If and imply. *Mind* 17: 453–5.

McCall, S. 1963. *Aristotle's modal syllogisms*. Amsterdam: North-Holland.

———. 1964. A new variety of implication. *Journal of Symbolic Logic* 29: 151–2.

———. 1966. Connexive implication. *Journal of Symbolic Logic* 31: 415–32.

———. 1967a. Connexive implication and the syllogism. *Mind* 76: 346–56.

———. 1967b. MacColl. In *The Encyclopedia of Philosophy*, edited by P. Edwards. London: Macmillan. 4: 545–6.

Patzig, G. 1969. *Die aristotelische Syllogistik. Logisch-philologische Untersuchungen über das Buch A der „Ersten Analytiken"*. 3rd ed. Göttingen: Vandenhoeck & Ruprecht.

Ploucquet, G. [1776] 1970. *Sammlung der Schriften, welche den logischen Kalkül des Herrn Prof. Ploucquet betreffen*. Frankfurt: A.F. Bök. Reprint, Stuttgart: Frommann-Holzboog.

Prior, A.N. 1949. Categoricals and hypotheticals in George Boole and his successors. *Australasian Journal of Philosophy* 27: 171–96.

Rahman, S. 1997a. Hugh MacColl—eine bibliographische Erschließung seiner Hauptwerke und Notizen zu ihrer Rezeptionsgeschichte. *History and Philosophy of Logic* 18: 165–83.

———. 1997b. *Die Logik der zusammenhängenden Behauptungen im frühen Werk von Hugh MacColl*. Basel: Birkhäuser, forthcoming.

———. 1999. Ways of understanding Hugh MacColl's concept of symbolic existence. *Nordic Journal of Philosophical Logic* 3: 35–58.

Rahman, S., and H. Rückert. 1998. Die Logik der Zusammenhängenden Aussagen: ein dialogischen Ansatz zur konnexen Logik. Fachrichtung 5.1 – Philosophie, Universität des Saarlandes, Memo no 28, September 1999.

Rahman, S., H. Rückert, and M. Fishmann. 2000. On dialogues and ontology. The dialogical approach to free logic. *Logique et Analyse*, forthcoming.

Sextus Empiricus. 1976. *Sextus Empiricus*, with an English translation by R.G. Bury. Vol. 1. Loeb Classical Library. Cambridge Mass.: Harvard University Press.

Stelzner, W. 1993a. *Philosophie und Logik, Frege-Kolloquien Jena 1989/1991*. Berlin: Walter de Gruyter.

———. 1993b. Hugh MacColl—Ein Klassiker der nicht-klassischen Logik. In Stelzner (1993a), 145–54.
Strawson, P. 1950. Truth II. In *Truth*, edited by George Pitcher. Englewood Cliffs: Prentice-Hall.
Venn, J. [1881] 1971. *Symbolic logic*. Reprint, New York: Chelsea (unaltered reprint of the 2nd ed. of 1894).

Universität des Saarlandes, Fachrichtung 5.1. – Philosophie, Postfach 151150, 66041 Saarbrücken, Germany

NICLA VASSALLO

PSYCHOLOGISM IN LOGIC: SOME SIMILARITIES BETWEEN BOOLE AND FREGE

This paper challenges the common view that Boole is a psychologistic philosopher and Frege an anti-psychologistic one. As there is no single agreed notion of what psychologism in (deductive) logic is, various characterizations given in the literature are employed to show that there are similarities between Boole and Frege whichever account is chosen. The emphasis on similarities is in no way meant to deny that there also exist differences between them; these however are not considered here.

1. INTRODUCTION

Boole's thought is not widely known or appreciated. This unsatisfactory situation is mainly due to two factors. First, most of us accept the immense superiority for contemporary philosophy and logic of Frege's ideas over Boole's, not least because it has been celebrated by one of today's finest minds, namely by Michael Dummett.

According to Dummett (1973), Frege was the first philosopher of language to whom the tradition of analytic philosophy owes an immense debt of gratitude. That tradition traces its very origins back to Frege.[1] Moreover he was 'the founder of modern mathematical logic' and the thinker who determined 'the sharp break between modern logic and the older logical tradition, to which George Boole and his followers still belonged'.[2] The exaltation of Frege's importance in the history of ideas is epitomized by Dummett's stressing on all possible occasions Frege's anti-psychologism together with its great influence and authority on seminal figures such as Husserl, Russell and Wittgenstein. Dummett can hardly do otherwise because one of the main traits he recognizes in analytical philosophy is its revolt against psychologism.[3] As long as Frege is to be closely linked to this strain of philosophy, he can scarcely be anything other than an anti-psychologistic thinker.

This account presents Frege as an innovator in more ways than one, and Boole as a figure linked to an old and moribund tradition. Our attachment to this view is strengthened by the second factor that encourages us to dismiss Boole's thought: it is often deemed psychologistic, and the term 'psychologism' has long been employed pejoratively to indicate an error of the past. When applied to the thought of a given philosopher, the term has been used as being roughly synonymous with 'bad thought', i.e. with thought that ought to be rejected; this indeed has been Boole's fate. By contrast, anti-psychologism has often been looked upon with approval as a legitimate and proper point of view, so Frege's philosophy is praised as intellectually virtuous.

J. Gasser (ed.), A Boole Anthology, 311–325.

My purpose here is not to engage in a large-scale comparison between Boole and Frege or in an examination of the role that they played in the history of philosophy and of logic. The idea of the insignificance of Boole for contemporary logic and philosophy has in any case already been contested by others. Goldfarb (1979), Grattan-Guinness (1988) and Putnam (1990)—among others—accord Boole an important place in the development of mathematical logic, and Bornet (1995) has linked him to the origins of analytical philosophy. What prompted me to write the present essay was the question of psychologism in connection with the observation that while the list of those who take Boole to be an aspiring or even an actual psychologistic philosopher is quite impressive,[4] it is rare to find a challenge to the view of Frege as anti-psychologistic, and when we do find one, as for example in Kitcher (1979), we soon enough find other authors ready to reassert that Frege's entire work was strongly anti-psychologistic.[5]

Should we accept this opposition of Boole and Frege on the subject of psychologism without further ado? I think not. Psychologism—at least in logic—is a tricky question, since no agreement exists as to what, precisely, it is. It is useful to consider definitions of psychologism as a way of coming to see the possibility, despite what we are told, of finding similarities between Boole and Frege.

In what follows I shall quote extensively both from Boole and Frege, rather than simply summarize their thoughts, lest it be objected that Frege (or Boole) did not say what I put into their mouths. I will try to be very simple, hoping thereby to convince even people unacquainted with the philosophy of these two men that they both can be placed in the same psychologistic boat. As I have just said, I shall describe this boat using definitions of psychologism that we find in the literature. I shall not question them here, but this does not mean that I think they are correct. My only purpose is to show that if psychologism is what is spelled out by the definition I consider, then we are compelled to say that both Boole and Frege are in the same boat. This result will lead me in the concluding section to some brief observations about the issue of psychologism that should encourage us to reassess Boole's thought.

2. LAWS OF THOUGHT[6]

Logic has to do with laws of thought is a definition of psychologism endorsed by Sober (Callebaut and Sober 1993, 98) who, asked by Callebaut about what psychologism is, said: 'Boole, for example, ... described logic as giving the laws of thought'. And Frege? According to Radnitzky (1976, 505):

> Thanks largely to the pioneering work of Frege and Husserl, psychologism in logic and metamathematics is largely a thing of the past: the attempt to reduce the norms of logic to laws of thought is now merely a historical curiosity.

So those who regard logic as linked to the laws of thought are psychologistic thinkers, and Frege is not to be counted among them. By contrast, Boole is certainly psychologistic. What evidence exists for this statement? For some[7] it is enough to refer to the title of his famous book, to wit, *An Investigation of the Laws of Thought on Which Are Founded the Mathematical Theories of Logic and Probabilities.*

Let me ask what can be meant by 'thought' here. If by 'thought' we signify something that belongs to the mind, then the above definition becomes: *logic has to do with something which is mental*. In particular, it can be claimed that *logic is the study of the mind*. If this is psychologism in virtue of the fact that 'psychological' may mean the mental as such, both Boole and Frege shared it. Boole (1854, 1) stated: 'The design of the following treatise is to investigate the fundamental laws of those operations of the mind by which reasoning is performed', and Frege (1977, 25) claimed: 'Neither logic nor mathematics has the task of investigating minds and contents of consciousness owned by individual men. Their task could perhaps be represented rather as the investigation of *the* mind; of *the* mind, not of minds'.

It can be promptly retorted that the similarity I am tracing between Boole and Frege is based on a misunderstanding, for while Frege explicitly assigns to logic the task of studying the mind, not minds, Boole makes no specific statement on the matter—in particular, then, he never specifies that the task of logic is not to deal with minds. I reply that if someone is willing to consider logic as the study of minds, he must surely think that the knowledge of their laws requires first of all an observation of many minds, and it is worth noticing that Boole explicitly denies this. In fact he affirms that 'the knowledge of the laws of the mind does not require as its basis any extensive collection of observations. The general truth is seen in the particular instance, and it is not confirmed by the repetition of instance ... [its] perception ... is not derived from an induction from many instances' (Boole 1854, 4).

Another way of interpreting 'thought' is to equate it with the process of thinking or reasoning. Since Boole wants to investigate the laws of those operations by which reasoning is conducted, he would mean by 'laws of thought' 'laws of thinking or reasoning'. The definition of psychologism then becomes: *logic has to do with the laws of thinking*. Why is it psychologism? According to Dummett (1983, xii) '... thinking (*das Denken*) ... is the proper subject of psychology'.[8] So if someone is interested in thinking he is doing psychology, and therefore if he says that logic has to do with thinking, he is psychologistic. Boole certainly is psychologistic in this sense. But so too is Frege. The full title of his logical masterpiece of 1879 is *Begriffsschrift, eine der arithmetischen nachgebildete Formelsprache des reinen Denkens*, i.e. *Begriffsschrift, a formalized language of pure thinking modelled upon the language of arithmetic*.[9]

Russell (1901, 366) suggested that equating the laws of logic and the laws of thinking means investigating how people actually reason. In fact he wrote that Boole 'was ... mistaken in supposing that he was dealing with the laws of thought: the question how people actually think was quite irrelevant to him'. But did Boole in fact suppose that?

3. STRONG PSYCHOLOGISM AND WEAK PSYCHOLOGISM

Haack (1978, 238) distinguishes three positions:

(i) logic is descriptive of ordinary mental processes.
(ii) logic is prescriptive of ordinary mental processes.
(iii) logic has nothing to do with ordinary mental processes.

She calls (i) strong psychologism, (ii) weak psychologism, and (iii) anti-psychologism, and claims that Frege holds (iii). If Russell had known of these distinctions, he would have said that Boole wanted to embrace (i). For three reasons I do not think Boole was so naive.

First, if someone tries to describe ordinary mental processes, then—even as a logician—he needs to refer to the findings of psychology, but Boole never did this. The psychology of his time was notoriously associationistic, which Boole refused to accord a place in logic:

> The ground of every system of notation employed in reasoning is the formal character of reasoning itself. If the process of inference is independent of the particular meaning of the concepts involved and depends only upon the general notion which those concepts manifest and thereupon only as it furnishes the basis of intellectual operations and of formal laws it is at once suggested to us that we express concepts not as in ordinary language by words the special meaning of which may through association of ideas interfere with our perception and application of the purely formal laws to which they are subject but by symbols in the employment of which the formal law and not the special meaning is present to the mind.[10]

Boole's general method in logic supplies the second reason for claiming that he did not seek to describe how people actually think. Since it involves uninterpretable symbols, Boole (1854, 68) insists that only the premises and the conclusions must be interpretable while the steps of inference must be carried out 'without regard to the question of the interpretability of the particular results obtained'. Therefore, the correspondence between processes of inference and their interpretation that exists in '*ordinary* applications of human reasoning',[11] where we need to see the meaning of every operation we perform, is not always present in Boole's general method. So, one important condition of our ordinary reasoning, i.e. its complete intelligibility, does not belong to Boole's logic. And this surely means that his logic was not designed to describe ordinary mental processes.

The third reason for saying that Boole is not rightly seen as a strongly psychologistic thinker is his conviction that '[Logic] investigates the form and expressions to which *correct* reasoning may be reduced and the laws upon which it is founded',[12] or, more briefly, that the laws of logic are 'the laws of *right* reasoning only'.[13] If the laws of logic were considered by Boole to be the laws of ordinary reasoning, it would be pointless to specify that they are the laws of right reasoning *only*. To say that the laws of logic are the laws of correct reasoning *only* implies that logic should recognize incorrect reasoning and judge it as incorrect, though it is not at all its task to describe it. If the laws of logic were considered the laws of ordinary reasoning, there would be no way to say that our ordinary reasoning is imperfect, when it happens to be so, and therefore no sense in speaking of 'right reasoning *only*'. If the laws of logic were like that, we could never make errors of reasoning. As Frege clearly stated: 'The laws in accordance with which we actually draw inferences are not to be identified with laws of valid inference; otherwise we could never draw a wrong inference'.[14] I will come back shortly with more details about Boole's and Frege's thought about our performance in our ordinary reasoning. We will see that their ideas are similar. But now, after having shown that we have no grounds to judge Boole as strongly psychologistic, I want to point out that he does present the traits of a weakly psychologistic thinker. What is more, so does Frege.

Grattan-Guinness (1997, xxxii) claims that 'Boole was psychologistic in that he saw his study as a normative account of the correct use of thinking'.[15] This is the position Haack calls weak psychologism. Boole (1854, 419) clearly subscribed to it, stating that 'it seems not improbable that [the laws of thought] should manifest their presence otherwise than by merely prescribing the condition of formal inference'. Though in some of his writings Frege was aware of the trouble that comes with defining the laws of logic as the laws of thought or thinking, in particular that it might lead one to suppose—as did Russell—that they are laws according to which we ordinarily reason,[16] he did not abandon calling them laws of thinking in virtue of their prescriptive character. And in fact, in the Introduction to his *Grundgesetze der Arithmetik*, he wrote:

> Any law asserting what is, can be conceived as prescribing that one ought to think in conformity with it, and is thus in that sense a law of thinking (ein Denkgesetz). This holds for laws of geometry and physics no less than for laws of logic. The latter have a special title to the name 'laws of thinking' (Denkgesetze) only if we mean to assert that they are the most general laws, which prescribe universally the way in which one ought to think if one is to think at all.[17]

Since both Boole and Frege share the conviction that logic is prescriptive of ordinary mental processes, they are both to be classified as weakly psychologistic philosophers.[18] So not only is Russell wrong in imagining that Boole supposed logic to describe thinking as it actually takes place, but Haack also is wrong in taking Frege to be anti-psychologistic since he clearly espoused the second of the positions she distinguishes—weak psychologism—and not the third, according to which logic has nothing to do with ordinary mental processes.

It can be retorted—as Husserl (1900) somehow did in his *Prolegomena*, like Johnson (1921) in his *Logic* and Carnap (1950) in his *Logical Foundations of Probabilities*—that to say that logic is normative is no more illuminating than to say the same of the laws of other sciences, e.g. 'than to say that mineralogy tells us how to think correctly about minerals'.[19] But, illuminating or not, if attributing to logic a normative character is psychologism, then both Boole and Frege embraced it. It can be further retorted that it is unfair or improper to say that the second position distinguished by Haack deserves to be called 'psychologism', though 'weak'. I do not propose to enter into this question. I am simply observing that if so-called 'weak psychologism' is, indeed, psychologism, then Frege is no less psychologistic than Boole.

4. THE MENTAL REALITY OF LOGIC

Until now I have taken it for granted that ordinary reasoning does not always coincide—though of course it can coincide—with correct reasoning. In doing so I have followed Frege in believing that we can draw erroneous inferences, i.e. that our ordinary reasoning can be mistaken from a logical point of view. One of the great merits of logic is precisely its usefulness as a device for detecting possible errors in our ordinary reasoning and arguments. Boole too observed that people make errors in their reasonings—an observation that would be quite difficult to accomodate, if he were strongly psychologistic—and offered us a good example of how logic can be applied to informal arguments, analyzing parts of Clarke's *Demonstration of the Being and Attributes of God* and of Spinoza's *Ethica ordine geometrico demonstrata*, and noticing that 'in what are regarded as the most rigorous examples of reasoning ... different trains of thought are blended together ... and, not infrequently, ... arguments, viewed by the strict laws of formal reasoning, are incorrect or inconclusive'.[20] If what are regarded as the most rigorous examples of reasoning are often incorrect, just imagine how little correct ordinary people's reasoning will be!

It is entirely possible simply to notice that our ordinary reasoning is not always correct, without going on to ask why. This however was not the route chosen by Boole and Frege. They both felt the necessity of asking 'how is it that we make mistakes in our reasoning', and they answered the question in a similar way. Frege (1989; letter dated 3 April 1920) maintained that our mistakes do not originate with logic, but come from the contaminations or interference to which logical activity is subjected. Boole (1854, 409) claimed that the phenomenon of incorrect reasoning is due 'to the interference of other laws with those laws of which *right* reasoning is the product'.

This explanation of errors derives from the conviction that logic has—among other things—some mental reality. These other things can interfere with our logical activity (Frege) or with the laws of which correct reasoning is the product (Boole), and this is why we err. But we do not only make errors; we also reason correctly. And to attribute to logic some mental reality has the aim of explaining our capacity for correct reasoning. In fact, if we were to succeed in avoiding what interferes in our mind with logic, then we should achieve correct reasoning and nothing else.

To show that Boole believed in the mental-reality-of-logic thesis is not difficult. The following quotations abundantly suffice. He claims that: '[logic] is based upon facts ... which have their abode in the constitution of the Mind';[21] '[there are] fixed principles in the right operations of human reason';[22] '[the laws of logic] have their ultimate abode in the human mind';[23] 'the essential standards of truth and correctness [are] ... deeply founded in the constitution of the human faculties'.[24]

In order to appreciate that Frege also embraced the mental-reality-of-logic thesis, it suffices to consider the following statement, bearing in mind that thinking can not be anything other than a mental activity: 'In the form in which thinking naturally develops the logical and the psychological are bound together'.[25] So something logical is in our mind, though bound to something psychological. And if Frege had not espoused the above thesis, it would be quite difficult to motivate his way of expressing the famous anti-psychologistic principle by the bizarre formula 'always to separate sharply the psychological from the logical',[26] and not more plainly as 'always to separate sharply psychology from logic'. It is precisely because he believed that ordinary thinking was composed of the logical and the psychological that he enjoins us to keep them apart so as to obtain a pure logical thinking, one cleansed of those errors that arise out of the interference of the psychological with the logical.

Sober (1978, 167) calls 'psychologism' *'the thesis that the laws of logic ... are (among other things) the laws of cognition'*,[27] and he specifies that 'this is the view that the rules of correct reasoning that logicians ... try to describe have *psychological reality*', denying that Frege endorsed this type of psychologism.[28] Of course Frege would have refused to claim that the laws of logic have psychological reality, because in his view they are part of mind and are intermingled with what is psychological, i.e. on his view what is mental does not coincide with what is psychological. Nevertheless, Frege would surely not have opposed the above thesis, which was once called the mental-reality-of-logic-thesis. In particular, he would not have opposed it if its claims were that the laws of logic are (among other things) the laws of cognition, or, better still, that—as Sober (1978, 175) puts it—'the laws of logic are constitutive of the laws of thought', i.e. of the laws of thinking or of the laws of mind. And if to subscribe to that thesis is to be psychologistic, then Frege is psychologistic, no less than Boole.[29]

5. THE APPEAL TO CONSCIOUSNESS

Musgrave (1972, 605) claims that Boole 'probably assumed that the normative laws of logic could also be drawn from a study of our consciousness' and judges *this appeal to consciousness* to be a type of psychologism from which Boole was able to free himself, pointing out that 'inferences drawn in accordance with [logical laws] are valid'. But it is still true that Boole never abandoned his conviction that we can know logic through our consciousness. Even in the *Laws of Thought,* where he extracts the laws of logic from language, thereby giving them a linguistic foundation, we read:

> For though in investigating the laws of signs, *a posteriori*, the immediate subject of examination is Language, with the rules which govern its use; while in making the internal processes of thought the direct object of inquiry, we appeal in a more immediate way to our personal consciousness,—it will be found that in both cases the results obtained are formally equivalent.[30]

So, though in the *Laws of Thought* Boole asks us to examine language and never to consult our consciousness as we determine the laws of logic, he still thinks that the results of the two methods would be identical.[31] This means that he does not give up the belief that our consciousness can yield the laws of logic. And this belief is evident also in his subsequent writings. Speaking of the principles of identity and excluded middle, he claims that 'our consciousness informs us' that they 'lie close to the very ground of the possibility of thought'.[32] Other illuminating passages are: '... the enquiry into the distinct operations of thought is entirely based upon an appeal to consciousness' (Boole 1997, 53), and '... we possess a ... test of the completeness of any account of laws of thought professedly based upon the testimony of consciousness in the completeness of the system of methods and processes into which the supposed laws are capable of being developed' (ibid.).[33]

If logic has some mental reality, it is quite natural to think that our consciousness can reveal logic to us. But it may be wondered how it can be that consciousness might reveal logic alone without also delivering to us those other things which interfere with logic in our mind and are responsible for errors in our reasoning. Of course this problem is by no means minor since we urgently need to avoid generating a logic with errors in it. As far as I know, Boole never explicitly raised this problem. This fact can be interpreted in two ways: first that he did not recognize the problem at all, second that he strongly believed in the force of logic and in the power of consciousness. I feel quite confident in excluding the first interpretation since he was quite aware of the interferences in our mind to which logic is subjected. The alternative is much more promising. Just after having told us that incorrect reasoning is due to the interference of other laws with the laws of right reasoning, Boole claims (1854, 409):

> Still it would remain that there exist among the intellectual laws a number marked out from the rest by this special character, viz., that every movement of the intellectual system which is accomplished solely under their direction is *right*, that every interference therewith by other laws is not interference only, but *violation*. It cannot but be felt that this circumstance would give to the laws in question a character of distinction and of predominance.

This passage reveals how certain Boole was of the force of laws of logic. Interference with them by other laws is neither natural nor obvious; indeed it is a violation. Laws of logic have a special character. They are distinct and predominant. So distinct and predominant—Boole must have thought—that they are immediately evident to our consciousness. This is further confirmed by the passage immediately after the above:

> Now a little consideration will show that there is nothing analogous to this in the government of the world by natural law. The realm of inorganic Nature admits neither of preference nor of distinctions. We cannot separate any portion of her laws from the rest, and pronounce them alone worthy of obedience,—alone charged with the fulfilment of her highest purpose.[34]

So, unlike the realm of inorganic nature, the realm of the mind admits of preference and of distinctions in virtue of those laws of logic that can be separated from the rest and it privileges them as the only ones worthy of obedience, as the only ones charged with the fulfilment of the mind's highest purpose. I think Boole did not see how our consciousness can fail to give us the sense of the importance of mind's highest purpose.

Still working under the hypothesis that an appeal to consciousness is a trait of psychologism, it is time now to look at Frege's thought. Frege (1950, x) strongly enjoined us 'always to separate sharply the psychological from the logical'. This anti-psychologistic principle finds an anticipation in Boole who believed that logic has to be separated from what interferes with it, and it can be so separated because it is so distinct and so predominant as to be immediately graspable by our consciousness. As that principle is not incompatible with espousing the mental-reality-of-logic thesis—indeed, as we have seen, albeit briefly, that it is perfectly consistent with this thesis—so it does admit an appeal to our consciousness. This is true not only for Boole but also for Frege.

Frege (1979, 2) speaks of 'isolating what is psychological, by consciously marking it off', and he claims (Frege 1979, 5) that 'we must reject all distinctions in logic that are made from a purely psychological standpoint'. He also allows (ibid.) that such a rejection 'does not mean that we want to banish any trace of what is psychological from thinking as it naturally takes place, which would be impossible; we only want to become aware of the logical justification for what we think'.[35] But what is being or becoming aware if not being or becoming conscious? In fact, repeating that 'in logic we must reject all distinctions that are made from a purely psychological point of view',[36] Frege (1979, 142) says that

> in human beings it is natural for thinking to be intermingled with having images and feeling. Logic has the task of isolating what is logical, not, to be sure, so that we should think without images, which is no doubt impossible, but so that we should consciously distinguish the logical from what is attached to it in the way of ideas and feeling.

The appeal to awareness or consciousness is due—as in Boole—to the acceptance of the mental-reality-of-logic thesis. This thesis clearly emerges also in the following statement: 'the required separation of the logical from the psychological is only a matter of distinguishing in our minds between them' (Frege 1979, 5–6). In our minds there is something logical and something psychological, and because of this fact the separation between them occurs in our mind.

It might be objected that Frege frees himself from psychologism as it arises out of the appeal to consciousness by proposing other ways of separating the logical from the psychological. For instance he writes (Frege 1979, 6):

> ... we can see the value of learning foreign language for one's logical education; when we see that the same thought can be worked in different ways, *our mind separated off the husk from the kernel*, though, in any given language, it appears as a natural and integral part of it. This is how the differences between languages can *facilitate our grasp of what is logical*. But still the difficulties are not wholly removed in this way and our logicians still keep dragging in a number of things which are really of no logical concern, though they belong to the grammar of languages akin to our own, if not to others. For this reason it is useful to be acquainted also with a means of expression of a quite different kind, such as we have, for instance, in the formula-language of algebra.[37]

I do not see how this passage can be taken as suggesting that Frege can rule out the appeal to consciousness. In fact, if our mind were unconsciously to separate off the husk from the kernel, it would then be difficult to understand how we can know the required separation has been performed, and Frege would surely not have been happy to hear that while we may perhaps succeed in keeping the logical and the psychological apart, we can not know whether we have succeeded or not. Moreover, learning a foreign language can facilitate our grasp of what is logical, but it does not guarantee us a faultless grasp. For this reason Frege recommends that we study formal languages. But how can they be free of what is psychological? Frege leaves us no other option but to say that the mathematicians or logicians who construct those languages have to keep the logical and the psychological separate from each other if there are to be languages free of what is psychological. And, since the study of natural foreign languages is not enough, it must also be true for mathematicians or logicians that 'the required separation of the logical from the psychological is only a matter of distinguishing in [their] minds between them',[38] of consciously distinguishing, that is.[39]

6. CONCLUSION

We have the following result: if we want to say that Boole was a psychologistic philosopher, we have to say the same of Frege. This result is not to be read as a desire on my part to attribute psychologistic theses to these authors at any price,[40] but as a protest at those readings that lead us to oppose Boole and Frege, and at the widespread practice of judging the former, but not the latter, psychologistic. It is also a protest at employing the word 'psychologistic' pejoratively to single out a terrible mistake of the past, thereby reinforcing the belief that Boole's philosophy is bad and outdated. Against those readings I am happy to claim that if Boole's philosophy must be considered bad and outdated because of its psychologism, then Frege's too deserves the same fate.

In any case I think it is improper to use 'psychologism' pejoratively. It is far better to employ it descriptively because only then will we analyze its strength and its weakness without prejudice. So the common traits between Boole and Frege that I have highlighted here are not to be rejected merely because they can be labelled psychologistic. But if they are rightly so labelled, we are not only compelled to review the history of philosophy and logic as it appears in those readings that make Boole psychologistic and Frege anti-psychologistic, but we should also reconsider Boole's and Frege's ideas in comparison with contemporary thought.[41] Though our century has been marked for a long period by its resolutely anti-psychologistic vein—and if Frege embraced some psychologism, his contribution to that vein as envisaged by Dummett must be revised—many thinkers now believe in what may be labelled 'naturalism' or 'psychologism', for example those who are proposing to reconsider logic in view of results from psychology.[42] Nothing of this kind was envisaged either by Boole or by Frege. Still others now defend what I called the 'mental-reality-of-logic thesis' with sophisticated arguments.[43] Of Boole and Frege, it was Boole who first embraced this thesis, and so his importance for our contemporary reflection cannot be easily dismissed when we come to write the history of ideas.

It might be retorted that I am right to say that the opposition between Boole and Frege on the question of psychologism makes no sense, but that I am on the wrong track. This opposition would not have been accepted even by Frege who never accused Boole of psychologism, while he did accuse others of it. Nevertheless, on more than one occasion he went to lengths to underline the differences between his own logic and that of Boole.[44] And since Frege surely considered himself anti-psychologistic and Boole as not committed to psychologism, I must be wrong in maintaining that they both subscribed to some forms of it, i.e. those we have seen here.

In reply I can allow that Frege is right and I am wrong if it be true that the common traits between Boole and Frege that have been emphasized here do not indicate psychologism. If, however, the common traits are indeed signs of psychologism, then Frege is wrong and I am right. Are they or are they not signs of it? For some authors they are. I do not propose here to agree or disagree. And in fact all my arguments depend on a conditional of the form: if psychologism is such-and-such, then we have to say that both Boole and Frege adhered to it. But what is certain is that if at this point we still want to have a justified belief in Frege's intense anti-psychologism, then the definitions of psychologism here employed must be wrong. What is also important is that this will force on us a reinterpretation of Boole's philosophical approach to logic—because what reasons will we then have to persist in attributing psychologism to him?

If this essay leaves you with the idea that psychologism is a tricky question, I shall welcome that. There is not any doubt about its trickiness. This is clear from the fact that in many cases the very same people who attacked psychologism have themselves subsequently been judged by others, or by others' standards, to be psychologistic.[45] This essay does nothing if not confirm how tricky the question is since Frege, the anti-psychologistic *par excellence*, cannot escape being viewed as psychologistic. The question is surely in need of a solution, especially now that psychologism or naturalism is again in fashion.[46] But while we await this solution, we must not remain mired in those standard readings that view Boole, but not Frege, as a partisan of psychologism. We must use the definitions of psychologism at our disposal and judge these authors by them without any cliché. But, of course, we can also reject those definitions. And if we reject those which appear in this essay, the similarities I envisaged between Boole and Frege vanish. This does not mean that others may not be found. That will depend on what other definitions of psychologism are adopted or coined.[47] I shall be delighted to see other definitions come under consideration—even if they lead to our coming up with great differences between Boole and Frege—because the ultimate aim of this paper is just that of being read—because it was so written—as a provocation. I would like to provoke an opening up of other lines of enquiry on psychologism in logic which can not but lead us to a reconsideration of Boole since he has been very often linked to this doctrine. These lines will surely help us to a deeper reading and assessment of Boole and his legacy.

ACKNOWLEDGEMENTS

This paper was originally given at the conference 'Boole 1997: One Hundred Fifty Years of Mathematical Analysis of Logic' held in Lausanne, September 1997. I thank the Swiss Society for Logic and Philosophy of Science for giving me the opportunity to try out the above ideas on that occasion. I am grateful to the audience for relevant comments. I should like to thank Anthony Savile for helpful suggestions before my writing the present version of the paper.

NOTES

[1] See also Dummett (1993).

[2] See Dummett (1978, 87). This is corroborated by Dummett (1978, 71) writing about Leibniz and Boole that 'the most important difference was surely that Boole published and Leibniz did not'.

[3] That trait is recognized not only by Dummett but also by Ryle. See e.g. his introduction to (1956).

[4] Being impressive, it is worth reporting: Russell (1901), Jourdain (1910), Jørgensen (1931), Cohen and Nagel (1934), Carnap (1950), Hesse (1952), Thomas (1955), Feys (1955), Bocheński (1956), Barone

(1961), Kneale and Kneale (1962), Mangione (1972), Musgrave (1972), Haack (1978), Dalla Chiara (1979), Laita (1980), Currie (1982), Grattan-Guinness (1982, 1997), Baker and Hacker (1984, 1989), Ferriani (1984), MacHale (1985), Cobb-Stevens (1990), Brockhaus (1991), Mangione and Bozzi (1993), Mugnai (1993, 1998), Callebaut and Sober (1993), Di Francesco (1996), Bornet (1997).

[5] In this specific case the reply is due to Weiner (1990) and to Picardi (1996).

[6] In this section, when I speak of 'the laws of thought' I am not referring to the laws of the content (*Gedanke*) that the subject thinks, simply because Frege did not have any expression for referring to them. In fact in German there is no expression such as 'laws of *Gedanke*', viz. '*Gedankengesetze*'. On content see the last footnote of this essay.

[7] See, for example, Carnap (1950, 40).

[8] My translation.

[9] See Frege (1952).

[10] See Boole (1997, 86). Cf. also the following passages: 'The validity ... of the process of *reasoning* depends not at all upon the pictorial element in our concepts whereby the images of sense are reproduced but only upon the formal laws and relations of those concepts' (Boole 1997, 70); '*When concepts are combined by composition whether immediately or mediately the order in which the several mental acts of composition are performed is indifferent*' (ibid., 78).

[11] See Boole (1854, 67).

[12] See Boole (1997, 1); my italics.

[13] See Boole (1854, 408). See also Boole (1952, 195): '... it is the office of the laws of reasoning to determine what is correct in the processes of thought'.

[14] See Frege (1979, 4). To be honest I should also draw attention to another passage from Boole (1997, 13): 'The object of Logic as a Science is to explain the laws of those mental operations by which ordinary Reasoning is conducted'. Taken in isolation, this statement would seem to credit the idea of Boole as a strongly psychologistic philosopher. But the essay in which it appears makes clear that he uses 'ordinary reasoning' just to oppose it to 'algebraic reasoning'—the former, but not the latter being applicable to our ordinary discourse.

[15] See also Bornet (1997) who agrees with Grattan-Guinness on this point.

[16] See, for example, Frege (1979, 4 and 145).

[17] See Frege (1964, 12).

[18] In order to underpin the prescriptive character of logic, both Boole and Frege assimilated the laws of logic with laws of ethics and distinguished them from the laws of nature. See for example Boole (1952, 195–7) and Frege (1979, 145).

[19] See Carnap (1950, 41).

[20] See Boole (1854, 186).

[21] See Boole (1952, 47).

[22] See Boole (1952, 194).

[23] See Boole (1952, 195).

[24] See Boole (1854, 2).

[25] See Frege (1979, 5). See also Frege (1979, 142): 'In human beings it is natural for thinking to be intermingled with having images and feeling. Logic has the task of isolating what is logical, not, to be sure, so that we should think without having images, which is no doubt impossible, but so that we should consciously distinguish the logical from what is attached to it in the way of ideas and feeling'; and Frege (1979, 269): 'Certainly the logical disposition of man *was* at work in the formation of language but equally alongside this many other dispositions—such as the poetic disposition. And so language is not constructed from a logical blueprint'.

[26] See Frege (1950, x).

[27] My italics.

[28] On Frege and the psychological reality thesis, see also Notturno (1982) according to whom Frege did not reject it.

[29] However, if I am right in maintaining that both Boole and Frege shared the conviction that logic has some mental reality, it is also clear why both thought that logic can study the mind, that is in so far as it is cleansed of the interference to which it is liable on account of its other processes or contents—called 'psychological' by Frege, and identified by Boole (1997, 179) as emotions and will. Logic can study the logical mind.

[30] See Boole (1854, 24–5).

[31] I have maintained in Vassallo (1997b) that the linguistic foundation Boole gives to logic in the *Laws of Thought* is a clear sign of anti-psychologism. This Boolean stance is less limpid in his subsequent writings.
[32] See Boole (1997, 51).
[33] Moreover he states (Boole, 1997, 53–4) that in order to 'construct the Science of Logic we must seek 1st. to distinguish the several operations of thought and determine their laws, consciousness here taking the place of experience. 2ndly. to construct upon the basis thus obtained the *general* methods and processes of the science. 3rdly. To ascertain whether there exist any valid processes of thought (consciousness being again the seat of appeal as to their validity) which are not included under the general methods and processes above referred to'. On consciousness see also Boole (1997, 62 and 179).
[34] See Boole (1854, 409).
[35] See Frege (1979, 5).
[36] See Frege (1979, 142).
[37] My italics. See also Frege (1979, 142, my italics): '... we think in some language or other and [...] grammar, which has a significance for language analogous to that which logic has for judgement, is a mixture of the logical and the psychological. If this were not so, all languages would necessarily have the same grammar. It is true that we can express the same thought in different languages; but the psychological trappings, the clothing of the thought, will often be different. This is why the learning of foreign languages is useful for one's logical education. Seeing that the same thought can be worked in different ways, we *learn better* to distinguish the verbal husk from the kernel with which, in any given language, it appears to be organically bound up. This is how the differences between languages can *facilitate our grasp of what is logical*. But still the difficulties are not wholly removed in this way, and our logic books still keep dragging in a number of things—subject and predicate, for example—which do not, strictly speaking, belong to logic. For this reason it is useful to be acquainted also with a means of expressing thoughts that is of a radically different nature, such as we have in the formula-language of arithmetic or in my concept-script'.
[38] See Frege (1979, 5–6).
[39] There are difficulties for Frege in his appeal to consciousness, not because this might be a trait of psychologism, but because of his characterization of objective and subjective. On these difficulties see, for example, Vassallo (1998a).
[40] In fact I believe that Boole's philosophy of logic is basically a-psychologistic; see Vassallo (1995).
[41] In Vassallo (1998b) I have tried to show that Boole's philosophy has great contemporary value.
[42] See for example Thagard (1982).
[43] See for example McGinn (1993).
[44] One of the main differences Frege envisages between his logic and Boole's was not the psychologistic flavour of the latter, but the fact that his concept-script was designed as the expression of a content, while Boole's logic was not. Frege (1979, 12) writes: 'In contrast we may now set out the aim of my concept-script. Right from the start I had in mind the *expression of a content*. What I am striving after is a *lingua characterica* in the first instance for mathematics, not a *calculus* restricted to pure logic'. It might be objected that two main essays in which Frege wrote about Boole—see Frege (1979, 9–46 and 47–52)—go back to a period when Frege had not yet taken a clear anti-psychologistic stance—see Vassallo (1997a) on this period—and this is why he did not accuse Boole of psychologism. This objection is weak, since after he adopted what most people consider to be that stance in espousing his anti-psychologistic principle, Frege wrote about Boole in numerous letters. Not only did he never accuse Boole of psychologism in his correspondence, but on at least one occasion he distinguished followers of Boole from psychologistic logicians—see his letter to Peano in Frege (1980), undated but probably written in 1891.
[45] See Mohanty (1989, 1), Seebohm (1991) and Vassallo (1995, 22–5) on this point.
[46] On this phenomenon, see Kitcher (1992), Villanueva (1993), French et al. (1994), and Agazzi and Vassallo (1998b).
[47] One position that can be labelled 'psychologistic' and that we did not look at in this essay is the following: *logical contents are psychological contents*. It is well known how much Frege fought against this; see, for example, Frege (1950, 37, n. 1). If I have not considered this form of psychologism here it is simply because the main aim of Boole's logic was not, like Frege's concept-script, as Frege himself (1979, 12) points out, the expression of a content. In any case I do not believe Boole embraced the above form of psychologism. This is evidenced by Boole (1997, 106) writing: 'To what extent can the content of a sign be suspended? For this All imaginations of qualities, impressions etc. physical or moral may be suspended—but the distinction between the signs of different things must be maintained'. So regarding

this form of psychologism, we can say that Boole and Frege agree at least in the sense that both rejected it, pointing out that logical contents are not psychological contents.

REFERENCES

Agazzi, E., and N. Vassallo, eds. 1998a. *George Boole: filosofia, logica, matematica*. Milano: Franco Angeli.

Agazzi, E., and N. Vassallo, eds. 1998b. *Introduzione al naturalismo filosofico contemporaneo*. Milano: Franco Angeli.

Baker, G.P., and P.M.S. Hacker. 1984. *Frege: Logical excavations*. New York: Oxford University Press.

———. 1989. Frege's anti-psychologism. In *Perspectives on psychologism*, edited by M.A. Notturno. Leiden: E.J. Brill.

Barone, F. 1961. 'Algebra della logica' e 'Leggi del pensiero'. *Filosofia* 12: 39–86.

Bocheński, I.M. 1956. *Formale Logik*. Freiburg/München: Karl Alber.

Boole, G. [1854] 1958. *An investigation of the laws of thought, on which are founded the mathematical theories of logic and probabilities*. Cambridge and London: Walton and Maberley. Reprint, New York: Dover.

———. 1952. *Studies in logic and probability*, edited by R. Rhees. London: Watts.

———. 1997. *Selected manuscripts on logic and its philosophy*, edited by I. Grattan-Guinness and G. Bornet. Basel: Birkhäuser.

Bornet, G. 1995. George Boole's linguistic turn and the origins of analytical philosophy. In *The British tradition in 20th century philosophy*, edited by J. Hintikka and K. Puhl. Wien: Holder-Pichler-Tempsky.

———. 1997. Boole's psychologism as a reception problem. In Boole (1997), xlvii–lviii.

Brockhaus, R.R. 1991. Realism and psychologism in 19th century logic. *Philosophy and Phenomenological Research* 51: 493–524.

Callebaut, W., and E. Sober. 1993. The return of psychologism. In *Taking the naturalistic turn or how real philosophy of science is done*, edited by W. Callebaut. Chicago: University of Chicago Press.

Carnap, R. 1950. *Logical foundations of probabilities*. Chicago: University of Chicago Press.

Cobb-Stevens, R. 1990. *Husserl and analytic philosophy*. Dordrecht: Kluwer.

Cohen, M.R., and E. Nagel. [1934] 1972. *Introduction to logic and scientific method*. New York: Harcourt Brace. Reprint, London: Routledge.

Currie, G. 1982. *Frege, an introduction to his philosophy*. Brighton: Harvester Press.

Dalla Chiara, M.L. 1979. *Logica*. Milano: Mondadori.

Di Francesco, M. 1996. *Introduzione alla filosofia della mente*. Roma: Laterza.

Dummett, M. 1973. *Frege. Philosophy of language*. London: Duckworth.

———. 1978. *Truth and other enigmas*. London: Duckworth.

———. 1983. 'Preface to the Italian edition' of Dummett (1973). Casale Monferrato: Marietti.

———. 1993. *Origins of analytical philosophy*. London: Duckworth.

Ferriani, M. 1984. Boole, Frege e la distinzione leibniziana *Lingua-Calculus*. *Annali di discipline filosofiche dell'Università di Bologna* 3: 145–71.

Feys, R. 1955. Boole as a logician. *Proceedings of the Royal Irish Academy* 57: 97–106.

Frege, G. 1950. *The foundations of arithmetic*. Oxford: Blackwell.

———. 1952. *Translations from the philosophical writing of Gottlob Frege*. Edited by P.T. Geach and M. Black. Oxford: Blackwell.

———. 1964. *The basic laws of arithmetic*. Berkeley: University of California Press.

———. 1977. *Logical investigations*. Edited by P.T. Geach. New Haven: Yale University Press.

———. 1979. *Posthumous writings*. Oxford: Blackwell.

———. 1980. *Philosophical and mathematical correspondence*. Oxford: Blackwell.

———. 1989.Briefe an Ludwig Wittgenstein. *Grazer Philosophische Studien* 33/34: 8–26.

French, P.A., T.E. Uehling, and H.K. Wettstein, eds. 1994. *Philosophical naturalism*. Midwest Studies in Philosophy no 19.

Goldfarb, W. 1979. Logic in the twenties: the nature of the quantifier. *Journal of Symbolic Logic* 44: 351–68.

Grattan-Guinness, I. 1982. Psychology in the foundation of logic and mathematics: the cases of Boole, Cantor and Brouwer. *History and Philosophy of Logic* 3: 33–53.

———. 1988. Living together and living apart. On the interactions between mathematics and logics from the French Revolution to the First World War. *South African Journal of Philosophy* 7: 73–82.
———. 1997. Boole's quest for the foundations of his logic. In Boole (1997), xiii–xlvii.
Haack, S. 1978. *Philosophy of logics*. New York: Cambridge University Press.
Hesse, M.B. 1952. Boole's philosophy of logic. *Annals of Science* 8: 61–81.
Husserl, E. 1900, 1913. *Logische Untersuchungen*. Vol. 1, *Prolegomena zur reinen Logik*. Tübingen: M. Niemeyer.
Johnson, W.E. 1921. *Logic, part 1*. Cambridge: Cambridge University Press.
Jørgensen, J. [1931] 1962. *A treatise of formal logic*. Oxford. Reprint, New York: Russell and Russell.
Jourdain, P.E.B. 1910. The development of the theories of mathematical logic and the principles of mathematics. *Quarterly Journal of Pure and Applied Mathematics* 41: 324–52.
Kitcher, P. 1979. Frege's epistemology. *Philosophical Review* 86: 235–62.
———. 1992. The naturalists return. *Philosophical Review* 101: 53–114.
Kneale, W., and M. Kneale. 1962. *The development of logic*. Oxford: Clarendon Press.
Laita, L.M. 1980. Boolean algebra and its extra-logical sources: the testimony of Mary Everest Boole. *History and Philosophy of Logic* 1: 37–60.
MacHale, D. 1985. *George Boole, his life and work*. Dublin: Boole Press.
Mangione, C. 1972. La svolta della logica nell'Ottocento. In *Storia del pensiero filosofico e scientifico*, edited by L. Geymonat. Milano: Garzanti.
Mangione, C., and S. Bozzi. 1993. *Storia della logica, da Boole ai nostri giorni*. Milano: Garzanti.
McGinn, C. 1993. Logic, mind and mathematics. *Philosophical Issues* 4: 101–27.
Mohanty, J.N. 1989. Psychologism. In *Perspectives on psychologism*, edited by M.A. Notturno. Leiden: E.J. Brill.
Mugnai, M. 1993. 'Introduction' to the Italian edition of G. Boole, *The mathematical analysis of logic* [in Boole (1952)]. Torino: Bollati e Boringhieri.
Mugnai, M. 1998. Boole e lo psicologismo: la caratterizzazione delle leggi logiche in *An investigation of the laws of thought*. In Agazzi and Vassallo (1998a), 111–130.
Musgrave, A. 1972. George Boole and psychologism. *Scientia* 107: 593–608.
Notturno, M.A. 1982. Frege and the psychological reality thesis. *Journal for the Theory of Social Behaviour* 12: 329–44.
Picardi, E. 1996. Frege's anti-psychologism. In *Frege: importance and legacy*, edited by M. Schirn. Berlin: Walter de Gruyter.
Putnam, H. 1990. Peirce the logician. In Putnam, H., *Realism with a human face*. Cambridge, Mass: Harvard University Press.
Radnitzky, G. 1976. Popperian philosophy of science as an antidote against relativism. In *Essays in memory of Imre Lakatos*, edited by R.S. Cohen et al. Dordrecht: Kluwer.
Russell, B. 1901. Recent work on the principles of mathematics. *International Monthly* 4: 83–101.
Ryle, G. 1956. *The revolution in philosophy*. London: Macmillan.
Seebohm, T.M. 1991. Psychologism revisited. In *Phenomenology and the formal sciences*, edited by T.M. Seebohm, D. Føllesdal, and J.N. Mohanty. Dordrecht: Kluwer.
Sober, E. 1978. Psychologism. *Journal for the Theory of Social Behaviour* 8: 165–91.
Thagard, P. 1982. From the descriptive to the normative in psychology and logic. *Philosophy of Science* 49: 24–42.
Thomas, I. 1955. Boole's concept of science. *Proceedings of the Royal Irish Academy* 57: 88–96.
Vassallo, N. 1995. *La depsicologizzazione della logica. Un confronto tra Boole e Frege*. Milano: Franco Angeli.
———. 1997a. On Dummett's early Frege and analytical philosophy. *Dialectica* 51: 171–87.
———. 1997b. Analysis versus laws: Boole's explanatory psychologism versus his explanatory anti-psychologism. *History and Philosophy of Logic* 18: 151–63.
———. 1998. Objective and subjective: some difficulties for Frege. In *La réalité et le signe*, edited by M. Negro. Fribourg: Editions Universitaires de Fribourg.
Villanueva, E., ed. 1993. Naturalism and normativity. *Philosophical Issues* 4.
Weiner, J. 1990. *Frege in perspective*. Ithaca: Cornell University Press.

University of Genoa, Dept. of Philosophy, via Balbi 4, I-16126 Genova, Italy

CONTRIBUTORS

Gérard Bornet read philosophy at the University of Bern, where he defended a dissertation on Bertrand Russell. He is currently Lecturer in Logic and Epistemology at the University of Fribourg (Switzerland). From 1991 to 1995, he carried out a research project supported by the Swiss National Science Foundation on the foundations of Boole's logic. He is co-editor (with Ivor Grattan-Guinness) of *George Boole, Selected Manuscripts on Logic and Its Philosophy* (Birkhäuser, 1997). He is currently mainly interested in semantics.

Marie-José Durand-Richard is Lecturer in the Mathematics Department of the University of Paris VIII (Vincennes-Saint Denis). Since defending a thesis in epistemology entitled 'George Peacock (1791–1858): La Synthèse Algébrique comme Loi Symbolique dans l'Angleterre des Réformes' at the Ecole des Hautes Etudes en Sciences Sociales, her research has concentrated on the historical contribution of the Symbolical Algebraic Network in England at that period. Specifically, she is interested in its numerous consequences in several different fields including algebra, logic, and symbolical calculus (operational calculus). She is currently preparing two collective works, one on the epistemological status of analogy in the scientific process, and the other on modelisations of the act of knowledge from Boole's Laws of Thought to constructivism.

Béatrice Godart-Wendling is Full Researcher of Linguistics at the CNRS (French National Center of Scientific Research) and teaches logic, history of language and epistemology at the University of Paris VII. Her main research interests concern the relations between linguistics and logic. In the field of semantics, she is the author of *La Vérité et le Menteur* (éditions du CNRS, Paris, 1990); her interest in syntax has resulted in the publication of several papers on the history of formal grammars.

Ivor Grattan-Guinness is Professor of the History of Mathematics and Logic at Middlesex University, England. From 1986 to 1988 he was the President of the British Society for the History of Mathematics. Editor of the history of science journal *Annals of Science* from 1974 to 1981, in 1979 he founded the journal *History and Philosophy of Logic*, editing it until 1992. He is an effective member of the *Académie Internationale d'Histoire des Sciences*. He edited a substantial *Companion Encyclopedia of the History and Philosophy of the Mathematical Sciences* (two volumes, 1994, London, Routledge). His books include *Convolutions in French Mathematics, 1800–1840*, three volumes (1990, Basel and Berlin, Birkhäuser); and *The Fontana* and *Norton History of the Mathematical Sciences. The Rainbow of Mathematics* (1997, London, Fontana; 1998, New York, Norton). To appear with Princeton University Press is *The Search for Mathematical Roots, 1870–1940. Logics, Set Theories and the Foundations of Mathematics from Cantor through Russell to Gödel*. He is the Associate Editor for mathematicians and statisticians for the *New Dictionary of National Biography*, to be published in 2004.

Theodore Hailperin is Emeritus Professor of Mathematics at Lehigh University. He is the author of *Boole's Logic and Probability* (North-Holland, Studies in Logic Vol. 85, 1986) and *Sentential Probability Logic* (Lehigh University Press and Associated University Presses, 1996). His current interests are in probabilistic inference and the philosophical foundations of logic.

Sriram Nambiar is studying philosophy at the State University of New York at Buffalo. He is finishing a dissertation on Boole's philosophy of logic under the supervision of Professor John Corcoran.

Maria Panteki is Lecturer of the History of Mathematics in the Mathematics Department of Aristotle University in Thessaloniki, Greece. Under the guidance of Professor I. Grattan-Guinness, she defended in 1992 her Ph.D. thesis on *Relationships between Algebra, Logic and Differential Equations in England (1800–1860).* Since then, she has been developing several issues raised in her thesis, especially that of the mathematical background of Boole's and De Morgan's work on logic. Other research interests include French enlightenment philosophy of education and celestial mechanics.

Volker Peckhaus is *Privatdozent* and *Oberassistent* at the Philosophy Department of the University of Erlangen-Nuremberg. He is author of *Hilbertprogramm und Kritische Philosophie* (1990); *Hermann Ulrici (1806–1884)* (1995); and *Logik, Mathesis universalis und allgemeine Wissenschaft* (1997). His research interests include the history and philosophy of logic and mathematics, philosophy of science, and methodological questions in humanities. He is a member of the editorial boards of *Modern Logic*, *Historia Mathematica*, and *Philosophia Scientiae*.

Shahid Rahman studied philosophy, mathematics, psychology and Spanish philology in Bahía Blanca (Argentina), Erlangen and Saarbrücken (Germany). Presently he is Visiting Professor at the University of Nancy 2 (France) and Privatdozent at the University of Saarbrücken. His main fields are philosophy and history of logic including free and paraconsistent logics, philosophy of science and language, philosophy of psychology and philosophy of literature. His books include *Über Dialoge, protologische Kategorien und andere Seltenheiten* (1993), *Philosophie pragmatique et les logiques non classiques* (1999), *Hugh MacColl on Connexive Logics and Symbolic Existence* (2000) and *New Perspectives in Dialogical Logic* (2000; with H. Rückert). He is also co-editor of *Philosophia Scientiae*.

Nicla Vassallo is a researcher in the Department of Philosophy at Genoa University. She is the author of *La Despicologizzazione della logica* (Franco Angeli, Milan, 1995), *La naturalizzazione dell'epistemologia* (Franco Angeli, Milan, 1997) and *Teorie della conoscenza filosofico-naturalistiche* (Franco Angeli, Milan, 1999). She has edited, in collaboration with Evandro Agazzi, *Introduzione al naturalismo filosofico contemporaneo* (Franco Angeli, Milan, 1998) and *George Boole* (Franco Angeli, Milan, 1998). Her main research interests lie in issues of epistemology, metaphysics and the philosophy of logic.

INDEX OF NAMES

INDEX OF SUBJECTS

SYNTHESE LIBRARY

1. J. M. Bochénski, *A Precis of Mathematical Logic*. Translated from French and German by O. Bird. 1959 ISBN 90-277-0073-7
2. P. Guiraud, *Problèmes et méthodes de la statistique linguistique*. 1959 ISBN 90-277-0025-7
3. H. Freudenthal (ed.), *The Concept and the Role of the Model in Mathematics and Natural and Social Sciences*. 1961 ISBN 90-277-0017-6
4. E. W. Beth, *Formal Methods*. An Introduction to Symbolic Logic and to the Study of Effective Operations in Arithmetic and Logic. 1962 ISBN 90-277-0069-9
5. B. H. Kazemier and D. Vuysje (eds.), *Logic and Language*. Studies dedicated to Professor Rudolf Carnap on the Occasion of His 70th Birthday. 1962 ISBN 90-277-0019-2
6. M. W. Wartofsky (ed.), *Proceedings of the Boston Colloquium for the Philosophy of Science, 1961–1962*. [Boston Studies in the Philosophy of Science, Vol. I] 1963 ISBN 90-277-0021-4
7. A. A. Zinov'ev, *Philosophical Problems of Many-valued Logic*. A revised edition, edited and translated (from Russian) by G. Küng and D.D. Comey. 1963 ISBN 90-277-0091-5
8. G. Gurvitch, *The Spectrum of Social Time*. Translated from French and edited by M. Korenbaum and P. Bosserman. 1964 ISBN 90-277-0006-0
9. P. Lorenzen, *Formal Logic*. Translated from German by F.J. Crosson. 1965 ISBN 90-277-0080-X
10. R. S. Cohen and M. W. Wartofsky (eds.), *Proceedings of the Boston Colloquium for the Philosophy of Science, 1962–1964*. In Honor of Philipp Frank. [Boston Studies in the Philosophy of Science, Vol. II] 1965 ISBN 90-277-9004-0
11. E. W. Beth, *Mathematical Thought*. An Introduction to the Philosophy of Mathematics. 1965 ISBN 90-277-0070-2
12. E. W. Beth and J. Piaget, *Mathematical Epistemology and Psychology*. Translated from French by W. Mays. 1966 ISBN 90-277-0071-0
13. G. Küng, *Ontology and the Logistic Analysis of Language*. An Enquiry into the Contemporary Views on Universals. Revised ed., translated from German. 1967 ISBN 90-277-0028-1
14. R. S. Cohen and M. W. Wartofsky (eds.), *Proceedings of the Boston Colloquium for the Philosophy of Sciences, 1964–1966*. In Memory of Norwood Russell Hanson. [Boston Studies in the Philosophy of Science, Vol. III] 1967 ISBN 90-277-0013-3
15. C. D. Broad, *Induction, Probability, and Causation*. Selected Papers. 1968 ISBN 90-277-0012-5
16. G. Patzig, *Aristotle's Theory of the Syllogism*. A Logical-philosophical Study of *Book A* of the *Prior Analytics*. Translated from German by J. Barnes. 1968 ISBN 90-277-0030-3
17. N. Rescher, *Topics in Philosophical Logic*. 1968 ISBN 90-277-0084-2
18. R. S. Cohen and M. W. Wartofsky (eds.), *Proceedings of the Boston Colloquium for the Philosophy of Science, 1966–1968, Part I*. [Boston Studies in the Philosophy of Science, Vol. IV] 1969 ISBN 90-277-0014-1
19. R. S. Cohen and M. W. Wartofsky (eds.), *Proceedings of the Boston Colloquium for the Philosophy of Science, 1966–1968, Part II*. [Boston Studies in the Philosophy of Science, Vol. V] 1969 ISBN 90-277-0015-X
20. J. W. Davis, D. J. Hockney and W. K. Wilson (eds.), *Philosophical Logic*. 1969 ISBN 90-277-0075-3
21. D. Davidson and J. Hintikka (eds.), *Words and Objections*. Essays on the Work of W. V. Quine. 1969, rev. ed. 1975 ISBN 90-277-0074-5; Pb 90-277-0602-6
22. P. Suppes, *Studies in the Methodology and Foundations of Science. Selected Papers from 1951 to 1969*. 1969 ISBN 90-277-0020-6
23. J. Hintikka, *Models for Modalities*. Selected Essays. 1969 ISBN 90-277-0078-8; Pb 90-277-0598-4

24. N. Rescher *et al.* (eds.), *Essays in Honor of Carl G. Hempel.* A Tribute on the Occasion of His 65th Birthday. 1969 ISBN 90-277-0085-0
25. P. V. Tavanec (ed.), *Problems of the Logic of Scientific Knowledge.* Translated from Russian. 1970 ISBN 90-277-0087-7
26. M. Swain (ed.), *Induction, Acceptance, and Rational Belief.* 1970 ISBN 90-277-0086-9
27. R. S. Cohen and R. J. Seeger (eds.), *Ernst Mach: Physicist and Philosopher.* [Boston Studies in the Philosophy of Science, Vol. VI]. 1970 ISBN 90-277-0016-8
28. J. Hintikka and P. Suppes, *Information and Inference.* 1970 ISBN 90-277-0155-5
29. K. Lambert, *Philosophical Problems in Logic.* Some Recent Developments. 1970 ISBN 90-277-0079-6
30. R. A. Eberle, *Nominalistic Systems.* 1970 ISBN 90-277-0161-X
31. P. Weingartner and G. Zecha (eds.), *Induction, Physics, and Ethics.* 1970 ISBN 90-277-0158-X
32. E. W. Beth, *Aspects of Modern Logic.* Translated from Dutch. 1970 ISBN 90-277-0173-3
33. R. Hilpinen (ed.), *Deontic Logic.* Introductory and Systematic Readings. 1971 *See also* No. 152. ISBN Pb (1981 rev.) 90-277-1302-2
34. J.-L. Krivine, *Introduction to Axiomatic Set Theory.* Translated from French. 1971 ISBN 90-277-0169-5; Pb 90-277-0411-2
35. J. D. Sneed, *The Logical Structure of Mathematical Physics.* 2nd rev. ed., 1979 ISBN 90-277-1056-2; Pb 90-277-1059-7
36. C. R. Kordig, *The Justification of Scientific Change.* 1971 ISBN 90-277-0181-4; Pb 90-277-0475-9
37. M. Čapek, *Bergson and Modern Physics.* A Reinterpretation and Re-evaluation. [Boston Studies in the Philosophy of Science, Vol. VII] 1971 ISBN 90-277-0186-5
38. N. R. Hanson, *What I Do Not Believe, and Other Essays.* Ed. by S. Toulmin and H. Woolf. 1971 ISBN 90-277-0191-1
39. R. C. Buck and R. S. Cohen (eds.), *PSA 1970.* Proceedings of the Second Biennial Meeting of the Philosophy of Science Association, Boston, Fall 1970. In Memory of Rudolf Carnap. [Boston Studies in the Philosophy of Science, Vol. VIII] 1971 ISBN 90-277-0187-3; Pb 90-277-0309-4
40. D. Davidson and G. Harman (eds.), *Semantics of Natural Language.* 1972 ISBN 90-277-0304-3; Pb 90-277-0310-8
41. Y. Bar-Hillel (ed.), *Pragmatics of Natural Languages.* 1971 ISBN 90-277-0194-6; Pb 90-277-0599-2
42. S. Stenlund, *Combinators, γ Terms and Proof Theory.* 1972 ISBN 90-277-0305-1
43. M. Strauss, *Modern Physics and Its Philosophy.* Selected Paper in the Logic, History, and Philosophy of Science. 1972 ISBN 90-277-0230-6
44. M. Bunge, *Method, Model and Matter.* 1973 ISBN 90-277-0252-7
45. M. Bunge, *Philosophy of Physics.* 1973 ISBN 90-277-0253-5
46. A. A. Zinov'ev, *Foundations of the Logical Theory of Scientific Knowledge (Complex Logic).* Revised and enlarged English edition with an appendix by G. A. Smirnov, E. A. Sidorenka, A. M. Fedina and L. A. Bobrova. [Boston Studies in the Philosophy of Science, Vol. IX] 1973 ISBN 90-277-0193-8; Pb 90-277-0324-8
47. L. Tondl, *Scientific Procedures.* A Contribution concerning the Methodological Problems of Scientific Concepts and Scientific Explanation. Translated from Czech by D. Short. Edited by R.S. Cohen and M.W. Wartofsky. [Boston Studies in the Philosophy of Science, Vol. X] 1973 ISBN 90-277-0147-4; Pb 90-277-0323-X
48. N. R. Hanson, *Constellations and Conjectures.* 1973 ISBN 90-277-0192-X

49. K. J. J. Hintikka, J. M. E. Moravcsik and P. Suppes (eds.), *Approaches to Natural Language.* 1973 ISBN 90-277-0220-9; Pb 90-277-0233-0
50. M. Bunge (ed.), *Exact Philosophy.* Problems, Tools and Goals. 1973 ISBN 90-277-0251-9
51. R. J. Bogdan and I. Niiniluoto (eds.), *Logic, Language and Probability.* 1973 ISBN 90-277-0312-4
52. G. Pearce and P. Maynard (eds.), *Conceptual Change.* 1973 ISBN 90-277-0287-X; Pb 90-277-0339-6
53. I. Niiniluoto and R. Tuomela, *Theoretical Concepts and Hypothetico-inductive Inference.* 1973 ISBN 90-277-0343-4
54. R. Fraissé, *Course of Mathematical Logic* – Volume 1: *Relation and Logical Formula.* Translated from French. 1973 ISBN 90-277-0268-3; Pb 90-277-0403-1
(For *Volume 2* see under No. 69).
55. A. Grünbaum, *Philosophical Problems of Space and Time.* Edited by R.S. Cohen and M.W. Wartofsky. 2nd enlarged ed. [Boston Studies in the Philosophy of Science, Vol. XII] 1973 ISBN 90-277-0357-4; Pb 90-277-0358-2
56. P. Suppes (ed.), *Space, Time and Geometry.* 1973 ISBN 90-277-0386-8; Pb 90-277-0442-2
57. H. Kelsen, *Essays in Legal and Moral Philosophy.* Selected and introduced by O. Weinberger. Translated from German by P. Heath. 1973 ISBN 90-277-0388-4
58. R. J. Seeger and R. S. Cohen (eds.), *Philosophical Foundations of Science.* [Boston Studies in the Philosophy of Science, Vol. XI] 1974 ISBN 90-277-0390-6; Pb 90-277-0376-0
59. R. S. Cohen and M. W. Wartofsky (eds.), *Logical and Epistemological Studies in Contemporary Physics.* [Boston Studies in the Philosophy of Science, Vol. XIII] 1973 ISBN 90-277-0391-4; Pb 90-277-0377-9
60. R. S. Cohen and M. W. Wartofsky (eds.), *Methodological and Historical Essays in the Natural and Social Sciences. Proceedings of the Boston Colloquium for the Philosophy of Science, 1969–1972.* [Boston Studies in the Philosophy of Science, Vol. XIV] 1974 ISBN 90-277-0392-2; Pb 90-277-0378-7
61. R. S. Cohen, J. J. Stachel and M. W. Wartofsky (eds.), *For Dirk Struik. Scientific, Historical and Political Essays.* [Boston Studies in the Philosophy of Science, Vol. XV] 1974 ISBN 90-277-0393-0; Pb 90-277-0379-5
62. K. Ajdukiewicz, *Pragmatic Logic.* Translated from Polish by O. Wojtasiewicz. 1974 ISBN 90-277-0326-4
63. S. Stenlund (ed.), *Logical Theory and Semantic Analysis.* Essays dedicated to Stig Kanger on His 50th Birthday. 1974 ISBN 90-277-0438-4
64. K. F. Schaffner and R. S. Cohen (eds.), *PSA 1972. Proceedings of the Third Biennial Meeting of the Philosophy of Science Association.* [Boston Studies in the Philosophy of Science, Vol. XX] 1974 ISBN 90-277-0408-2; Pb 90-277-0409-0
65. H. E. Kyburg, Jr., *The Logical Foundations of Statistical Inference.* 1974 ISBN 90-277-0330-2; Pb 90-277-0430-9
66. M. Grene, *The Understanding of Nature.* Essays in the Philosophy of Biology. [Boston Studies in the Philosophy of Science, Vol. XXIII] 1974 ISBN 90-277-0462-7; Pb 90-277-0463-5
67. J. M. Broekman, *Structuralism: Moscow, Prague, Paris.* Translated from German. 1974 ISBN 90-277-0478-3
68. N. Geschwind, *Selected Papers on Language and the Brain.* [Boston Studies in the Philosophy of Science, Vol. XVI] 1974 ISBN 90-277-0262-4; Pb 90-277-0263-2
69. R. Fraissé, *Course of Mathematical Logic* – Volume 2: *Model Theory.* Translated from French. 1974 ISBN 90-277-0269-1; Pb 90-277-0510-0
(For *Volume 1* see under No. 54)

70. A. Grzegorczyk, *An Outline of Mathematical Logic*. Fundamental Results and Notions explained with all Details. Translated from Polish. 1974 ISBN 90-277-0359-0; Pb 90-277-0447-3
71. F. von Kutschera, *Philosophy of Language*. 1975 ISBN 90-277-0591-7
72. J. Manninen and R. Tuomela (eds.), *Essays on Explanation and Understanding*. Studies in the Foundations of Humanities and Social Sciences. 1976 ISBN 90-277-0592-5
73. J. Hintikka (ed.), *Rudolf Carnap, Logical Empiricist*. Materials and Perspectives. 1975 ISBN 90-277-0583-6
74. M. Čapek (ed.), *The Concepts of Space and Time*. Their Structure and Their Development. [Boston Studies in the Philosophy of Science, Vol. XXII] 1976 ISBN 90-277-0355-8; Pb 90-277-0375-2
75. J. Hintikka and U. Remes, *The Method of Analysis*. Its Geometrical Origin and Its General Significance. [Boston Studies in the Philosophy of Science, Vol. XXV] 1974 ISBN 90-277-0532-1; Pb 90-277-0543-7
76. J. E. Murdoch and E. D. Sylla (eds.), *The Cultural Context of Medieval Learning*. [Boston Studies in the Philosophy of Science, Vol. XXVI] 1975 ISBN 90-277-0560-7; Pb 90-277-0587-9
77. S. Amsterdamski, *Between Experience and Metaphysics*. Philosophical Problems of the Evolution of Science. [Boston Studies in the Philosophy of Science, Vol. XXXV] 1975 ISBN 90-277-0568-2; Pb 90-277-0580-1
78. P. Suppes (ed.), *Logic and Probability in Quantum Mechanics*. 1976 ISBN 90-277-0570-4; Pb 90-277-1200-X
79. H. von Helmholtz: *Epistemological Writings. The Paul Hertz / Moritz Schlick Centenary Edition of 1921 with Notes and Commentary by the Editors*. Newly translated from German by M. F. Lowe. Edited, with an Introduction and Bibliography, by R. S. Cohen and Y. Elkana. [Boston Studies in the Philosophy of Science, Vol. XXXVII] 1975 ISBN 90-277-0290-X; Pb 90-277-0582-8
80. J. Agassi, *Science in Flux*. [Boston Studies in the Philosophy of Science, Vol. XXVIII] 1975 ISBN 90-277-0584-4; Pb 90-277-0612-2
81. S. G. Harding (ed.), *Can Theories Be Refuted?* Essays on the Duhem-Quine Thesis. 1976 ISBN 90-277-0629-8; Pb 90-277-0630-1
82. S. Nowak, *Methodology of Sociological Research*. General Problems. 1977 ISBN 90-277-0486-4
83. J. Piaget, J.-B. Grize, A. Szeminśska and V. Bang, *Epistemology and Psychology of Functions*. Translated from French. 1977 ISBN 90-277-0804-5
84. M. Grene and E. Mendelsohn (eds.), *Topics in the Philosophy of Biology*. [Boston Studies in the Philosophy of Science, Vol. XXVII] 1976 ISBN 90-277-0595-X; Pb 90-277-0596-8
85. E. Fischbein, *The Intuitive Sources of Probabilistic Thinking in Children*. 1975 ISBN 90-277-0626-3; Pb 90-277-1190-9
86. E. W. Adams, *The Logic of Conditionals*. An Application of Probability to Deductive Logic. 1975 ISBN 90-277-0631-X
87. M. Przełęcki and R. Wójcicki (eds.), *Twenty-Five Years of Logical Methodology in Poland*. Translated from Polish. 1976 ISBN 90-277-0601-8
88. J. Topolski, *The Methodology of History*. Translated from Polish by O. Wojtasiewicz. 1976 ISBN 90-277-0550-X
89. A. Kasher (ed.), *Language in Focus: Foundations, Methods and Systems*. Essays dedicated to Yehoshua Bar-Hillel. [Boston Studies in the Philosophy of Science, Vol. XLIII] 1976 ISBN 90-277-0644-1; Pb 90-277-0645-X

90. J. Hintikka, *The Intentions of Intentionality and Other New Models for Modalities.* 1975 ISBN 90-277-0633-6; Pb 90-277-0634-4
91. W. Stegmüller, *Collected Papers on Epistemology, Philosophy of Science and History of Philosophy.* 2 Volumes. 1977 Set ISBN 90-277-0767-7
92. D. M. Gabbay, *Investigations in Modal and Tense Logics with Applications to Problems in Philosophy and Linguistics.* 1976 ISBN 90-277-0656-5
93. R. J. Bogdan, *Local Induction.* 1976 ISBN 90-277-0649-2
94. S. Nowak, *Understanding and Prediction.* Essays in the Methodology of Social and Behavioral Theories. 1976 ISBN 90-277-0558-5; Pb 90-277-1199-2
95. P. Mittelstaedt, *Philosophical Problems of Modern Physics.* [Boston Studies in the Philosophy of Science, Vol. XVIII] 1976 ISBN 90-277-0285-3; Pb 90-277-0506-2
96. G. Holton and W. A. Blanpied (eds.), *Science and Its Public: The Changing Relationship.* [Boston Studies in the Philosophy of Science, Vol. XXXIII] 1976 ISBN 90-277-0657-3; Pb 90-277-0658-1
97. M. Brand and D. Walton (eds.), *Action Theory.* 1976 ISBN 90-277-0671-9
98. P. Gochet, *Outline of a Nominalist Theory of Propositions.* An Essay in the Theory of Meaning and in the Philosophy of Logic. 1980 ISBN 90-277-1031-7
99. R. S. Cohen, P. K. Feyerabend, and M. W. Wartofsky (eds.), *Essays in Memory of Imre Lakatos.* [Boston Studies in the Philosophy of Science, Vol. XXXIX] 1976 ISBN 90-277-0654-9; Pb 90-277-0655-7
100. R. S. Cohen and J. J. Stachel (eds.), *Selected Papers of Léon Rosenfield.* [Boston Studies in the Philosophy of Science, Vol. XXI] 1979 ISBN 90-277-0651-4; Pb 90-277-0652-2
101. R. S. Cohen, C. A. Hooker, A. C. Michalos and J. W. van Evra (eds.), *PSA 1974. Proceedings of the 1974 Biennial Meeting of the Philosophy of Science Association.* [Boston Studies in the Philosophy of Science, Vol. XXXII] 1976 ISBN 90-277-0647-6; Pb 90-277-0648-4
102. Y. Fried and J. Agassi, *Paranoia.* A Study in Diagnosis. [Boston Studies in the Philosophy of Science, Vol. L] 1976 ISBN 90-277-0704-9; Pb 90-277-0705-7
103. M. Przełęcki, K. Szaniawski and R. Wójcicki (eds.), *Formal Methods in the Methodology of Empirical Sciences.* 1976 ISBN 90-277-0698-0
104. J. M. Vickers, *Belief and Probability.* 1976 ISBN 90-277-0744-8
105. K. H. Wolff, *Surrender and Catch.* Experience and Inquiry Today. [Boston Studies in the Philosophy of Science, Vol. LI] 1976 ISBN 90-277-0758-8; Pb 90-277-0765-0
106. K. Kosík, *Dialectics of the Concrete.* A Study on Problems of Man and World. [Boston Studies in the Philosophy of Science, Vol. LII] 1976 ISBN 90-277-0761-8; Pb 90-277-0764-2
107. N. Goodman, *The Structure of Appearance.* 3rd ed. with an Introduction by G. Hellman. [Boston Studies in the Philosophy of Science, Vol. LIII] 1977 ISBN 90-277-0773-1; Pb 90-277-0774-X
108. K. Ajdukiewicz, *The Scientific World-Perspective and Other Essays, 1931-1963.* Translated from Polish. Edited and with an Introduction by J. Giedymin. 1978 ISBN 90-277-0527-5
109. R. L. Causey, *Unity of Science.* 1977 ISBN 90-277-0779-0
110. R. E. Grandy, *Advanced Logic for Applications.* 1977 ISBN 90-277-0781-2
111. R. P. McArthur, *Tense Logic.* 1976 ISBN 90-277-0697-2
112. L. Lindahl, *Position and Change.* A Study in Law and Logic. Translated from Swedish by P. Needham. 1977 ISBN 90-277-0787-1
113. R. Tuomela, *Dispositions.* 1978 ISBN 90-277-0810-X
114. H. A. Simon, *Models of Discovery and Other Topics in the Methods of Science.* [Boston Studies in the Philosophy of Science, Vol. LIV] 1977 ISBN 90-277-0812-6; Pb 90-277-0858-4

115. R. D. Rosenkrantz, *Inference, Method and Decision.* Towards a Bayesian Philosophy of Science. 1977 ISBN 90-277-0817-7; Pb 90-277-0818-5
116. R. Tuomela, *Human Action and Its Explanation.* A Study on the Philosophical Foundations of Psychology. 1977 ISBN 90-277-0824-X
117. M. Lazerowitz, *The Language of Philosophy.* Freud and Wittgenstein. [Boston Studies in the Philosophy of Science, Vol. LV] 1977 ISBN 90-277-0826-6; Pb 90-277-0862-2
118. Not published 119. J. Pelc (ed.), *Semiotics in Poland, 1894–1969.* Translated from Polish. 1979 ISBN 90-277-0811-8
120. I. Pörn, *Action Theory and Social Science.* Some Formal Models. 1977 ISBN 90-277-0846-0
121. J. Margolis, *Persons and Mind.* The Prospects of Nonreductive Materialism. [Boston Studies in the Philosophy of Science, Vol. LVII] 1977 ISBN 90-277-0854-1; Pb 90-277-0863-0
122. J. Hintikka, I. Niiniluoto, and E. Saarinen (eds.), *Essays on Mathematical and Philosophical Logic.* 1979 ISBN 90-277-0879-7
123. T. A. F. Kuipers, *Studies in Inductive Probability and Rational Expectation.* 1978 ISBN 90-277-0882-7
124. E. Saarinen, R. Hilpinen, I. Niiniluoto and M. P. Hintikka (eds.), *Essays in Honour of Jaakko Hintikka on the Occasion of His 50th Birthday.* 1979 ISBN 90-277-0916-5
125. G. Radnitzky and G. Andersson (eds.), *Progress and Rationality in Science.* [Boston Studies in the Philosophy of Science, Vol. LVIII] 1978 ISBN 90-277-0921-1; Pb 90-277-0922-X
126. P. Mittelstaedt, *Quantum Logic.* 1978 ISBN 90-277-0925-4
127. K. A. Bowen, *Model Theory for Modal Logic.* Kripke Models for Modal Predicate Calculi. 1979 ISBN 90-277-0929-7
128. H. A. Bursen, *Dismantling the Memory Machine.* A Philosophical Investigation of Machine Theories of Memory. 1978 ISBN 90-277-0933-5
129. M. W. Wartofsky, *Models.* Representation and the Scientific Understanding. [Boston Studies in the Philosophy of Science, Vol. XLVIII] 1979 ISBN 90-277-0736-7; Pb 90-277-0947-5
130. D. Ihde, *Technics and Praxis.* A Philosophy of Technology. [Boston Studies in the Philosophy of Science, Vol. XXIV] 1979 ISBN 90-277-0953-X; Pb 90-277-0954-8
131. J. J. Wiatr (ed.), *Polish Essays in the Methodology of the Social Sciences.* [Boston Studies in the Philosophy of Science, Vol. XXIX] 1979 ISBN 90-277-0723-5; Pb 90-277-0956-4
132. W. C. Salmon (ed.), *Hans Reichenbach: Logical Empiricist.* 1979 ISBN 90-277-0958-0
133. P. Bieri, R.-P. Horstmann and L. Krüger (eds.), *Transcendental Arguments in Science.* Essays in Epistemology. 1979 ISBN 90-277-0963-7; Pb 90-277-0964-5
134. M. Marković and G. Petrović (eds.), *Praxis.* Yugoslav Essays in the Philosophy and Methodology of the Social Sciences. [Boston Studies in the Philosophy of Science, Vol. XXXVI] 1979 ISBN 90-277-0727-8; Pb 90-277-0968-8
135. R. Wójcicki, *Topics in the Formal Methodology of Empirical Sciences.* Translated from Polish. 1979 ISBN 90-277-1004-X
136. G. Radnitzky and G. Andersson (eds.), *The Structure and Development of Science.* [Boston Studies in the Philosophy of Science, Vol. LIX] 1979 ISBN 90-277-0994-7; Pb 90-277-0995-5
137. J. C. Webb, *Mechanism, Mentalism and Metamathematics.* An Essay on Finitism. 1980 ISBN 90-277-1046-5
138. D. F. Gustafson and B. L. Tapscott (eds.), *Body, Mind and Method.* Essays in Honor of Virgil C. Aldrich. 1979 ISBN 90-277-1013-9
139. L. Nowak, *The Structure of Idealization.* Towards a Systematic Interpretation of the Marxian Idea of Science. 1980 ISBN 90-277-1014-7

140. C. Perelman, *The New Rhetoric and the Humanities.* Essays on Rhetoric and Its Applications. Translated from French and German. With an Introduction by H. Zyskind. 1979
ISBN 90-277-1018-X; Pb 90-277-1019-8
141. W. Rabinowicz, *Universalizability.* A Study in Morals and Metaphysics. 1979
ISBN 90-277-1020-2
142. C. Perelman, *Justice, Law and Argument.* Essays on Moral and Legal Reasoning. Translated from French and German. With an Introduction by H.J. Berman. 1980
ISBN 90-277-1089-9; Pb 90-277-1090-2
143. S. Kanger and S. Öhman (eds.), *Philosophy and Grammar.* Papers on the Occasion of the Quincentennial of Uppsala University. 1981 ISBN 90-277-1091-0
144. T. Pawlowski, *Concept Formation in the Humanities and the Social Sciences.* 1980
ISBN 90-277-1096-1
145. J. Hintikka, D. Gruender and E. Agazzi (eds.), *Theory Change, Ancient Axiomatics and Galileo's Methodology.* Proceedings of the 1978 Pisa Conference on the History and Philosophy of Science, Volume I. 1981 ISBN 90-277-1126-7
146. J. Hintikka, D. Gruender and E. Agazzi (eds.), *Probabilistic Thinking, Thermodynamics, and the Interaction of the History and Philosophy of Science.* Proceedings of the 1978 Pisa Conference on the History and Philosophy of Science, Volume II. 1981 ISBN 90-277-1127-5
147. U. Mönnich (ed.), *Aspects of Philosophical Logic.* Some Logical Forays into Central Notions of Linguistics and Philosophy. 1981 ISBN 90-277-1201-8
148. D. M. Gabbay, *Semantical Investigations in Heyting's Intuitionistic Logic.* 1981
ISBN 90-277-1202-6
149. E. Agazzi (ed.), *Modern Logic – A Survey.* Historical, Philosophical, and Mathematical Aspects of Modern Logic and Its Applications. 1981 ISBN 90-277-1137-2
150. A. F. Parker Rhodes, *The Theory of Indistinguishables.* A Search for Explanatory Principles below the Level of Physics. 1981 ISBN 90-277-1214-X
151. J. C. Pitt, *Pictures, Images, and Conceptual Change.* An Analysis of Wilfrid Sellars' Philosophy of Science. 1981 ISBN 90-277-1276-X; Pb 90-277-1277-8
152. R. Hilpinen (ed.), *New Studies in Deontic Logic.* Norms, Actions, and the Foundations of Ethics. 1981 ISBN 90-277-1278-6; Pb 90-277-1346-4
153. C. Dilworth, *Scientific Progress.* A Study Concerning the Nature of the Relation between Successive Scientific Theories. 3rd rev. ed., 1994 ISBN 0-7923-2487-0; Pb 0-7923-2488-9
154. D. Woodruff Smith and R. McIntyre, *Husserl and Intentionality.* A Study of Mind, Meaning, and Language. 1982 ISBN 90-277-1392-8; Pb 90-277-1730-3
155. R. J. Nelson, *The Logic of Mind.* 2nd. ed., 1989 ISBN 90-277-2819-4; Pb 90-277-2822-4
156. J. F. A. K. van Benthem, *The Logic of Time.* A Model-Theoretic Investigation into the Varieties of Temporal Ontology, and Temporal Discourse. 1983; 2nd ed., 1991 ISBN 0-7923-1081-0
157. R. Swinburne (ed.), *Space, Time and Causality.* 1983 ISBN 90-277-1437-1
158. E. T. Jaynes, *Papers on Probability, Statistics and Statistical Physics.* Ed. by R. D. Rozenkrantz. 1983 ISBN 90-277-1448-7; Pb (1989) 0-7923-0213-3
159. T. Chapman, *Time: A Philosophical Analysis.* 1982 ISBN 90-277-1465-7
160. E. N. Zalta, *Abstract Objects.* An Introduction to Axiomatic Metaphysics. 1983
ISBN 90-277-1474-6
161. S. Harding and M. B. Hintikka (eds.), *Discovering Reality.* Feminist Perspectives on Epistemology, Metaphysics, Methodology, and Philosophy of Science. 1983
ISBN 90-277-1496-7; Pb 90-277-1538-6
162. M. A. Stewart (ed.), *Law, Morality and Rights.* 1983 ISBN 90-277-1519-X

163. D. Mayr and G. Süssmann (eds.), *Space, Time, and Mechanics.* Basic Structures of a Physical Theory. 1983 ISBN 90-277-1525-4
164. D. Gabbay and F. Guenthner (eds.), *Handbook of Philosophical Logic.* Vol. I: Elements of Classical Logic. 1983 ISBN 90-277-1542-4
165. D. Gabbay and F. Guenthner (eds.), *Handbook of Philosophical Logic.* Vol. II: Extensions of Classical Logic. 1984 ISBN 90-277-1604-8
166. D. Gabbay and F. Guenthner (eds.), *Handbook of Philosophical Logic.* Vol. III: Alternative to Classical Logic. 1986 ISBN 90-277-1605-6
167. D. Gabbay and F. Guenthner (eds.), *Handbook of Philosophical Logic.* Vol. IV: Topics in the Philosophy of Language. 1989 ISBN 90-277-1606-4
168. A. J. I. Jones, *Communication and Meaning.* An Essay in Applied Modal Logic. 1983 ISBN 90-277-1543-2
169. M. Fitting, *Proof Methods for Modal and Intuitionistic Logics.* 1983 ISBN 90-277-1573-4
170. J. Margolis, *Culture and Cultural Entities.* Toward a New Unity of Science. 1984 ISBN 90-277-1574-2
171. R. Tuomela, *A Theory of Social Action.* 1984 ISBN 90-277-1703-6
172. J. J. E. Gracia, E. Rabossi, E. Villanueva and M. Dascal (eds.), *Philosophical Analysis in Latin America.* 1984 ISBN 90-277-1749-4
173. P. Ziff, *Epistemic Analysis.* A Coherence Theory of Knowledge. 1984 ISBN 90-277-1751-7
174. P. Ziff, *Antiaesthetics.* An Appreciation of the Cow with the Subtile Nose. 1984 ISBN 90-277-1773-7
175. W. Balzer, D. A. Pearce, and H.-J. Schmidt (eds.), *Reduction in Science.* Structure, Examples, Philosophical Problems. 1984 ISBN 90-277-1811-3
176. A. Peczenik, L. Lindahl and B. van Roermund (eds.), *Theory of Legal Science.* Proceedings of the Conference on Legal Theory and Philosophy of Science (Lund, Sweden, December 1983). 1984 ISBN 90-277-1834-2
177. I. Niiniluoto, *Is Science Progressive?* 1984 ISBN 90-277-1835-0
178. B. K. Matilal and J. L. Shaw (eds.), *Analytical Philosophy in Comparative Perspective.* Exploratory Essays in Current Theories and Classical Indian Theories of Meaning and Reference. 1985 ISBN 90-277-1870-9
179. P. Kroes, *Time: Its Structure and Role in Physical Theories.* 1985 ISBN 90-277-1894-6
180. J. H. Fetzer, *Sociobiology and Epistemology.* 1985 ISBN 90-277-2005-3; Pb 90-277-2006-1
181. L. Haaparanta and J. Hintikka (eds.), *Frege Synthesized.* Essays on the Philosophical and Foundational Work of Gottlob Frege. 1986 ISBN 90-277-2126-2
182. M. Detlefsen, *Hilbert's Program.* An Essay on Mathematical Instrumentalism. 1986 ISBN 90-277-2151-3
183. J. L. Golden and J. J. Pilotta (eds.), *Practical Reasoning in Human Affairs.* Studies in Honor of Chaim Perelman. 1986 ISBN 90-277-2255-2
184. H. Zandvoort, *Models of Scientific Development and the Case of Nuclear Magnetic Resonance.* 1986 ISBN 90-277-2351-6
185. I. Niiniluoto, *Truthlikeness.* 1987 ISBN 90-277-2354-0
186. W. Balzer, C. U. Moulines and J. D. Sneed, *An Architectonic for Science.* The Structuralist Program. 1987 ISBN 90-277-2403-2
187. D. Pearce, *Roads to Commensurability.* 1987 ISBN 90-277-2414-8
188. L. M. Vaina (ed.), *Matters of Intelligence.* Conceptual Structures in Cognitive Neuroscience. 1987 ISBN 90-277-2460-1

189. H. Siegel, *Relativism Refuted*. A Critique of Contemporary Epistemological Relativism. 1987 ISBN 90-277-2469-5
190. W. Callebaut and R. Pinxten, *Evolutionary Epistemology*. A Multiparadigm Program, with a Complete Evolutionary Epistemology Bibliograph. 1987 ISBN 90-277-2582-9
191. J. Kmita, *Problems in Historical Epistemology*. 1988 ISBN 90-277-2199-8
192. J. H. Fetzer (ed.), *Probability and Causality*. Essays in Honor of Wesley C. Salmon, with an Annotated Bibliography. 1988 ISBN 90-277-2607-8; Pb 1-5560-8052-2
193. A. Donovan, L. Laudan and R. Laudan (eds.), *Scrutinizing Science*. Empirical Studies of Scientific Change. 1988 ISBN 90-277-2608-6
194. H.R. Otto and J.A. Tuedio (eds.), *Perspectives on Mind*. 1988 ISBN 90-277-2640-X
195. D. Batens and J.P. van Bendegem (eds.), *Theory and Experiment*. Recent Insights and New Perspectives on Their Relation. 1988 ISBN 90-277-2645-0
196. J. Österberg, *Self and Others*. A Study of Ethical Egoism. 1988 ISBN 90-277-2648-5
197. D.H. Helman (ed.), *Analogical Reasoning*. Perspectives of Artificial Intelligence, Cognitive Science, and Philosophy. 1988 ISBN 90-277-2711-2
198. J. Woleński, *Logic and Philosophy in the Lvov-Warsaw School*. 1989 ISBN 90-277-2749-X
199. R. Wójcicki, *Theory of Logical Calculi*. Basic Theory of Consequence Operations. 1988 ISBN 90-277-2785-6
200. J. Hintikka and M.B. Hintikka, *The Logic of Epistemology and the Epistemology of Logic*. Selected Essays. 1989 ISBN 0-7923-0040-8; Pb 0-7923-0041-6
201. E. Agazzi (ed.), *Probability in the Sciences*. 1988 ISBN 90-277-2808-9
202. M. Meyer (ed.), *From Metaphysics to Rhetoric*. 1989 ISBN 90-277-2814-3
203. R.L. Tieszen, *Mathematical Intuition*. Phenomenology and Mathematical Knowledge. 1989 ISBN 0-7923-0131-5
204. A. Melnick, *Space, Time, and Thought in Kant*. 1989 ISBN 0-7923-0135-8
205. D.W. Smith, *The Circle of Acquaintance*. Perception, Consciousness, and Empathy. 1989 ISBN 0-7923-0252-4
206. M.H. Salmon (ed.), *The Philosophy of Logical Mechanism*. Essays in Honor of Arthur W. Burks. With his Responses, and with a Bibliography of Burk's Work. 1990 ISBN 0-7923-0325-3
207. M. Kusch, *Language as Calculus vs. Language as Universal Medium*. A Study in Husserl, Heidegger, and Gadamer. 1989 ISBN 0-7923-0333-4
208. T.C. Meyering, *Historical Roots of Cognitive Science*. The Rise of a Cognitive Theory of Perception from Antiquity to the Nineteenth Century. 1989 ISBN 0-7923-0349-0
209. P. Kosso, *Observability and Observation in Physical Science*. 1989 ISBN 0-7923-0389-X
210. J. Kmita, *Essays on the Theory of Scientific Cognition*. 1990 ISBN 0-7923-0441-1
211. W. Sieg (ed.), *Acting and Reflecting*. The Interdisciplinary Turn in Philosophy. 1990 ISBN 0-7923-0512-4
212. J. Karpiński, *Causality in Sociological Research*. 1990 ISBN 0-7923-0546-9
213. H.A. Lewis (ed.), *Peter Geach: Philosophical Encounters*. 1991 ISBN 0-7923-0823-9
214. M. Ter Hark, *Beyond the Inner and the Outer*. Wittgenstein's Philosophy of Psychology. 1990 ISBN 0-7923-0850-6
215. M. Gosselin, *Nominalism and Contemporary Nominalism*. Ontological and Epistemological Implications of the Work of W.V.O. Quine and of N. Goodman. 1990 ISBN 0-7923-0904-9
216. J.H. Fetzer, D. Shatz and G. Schlesinger (eds.), *Definitions and Definability*. Philosophical Perspectives. 1991 ISBN 0-7923-1046-2
217. E. Agazzi and A. Cordero (eds.), *Philosophy and the Origin and Evolution of the Universe*. 1991 ISBN 0-7923-1322-4

218. M. Kusch, *Foucault's Strata and Fields.* An Investigation into Archaeological and Genealogical Science Studies. 1991 ISBN 0-7923-1462-X
219. C.J. Posy, *Kant's Philosophy of Mathematics.* Modern Essays. 1992 ISBN 0-7923-1495-6
220. G. Van de Vijver, *New Perspectives on Cybernetics.* Self-Organization, Autonomy and Connectionism. 1992 ISBN 0-7923-1519-7
221. J.C. Nyíri, *Tradition and Individuality.* Essays. 1992 ISBN 0-7923-1566-9
222. R. Howell, *Kant's Transcendental Deduction.* An Analysis of Main Themes in His Critical Philosophy. 1992 ISBN 0-7923-1571-5
223. A. García de la Sienra, *The Logical Foundations of the Marxian Theory of Value.* 1992 ISBN 0-7923-1778-5
224. D.S. Shwayder, *Statement and Referent.* An Inquiry into the Foundations of Our Conceptual Order. 1992 ISBN 0-7923-1803-X
225. M. Rosen, *Problems of the Hegelian Dialectic.* Dialectic Reconstructed as a Logic of Human Reality. 1993 ISBN 0-7923-2047-6
226. P. Suppes, *Models and Methods in the Philosophy of Science: Selected Essays.* 1993 ISBN 0-7923-2211-8
227. R. M. Dancy (ed.), *Kant and Critique: New Essays in Honor of W. H. Werkmeister.* 1993 ISBN 0-7923-2244-4
228. J. Woleński (ed.), *Philosophical Logic in Poland.* 1993 ISBN 0-7923-2293-2
229. M. De Rijke (ed.), *Diamonds and Defaults.* Studies in Pure and Applied Intensional Logic. 1993 ISBN 0-7923-2342-4
230. B.K. Matilal and A. Chakrabarti (eds.), *Knowing from Words.* Western and Indian Philosophical Analysis of Understanding and Testimony. 1994 ISBN 0-7923-2345-9
231. S.A. Kleiner, *The Logic of Discovery.* A Theory of the Rationality of Scientific Research. 1993 ISBN 0-7923-2371-8
232. R. Festa, *Optimum Inductive Methods.* A Study in Inductive Probability, Bayesian Statistics, and Verisimilitude. 1993 ISBN 0-7923-2460-9
233. P. Humphreys (ed.), *Patrick Suppes: Scientific Philosopher.* Vol. 1: Probability and Probabilistic Causality. 1994 ISBN 0-7923-2552-4
234. P. Humphreys (ed.), *Patrick Suppes: Scientific Philosopher.* Vol. 2: Philosophy of Physics, Theory Structure, and Measurement Theory. 1994 ISBN 0-7923-2553-2
235. P. Humphreys (ed.), *Patrick Suppes: Scientific Philosopher.* Vol. 3: Language, Logic, and Psychology. 1994 ISBN 0-7923-2862-0
 Set ISBN (Vols 233–235) 0-7923-2554-0
236. D. Prawitz and D. Westerståhl (eds.), *Logic and Philosophy of Science in Uppsala.* Papers from the 9th International Congress of Logic, Methodology, and Philosophy of Science. 1994 ISBN 0-7923-2702-0
237. L. Haaparanta (ed.), *Mind, Meaning and Mathematics.* Essays on the Philosophical Views of Husserl and Frege. 1994 ISBN 0-7923-2703-9
238. J. Hintikka (ed.), *Aspects of Metaphor.* 1994 ISBN 0-7923-2786-1
239. B. McGuinness and G. Oliveri (eds.), *The Philosophy of Michael Dummett.* With Replies from Michael Dummett. 1994 ISBN 0-7923-2804-3
240. D. Jamieson (ed.), *Language, Mind, and Art.* Essays in Appreciation and Analysis, In Honor of Paul Ziff. 1994 ISBN 0-7923-2810-8
241. G. Preyer, F. Siebelt and A. Ulfig (eds.), *Language, Mind and Epistemology.* On Donald Davidson's Philosophy. 1994 ISBN 0-7923-2811-6
242. P. Ehrlich (ed.), *Real Numbers, Generalizations of the Reals, and Theories of Continua.* 1994 ISBN 0-7923-2689-X

SYNTHESE LIBRARY

243. G. Debrock and M. Hulswit (eds.), *Living Doubt*. Essays concerning the epistemology of Charles Sanders Peirce. 1994 ISBN 0-7923-2898-1
244. J. Srzednicki, *To Know or Not to Know*. Beyond Realism and Anti-Realism. 1994 ISBN 0-7923-2909-0
245. R. Egidi (ed.), *Wittgenstein: Mind and Language*. 1995 ISBN 0-7923-3171-0
246. A. Hyslop, *Other Minds*. 1995 ISBN 0-7923-3245-8
247. L. Pólos and M. Masuch (eds.), *Applied Logic: How, What and Why*. Logical Approaches to Natural Language. 1995 ISBN 0-7923-3432-9
248. M. Krynicki, M. Mostowski and L.M. Szczerba (eds.), *Quantifiers: Logics, Models and Computation*. Volume One: Surveys. 1995 ISBN 0-7923-3448-5
249. M. Krynicki, M. Mostowski and L.M. Szczerba (eds.), *Quantifiers: Logics, Models and Computation*. Volume Two: Contributions. 1995 ISBN 0-7923-3449-3
Set ISBN (Vols 248 + 249) 0-7923-3450-7
250. R.A. Watson, *Representational Ideas from Plato to Patricia Churchland*. 1995 ISBN 0-7923-3453-1
251. J. Hintikka (ed.), *From Dedekind to Gödel*. Essays on the Development of the Foundations of Mathematics. 1995 ISBN 0-7923-3484-1
252. A. Wiśniewski, *The Posing of Questions*. Logical Foundations of Erotetic Inferences. 1995 ISBN 0-7923-3637-2
253. J. Peregrin, *Doing Worlds with Words*. Formal Semantics without Formal Metaphysics. 1995 ISBN 0-7923-3742-5
254. I.A. Kieseppä, *Truthlikeness for Multidimensional, Quantitative Cognitive Problems*. 1996 ISBN 0-7923-4005-1
255. P. Hugly and C. Sayward: *Intensionality and Truth*. An Essay on the Philosophy of A.N. Prior. 1996 ISBN 0-7923-4119-8
256. L. Hankinson Nelson and J. Nelson (eds.): *Feminism, Science, and the Philosophy of Science*. 1997 ISBN 0-7923-4162-7
257. P.I. Bystrov and V.N. Sadovsky (eds.): *Philosophical Logic and Logical Philosophy*. Essays in Honour of Vladimir A. Smirnov. 1996 ISBN 0-7923-4270-4
258. Å.E. Andersson and N-E. Sahlin (eds.): *The Complexity of Creativity*. 1996 ISBN 0-7923-4346-8
259. M.L. Dalla Chiara, K. Doets, D. Mundici and J. van Benthem (eds.): *Logic and Scientific Methods*. Volume One of the Tenth International Congress of Logic, Methodology and Philosophy of Science, Florence, August 1995. 1997 ISBN 0-7923-4383-2
260. M.L. Dalla Chiara, K. Doets, D. Mundici and J. van Benthem (eds.): *Structures and Norms in Science*. Volume Two of the Tenth International Congress of Logic, Methodology and Philosophy of Science, Florence, August 1995. 1997 ISBN 0-7923-4384-0
Set ISBN (Vols 259 + 260) 0-7923-4385-9
261. A. Chakrabarti: *Denying Existence*. The Logic, Epistemology and Pragmatics of Negative Existentials and Fictional Discourse. 1997 ISBN 0-7923-4388-3
262. A. Biletzki: *Talking Wolves*. Thomas Hobbes on the Language of Politics and the Politics of Language. 1997 ISBN 0-7923-4425-1
263. D. Nute (ed.): *Defeasible Deontic Logic*. 1997 ISBN 0-7923-4630-0
264. U. Meixner: *Axiomatic Formal Ontology*. 1997 ISBN 0-7923-4747-X
265. I. Brinck: *The Indexical 'I'*. The First Person in Thought and Language. 1997 ISBN 0-7923-4741-2
266. G. Hölmström-Hintikka and R. Tuomela (eds.): *Contemporary Action Theory*. Volume 1: Individual Action. 1997 ISBN 0-7923-4753-6; Set: 0-7923-4754-4

267. G. Hölmström-Hintikka and R. Tuomela (eds.): *Contemporary Action Theory.* Volume 2: Social Action. 1997 ISBN 0-7923-4752-8; Set: 0-7923-4754-4
268. B.-C. Park: *Phenomenological Aspects of Wittgenstein's Philosophy.* 1998 ISBN 0-7923-4813-3
269. J. Paśniczek: *The Logic of Intentional Objects.* A Meinongian Version of Classical Logic. 1998 Hb ISBN 0-7923-4880-X; Pb ISBN 0-7923-5578-4
270. P.W. Humphreys and J.H. Fetzer (eds.): *The New Theory of Reference.* Kripke, Marcus, and Its Origins. 1998 ISBN 0-7923-4898-2
271. K. Szaniawski, A. Chmielewski and J. Woleński (eds.): *On Science, Inference, Information and Decision Making.* Selected Essays in the Philosophy of Science. 1998 ISBN 0-7923-4922-9
272. G.H. von Wright: *In the Shadow of Descartes.* Essays in the Philosophy of Mind. 1998 ISBN 0-7923-4992-X
273. K. Kijania-Placek and J. Woleński (eds.): *The Lvov–Warsaw School and Contemporary Philosophy.* 1998 ISBN 0-7923-5105-3
274. D. Dedrick: *Naming the Rainbow.* Colour Language, Colour Science, and Culture. 1998 ISBN 0-7923-5239-4
275. L. Albertazzi (ed.): *Shapes of Forms.* From Gestalt Psychology and Phenomenology to Ontology and Mathematics. 1999 ISBN 0-7923-5246-7
276. P. Fletcher: *Truth, Proof and Infinity.* A Theory of Constructions and Constructive Reasoning. 1998 ISBN 0-7923-5262-9
277. M. Fitting and R.L. Mendelsohn (eds.): *First-Order Modal Logic.* 1998 Hb ISBN 0-7923-5334-X; Pb ISBN 0-7923-5335-8
278. J.N. Mohanty: *Logic, Truth and the Modalities from a Phenomenological Perspective.* 1999 ISBN 0-7923-5550-4
279. T. Placek: *Mathematical Intiutionism and Intersubjectivity.* A Critical Exposition of Arguments for Intuitionism. 1999 ISBN 0-7923-5630-6
280. A. Cantini, E. Casari and P. Minari (eds.): *Logic and Foundations of Mathematics.* 1999 ISBN 0-7923-5659-4 set ISBN 0-7923-5867-8
281. M.L. Dalla Chiara, R. Giuntini and F. Laudisa (eds.): *Language, Quantum, Music.* 1999 ISBN 0-7923-5727-2; set ISBN 0-7923-5867-8
282. R. Egidi (ed.): *In Search of a New Humanism.* The Philosophy of Georg Hendrik von Wright. 1999 ISBN 0-7923-5810-4
283. F. Vollmer: *Agent Causality.* 1999 ISBN 0-7923-5848-1
284. J. Peregrin (ed.): *Truth and Its Nature (if Any).* 1999 ISBN 0-7923-5865-1
285. M. De Caro (ed.): *Interpretations and Causes.* New Perspectives on Donald Davidson's Philosophy. 1999 ISBN 0-7923-5869-4
286. R. Murawski: *Recursive Functions and Metamathematics.* Problems of Completeness and Decidability, Gödel's Theorems. 1999 ISBN 0-7923-5904-6
287. T.A.F. Kuipers: *From Instrumentalism to Constructive Realism.* On Some Relations between Confirmation, Empirical Progress, and Truth Approximation. 2000 ISBN 0-7923-6086-9
288. G. Holmström-Hintikka (ed.): *Medieval Philosophy and Modern Times.* 2000 ISBN 0-7923-6102-4
289. E. Grosholz and H. Breger (eds.): *The Growth of Mathematical Knowledge.* 2000 ISBN 0-7923-6151-2

290. G. Sommaruga: *History and Philosophy of Constructive Type Theory*. 2000 ISBN 0-7923-6180-6
291. J. Gasser (ed.): *A Boole Anthology*. Recent and Classical Studies in the Logic of George Boole. 2000 ISBN 0-7923-6380-9
292. V.F. Hendricks, S.A. Pedersen and K.F. Jørgensen (eds.): *Proof Theory*. History and Philosophical Significance. 2000 ISBN 0-7923-6544-5

Previous volumes are still available.

KLUWER ACADEMIC PUBLISHERS – DORDRECHT / BOSTON / LONDON